METHODS IN MOLECULAR BIOLOGY™

Series Editor
John M. Walker
School of Life Sciences
University of Hertfordshire
Hatfield, Hertfordshire, AL10 9AB, UK

For further volumes:
http://www.springer.com/series/7651

Neural Development

Methods and Protocols

Edited by

Renping Zhou

Department of Chemical Biology, Susan Lehman Cullman Laboratory for Cancer Research, Ernest Mario School of Pharmacy, Rutgers University, Piscataway, NJ, USA

Lin Mei

Institute of Molecular Medicine and Genetics, Georgia Regents University, Augusta, GA, USA

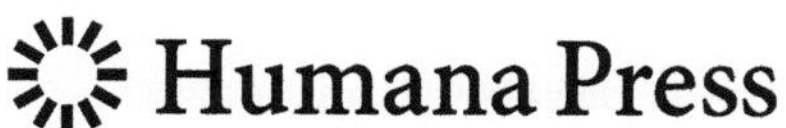

Editors
Renping Zhou
Department of Chemical Biology
Susan Lehman Cullman Laboratory
for Cancer Research
Ernest Mario School of Pharmacy
Rutgers University
Piscataway, NJ, USA

Lin Mei
Institute of Molecular Medicine and Genetics
Georgia Regents University
Augusta, GA, USA

ISSN 1064-3745 ISSN 1940-6029 (electronic)
ISBN 978-1-62703-443-2 ISBN 978-1-62703-444-9 (eBook)
DOI 10.1007/978-1-62703-444-9
Springer New York Heidelberg Dordrecht London

Library of Congress Control Number: 2013937290

Printed on acid-free paper

Humana Press is a brand of Springer
Springer is part of Springer Science+Business Media (www.springer.com)

Preface

The brain is a remarkable organ that coordinates all the activities necessary for survival. The human brain consists of many different types of cells. There is an astronomically large number of neurons, 14 times (!) more than that of human beings on Earth. They are located in strategic regions in the brain including different layers in the cortex and nuclei deep in the brain. Each neuron could make 1,000 or more synapses with other neurons in the brain or target cells in tissues or organs outside the brain. The contacts of a given neuron are estimated to be more than the number of friends an average Facebook user has. Neurons perceive signals from the environment, integrate them, and govern the body's responses. In between neurons are glial cells that are thought to represent 80 % of cells in the brain and outnumber neurons by 10–50 times. They support neurons, insulate nerve tracks, and regulate neural development and synaptic transmission.

How the myriad types of neurons and glia are generated and wired to form a functional nervous system is the central question of developmental neurobiology. Neural inductive activity was first discovered by Spemann and colleagues in the 1920s, and the growth cone and its chemotactic ability were proposed by Cajal in late 1900. Since these landmark discoveries, research in developmental neuroscience has made tremendous progress. A family of inhibitors for the bone morphogenic protein (BMP) pathway has been identified as the Spemann Organizer activity for neural induction. Transcription factor combinations that determine different neuronal fate are being defined. With these advances, we are now on the verge of generating neurons of specific types from many nonneural tissues, providing hope that one day degenerated dopaminergic neurons in Parkinson's disease or damaged spinal cords can be repaired with the specific neurons generated from the skin fibroblasts of the patients. Research in the last 20 years also identified numerous proteins that regulate directional migration of growth cones, and the guidance functions of these factors have been confirmed in vivo in many cases. Current efforts are being directed toward elucidating the intricate interactions among the guidance signals in the specification of neural circuits that control particular behavior. Understanding the molecular and cellular mechanisms underlying the development of specific neural circuits is not just an intellectual curiosity but also central to our ability to develop therapeutic approaches to repair damaged pathways in the future.

Scientific progress depends in a large degree on the invention of new assays and the deployment of new technology. Techniques and methods for current developmental studies are likely to be used in the future. However, many have only been published in research papers and often short of details because of the limited space of journals. We collect here commonly used protocols to facilitate future research in developmental neuroscience. Loosely reflecting the developmental sequence of the nervous system, this book consists of 4 parts. The first part is on techniques of culturing neurons and glia as well as their growth and differentiation. Modern neurobiological research depends critically on the ability to manipulate expression of specific genes in neurons. Methods of gene delivery and down-regulation are presented in the second part. The third part focuses on protocols for analyzing axon growth and guidance and synapse formation. The final fourth part includes basic methods to analyze brain morphology and axon pathways in developing animals. The protocols presented here are written in enough detail to provide useful guidance for students and postdoctoral fellows new to developmental neurobiology.

We are deeply indebted to the authors of the chapters in this book. We would also like to thank the series editor, Dr. John M. Walker, and the Humana Press staff, especially David Casey, for valuable suggestions and patience, and Ms. Erica DiPaola for editorial assistance.

Piscataway, NJ, USA — *Renping Zhou*
Augusta, GA, USA — *Lin Mei*

Contents

Part III Axon Pathways and Synapses

Contributors

YOSHITOSHI ATOBE • *Department of Neuroanatomy, Yokohama City University Graduate School of Medicine, Yokohama, Japan*

MARTIN BASTMEYER • *Department of Cell and Neurobiology, Zoological Institute, Karlsruhe Institute of Technology, Karlsruhe, Germany; DFG-Center for Functional Nanostructures, Karlsruhe Institute of Technology, Karlsruhe, Germany*

KANG-YELL CHOI • *Translational Research Center for Protein Function Control, College of Life Science and Biotechnology, Yonsei University, Seoul, South Korea*

WON-SEOK CHOI • *School of Biological Sciences and Technology, College of Natural Sciences, College of Medicine, Chonnam National University, Gwangju, South Korea; Department of Environmental and Occupational Health Sciences, University of Washington, Seattle, WA, USA*

MARGARET A. COOPER • *Department of Molecular Biophysics and Biochemistry, Yale University, New Haven, CT, USA*

GITANJALI DAS • *Department of Chemical Biology, Susan Lehman Cullman Laboratory for Cancer Research, Ernest Mario School of Pharmacy, Rutgers University, Piscataway, NJ, USA*

ERIK MICHAEL DEBOER • *Department of Neuroscience and Cell Biology, UMDNJ-Robert Wood Johnson Medical School, Piscataway, NJ, USA*

BAOJIN DING • *Department of Microbiology and Physiological Systems, University of Massachusetts Medical School, Worcester, MA, USA; Program in Neuroscience, University of Massachusetts Medical School, Worcester, MA, USA*

DAISUKE DOI • *Department of Clinical Application, Center for iPS Cell, Kyoto University, Kyoto, Japan*

CHERYL F. DREYFUS • *Department of Neuroscience and Cell Biology, UMDNJ-Robert Wood Johnson Medical School, Piscataway, NJ, USA*

ANNA DUNAEVSKY • *Developmental Neuroscience, Munroe-Meyer Institute, University of Nebraska Medical Center, Omaha, NE, USA*

EMMA K. FARLEY • *Division of Genetics, Genomics and Development Center for Integrative Genomics, Department of Molecular and Cell Biology, University of California, Berkeley, CA, USA*

BONNIE L. FIRESTEIN • *Department of Cell Biology and Neuroscience, Rutgers University, Piscataway, NJ, USA; Department of Biomedical Engineering, Rutgers University, Piscataway, NJ, USA*

MARTIN FRITZ • *Department of Cell and Neurobiology, Zoological Institute, Karlsruhe Institute of Technology, Karlsruhe, Germany*

CLIFTON G. FULMER • *Department of Neuroscience and Cell Biology, Graduate School of Biochemical Sciences, UMDNJ-Robert Wood Johnson Medical School, Piscataway, NJ, USA*

KENGO FUNAKOSHI • *Department of Neuroanatomy, Yokohama City University Graduate School of Medicine, Yokohama, Japan*

YANGYANG HUANG • *Department of Neuroscience and Cell Biology, UMDNJ-Robert Wood Johnson Medical School, Piscataway, NJ, USA*
DANIEL L. KILPATRICK • *Department of Microbiology and Physiological Systems, University of Massachusetts Medical School, Worcester, MA, USA; Program in Neuroscience, University of Massachusetts Medical School, Worcester, MA, USA*
HYUNG-WOOK KIM • *Department of Environmental and Occupational Health Sciences, University of Washington, Seattle, WA, USA*
MI-YEON KIM • *Translational Research Center for Protein Function Control, College of Life Science and Biotechnology, Yonsei University, Seoul, South Korea*
BERND KNÖLL • *Institute for Physiological Chemistry, Ulm University, Ulm, Germany*
MELINDA K. KUTZING • *Department of Cell Biology and Neuroscience, Graduate Program in Biomedical Engineering, Rutgers University, Piscataway, NJ, USA*
MUNJIN KWON • *Department of Cell Biology and Neuroscience, Rutgers University, Piscataway, NJ, USA*
CHRIS L. LANGHAMMER • *Department of Cell Biology and Neuroscience, Graduate Program in Biomedical Engineering, Rutgers University, Piscataway, NJ, USA*
NIKOLAI E. LAZAROV • *Department of Anatomy and Histology, Medical University—Sofia, Sofia, Bulgaria; Institute of Neurobiology, Bulgarian Academy of Sciences, Sofia, Bulgaria*
DAEHOON LEE • *Institute of Molecular Medicine and Genetics, Medical College of Georgia, Georgia Regents University, Augusta, GA, USA; Department of Neurology, Medical College of Georgia, Georgia Regents University, Augusta, GA, USA*
BO LI • *Cold Spring Harbor Laboratory, Cold Spring Harbor, NY, USA*
HIROSHI MAKINO • *Section of Neurobiology, Division of Biological Sciences, Center for Neural Circuits and Behavior, University of California, San Diego, La Jolla, CA, USA; Department of Neurosciences, Center for Neural Circuits and Behavior, University of California, San Diego, La Jolla, CA, USA*
MICHAEL P. MATISE • *Department of Neuroscience and Cell Biology, UMDNJ-Robert Wood Johnson Medical School, Piscataway, NJ, USA*
W. GEOFFREY MCAULIFFE • *Department of Neuroscience and Cell Biology, UMDNJ-Robert Wood Johnson Medical School, Piscataway, NJ, USA*
LIN MEI • *Institute of Molecular Medicine and Genetics, Georgia Regents University, Augusta, GA, USA*
BYOUNG-SAN MOON • *Translational Research Center for Protein Function Control, College of Life Science and Biotechnology, Yonsei University, Seoul, South Korea*
ASUKA MORIZANE • *Department of Clinical Application, Center for iPS Cell, Kyoto University, Kyoto, Japan*
YUN PENG • *Institute of Molecular Medicine and Genetics, Medical College of Georgia, Georgia Regents University, Augusta, GA, USA; Department of Neurology, Medical College of Georgia, Georgia Regents University, Augusta, GA, USA*
MLADEN-ROKO RASIN • *Department of Neuroscience and Cell Biology, UMDNJ-Robert Wood Johnson Medical School, Piscataway, NJ, USA*
KENNETH REUHL • *Neurotoxicology Laboratory, Department of Pharmacology and Toxicology, Rutgers University, Piscataway, NJ, USA*
THARAKESWARI SELVAKUMAR • *Department of Microbiology and Physiological Systems, University of Massachusetts Medical School, Worcester, MA, USA; Program in Neuroscience, University of Massachusetts Medical School, Worcester, MA, USA*

KATIE SOKOLOWSKI • *Center for Neuroscience Research, Children's National Medical Center, Washington, DC, USA*
ALEXANDER I. SON • *Department of Chemical Biology, Susan Lehman Cullman Laboratory for Cancer Research, Ernest Mario School of Pharmacy, Rutgers University, Piscataway, NJ, USA*
ERIC S. SWEET • *Neuroscience Graduate Program, Department of Cell Biology and Neuroscience, Rutgers University, Piscataway, NJ, USA*
JUN TAKAHASHI • *Department of Biological Repair, Field of Clinical Application, Institute for Frontier Medical Sciences, Department of Clinical Application Center for iPS Cell Research and Application, Kyoto University, Kyoto, Japan*
YANMEI TAO • *Institute of Development and Regenerative Biology, Hangzhou Normal University, Hangzhou, Zhejiang, China*
CHUNLEI WANG • *Institute of Molecular Medicine and Genetics, Georgia Regents University, Augusta, GA, USA*
HUI WANG • *Department of Neuroscience and Cell Biology, UMDNJ-Robert Wood Johnson Medical School, Piscataway, NJ, USA*
MARKUS WESCHENFELDER • *Department of Cell and Neurobiology, Zoological Institute, Karlsruhe Institute of Technology, Karlsruhe, Germany*
FRANCO WETH • *Department of Cell and Neurobiology, Zoological Institute, Karlsruhe Institute of Technology, Karlsruhe, Germany*
HAITAO WU • *Department of Cognitive Sciences, Beijing Institute of Basic Medical Sciences, Beijing, China*
ZHENGUI XIA • *Department of Environmental and Occupational Health Sciences, University of Washington, Seattle, WA, USA*
SHAN XIONG • *Institute of Molecular Medicine and Genetics, Medical College of Georgia, Georgia Regents University, Augusta, GA, USA; Department of Neurology, Medical College of Georgia, Georgia Regents University, Augusta, GA, USA*
WEN-CHENG XIONG • *Institute of Molecular Medicine and Genetics, Medical College of Georgia, Georgia Regents University, Augusta, GA, USA; Department of Neurology, Medical College of Georgia, Georgia Regents University, Augusta, GA, USA*
AKIRA YOSHIKAWA • *Department of Neuroanatomy, Yokohama City University Graduate School of Medicine, Yokohama, Japan*
XIN YUE • *Department of Chemical Biology, Susan Lehman Cullman Laboratory for Cancer Research, Ernest Mario School of Pharmacy, Rutgers University, Piscataway, NJ, USA*
RENPING ZHOU • *Department of Chemical Biology, Susan Lehman Cullman Laboratory for Cancer Research, Ernest Mario School of Pharmacy, Rutgers University, Piscataway, NJ, USA*

Part I

Neuron and Glial Cell Culture

Chapter 1

Isolation and Maintenance of Cortical Neural Progenitor Cells In Vitro

Mi-Yeon Kim, Byoung-San Moon, and Kang-Yell Choi

Abstract

Neural progenitor cells (NPCs) or neural stem cells are important tools for investigating central nervous system (CNS) development. NPCs can be used in therapeutic strategies and for characterizing differentiation mechanisms. Here, we describe methods for isolating and culturing embryonic NPCs.

Key words Multipotent stem cell, Neural progenitor cells, Central nervous system, Development

1 Introduction

Neural progenitor cells (NPCs), first defined by Reynolds and colleagues [1], are self-renewing, multipotent cells that can differentiate into neurons, astrocytes, and oligodendrocytes, all of which are constitutive central nervous system (CNS) cells [2]. During neural development, the rapidly dividing NPCs are located in the ventricular zone of the brain [3]. NPCs migrate and subsequently differentiate into functional cells in the cortical plate of the brain cortex [4], indicating that NPCs are physically and functionally important in brain developmental processes including the sequential processes involved in brain maturation.

NPCs are commonly used to investigate the formation and regulation of the nervous system; these NPC studies can provide data to support therapeutic strategies for the treatment of brain diseases and injuries [5]. NPCs can also be differentiated into cells that could be used for therapeutic strategies such as implanting dopaminergic neurons to treat Parkinson's disease [6] or oligodendrocytes to treat demyelination disease [7]. Finally, NPCs can be used to identify neural or other types of cell differentiation mechanisms.

It is important to isolate NPCs during the proper developmental stage and from specific areas of the brain to obtain high-quality

Renping Zhou and Lin Mei (eds.), *Neural Development: Methods and Protocols*, Methods in Molecular Biology, vol. 1018, DOI 10.1007/978-1-62703-444-9_1,

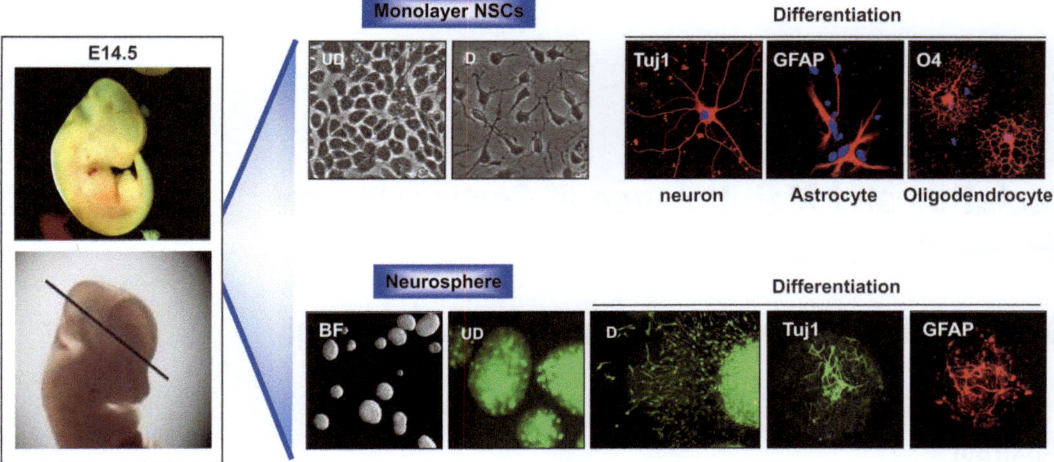

Fig. 1 Overview of the NPC culture. From E14.5 embryo brain, NPCs can be isolated. NPCs can be plated monolayer on poly-L-ornithine/fibronectin-coated dishes and analyzed undifferentiation and differentiation status. NPCs can also form free-floating clusters which are named neurospheres. Neurospheres can be cultured several times and be preserved

primary cells. NPCs are located in the cerebral cortex, hippocampus, basal forebrain, and cerebellum at embryonic stages. In the subventricular zone of rodent brain development, embryonic (E) day 8–14 NPCs are in an expansion phase [3] and can therefore, be easily obtained at early developmental stages prior to progression to neurogenesis.

Isolated NPCs can be cultured in either monolayer or neurospheres. In monolayer culture, fibronectin coating promotes survival and migration of NPCs [8] and can act as scaffolding [9]. Neurospheres, non-adherent spheroid clusters of NPCs [10], form from NPCs that retain self-renewal potential. The formation of neurospheres is affected by various factors including density and cell passage number. Here, we describe efficient methods for isolating and maintaining cortical NPCs as well as for inducing formation of neurospheres (*see* Fig. 1).

2 Materials

2.1 Animal

E14.5 impregnated Sprague–Dawley rats.

2.2 Instruments

1. Fine scissors.
2. Micro scissors.
3. Fine forceps.
4. Needles.

5. 100-mm petri dish.
6. 6-Well plate.
7. 15-ml sterile conical tube.
8. Stereomicroscope.
9. T25 flask.

2.3 Solutions and Medium

1. 10× Hanks balanced salt solution (HBSS).
   Mix 4 g KCl, 0.6 g KH_2PO_4, 80 g NaCl, and 0.9 g $Na_2HPO_4{\cdot}7H_2O$ into a 1 L graduated cylinder containing approximately 700 ml distilled water. After stirring, bring the volume up to 1 L using distilled water and autoclave the solution. Prior to use, dilute 50 ml of this solution with 450 ml of distilled, deionized water.
2. NPC basal medium (N2 medium).
   NPCs were cultured in N2 medium; 486 ml Dulbecco's modified Eagle medium: Nutrient Mixture F-12 (DMEM/F12) with 50 μl 1 M putrescine, 50 μl 0.5 mM selenite, 10 μl 1 mM progesterone, 1 ml 50 mg/ml transferrin, 3.13 ml 4 mg/ml insulin, 1.25 ml 200 mM Glutamax, 0.775 g D(+)-glucose, and 5 ml 100× penicillin/streptomycin (*see* **Note 1**).
3. Poly-L-ornithine/fibronectin-coated culture dishes.
   (a) Dilute 50 mg/ml poly-L-ornithine in distilled, deionized water to a final concentration of 15 μg/ml. Store the stock solution at −20 °C.
   (b) Dilute 5 mg/ml fibronectin in distilled, deionized water to a final concentration of 10 μg/ml. Store the stock solution at −20 °C.
   (c) Dilute the 50 mg/ml poly-L-ornithine stock solution to 15 μg/ml in phosphate-buffered saline (PBS), and coat the surface of each well or dish by placing enough solution followed by incubation of the plates and dishes overnight at 37 °C. After incubation, aspirate the coating solution and wash the plates or dishes three times with PBS to remove excess poly-L-ornithine. For fibronectin coating, dilute the 5 mg/ml fibronectin solution to 10 μg/ml with PBS, and plates or dishes filled with the solution were incubated at 37 °C for at least 2 h.
4. Growth factors.
   (a) Prepare a 10 ng/ml human basic fibroblast growth factor (FGF2) (Peprotech, NJ) by diluting 10 mg/ml FGF2 with PBS.
   (b) Prepare a 20 ng/ml human epidermal growth factor (EGF) (Peprotech) by diluting a 20 mg/ml EGF with PBS.

5. Solutions for NPC subculture.
   (a) N2 medium (Gibco, NY).
   (b) Trypsin–EDTA solution: 1× TrypLE™ express (Gibco).
   (c) PBS.

3 Methods

3.1 Monolayer Culture

Isolate the NPCs from E14.5 rat brain cortices and plate onto poly-L-ornithine/fibronectin coated dishes.

3.1.1 Isolation of NPCs

Dissection of the Embryo Brain (*See* Fig. 2)

1. Euthanize a pregnant (E14.5) rat in a CO_2 chamber.
2. Cut an opening from the abdominal region to the chest.
3. Remove the uterus and place it in a 100-mm petri dish containing ice-cold 1× HBSS.
4. Incise the amnion with micro spring scissors.
5. Remove the embryos from the amnion.
6. Transfer the embryos from the uterus into the petri dish containing fresh, ice-cold 1× HBSS.

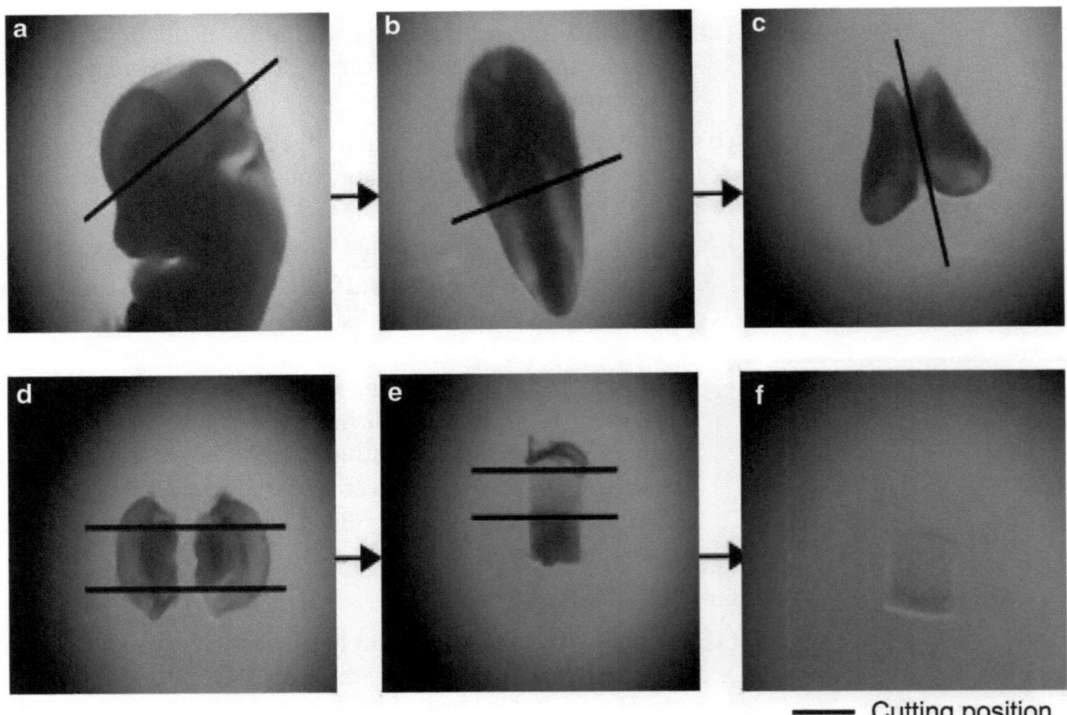

Fig. 2 Isolation of the cortex from E14. 5 rat embryo brain. (**a**) Cut out the E14.5 embryo brain with needle. (**b**) Cut out the forebrain. (**c**) Divide each hemispheres. (**d**) Make a scratch on edge of each hemispheres. (**e**) Open the ventricle and make one layer. (**f**) Cut out the edge region and take the cortex tissue

7. Stabilize the embryo with forceps and remove the head with a bent (90°) needle. Remove the ectomeninx.
8. Divide the hemispheres, and cut out the cortex region.
9. Transfer the cortices, and add 1 ml 1× HBSS solution to a 15-ml conical tube using a 1 ml pipette tip with the end cut-off.

Isolation of NPCs from Brain Tissue

1. All the procedures for isolation of NPCs should perform on a clean bench. Place the 15 ml tube containing cortices in 1× HBSS for approximately 5 min.
2. Aspirate the 1× HBSS solution and add 3 ml N2 medium.
3. Shake the tube lightly, set it down gently, and allow it to sit for 2 min (*see* **Note 2**).
4. Aspirate the N2 medium and add 1 ml fresh N2 medium.
5. Using only a 1 ml pipette tip, dissociate the cortices in the N2 medium to create a single-cell suspension (*see* **Note 3**).
6. Allow the cell suspension to sit for 1 min to allow other types of cells, such as red blood cells, to settle down to the bottom of the tube.
7. Transfer the supernatant to a new 15 ml conical tube.
8. Count the number of cells.
9. NPCs can be seeded onto fibronectin-coated dishes by aspirating the fibronectin solution from the 100 mm dish and then seeding the NPCs at a density of 2×10^6 cells (*see* Fig. 3).
10. Add N2 medium supplemented with 10 ng/ml FGF2. Incubate the cells at 37 °C in 5 % CO_2 incubator for 4 days (*see* **Note 4**).

3.1.2 Passaging the NPCs

1. Prepare poly-L-ornithine/fibronectin-coated plates before detaching the cultured NPCs (*see* the Subheading 2.3, **item 3**).
2. Aspirate the medium from the cultured NPCs.

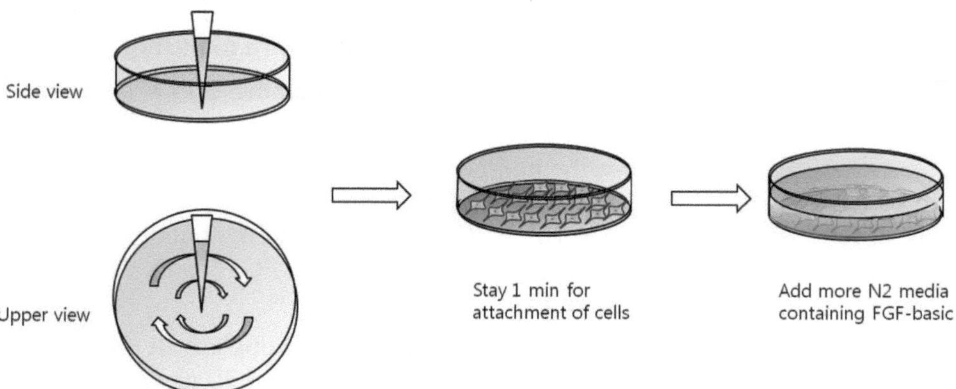

Fig. 3 Seeding NPCs on coated dish. After suction of fibronectin coating solution, counted NPCs are plated on dish in a circular swirl motion. Set it for 1 min for attachment of cells and add N2 media containing FGF-basic

3. Add 1 ml 1× TrypLE solution to the NPCs and swirl to distribute the cells evenly.
4. Aspirate the 1× TrypLE solution and incubate the cells at 37 °C for 1 min.
5. Suspend cells in 2 ml N2 medium and harvest the detached cells.
6. Centrifuge the cells at 800 × *g* for 2 min at room temperature.
7. Resuspend the cells in 1 ml N2 medium.
8. Add an additional 4 ml N2 medium and count the number of cells.
9. NPCs can be seeded at a density of 2×10^5 cells onto 6-well fibronectin-coated dishes after aspirating the fibronectin solution (*see* the Subheading 3.1.2, **step 1**).
10. To maintain cells in an undifferentiated state, culture the NPCs in N2 medium containing 10 ng/ml FGF2. Replace the medium with fresh N2 medium supplemented with 10 ng/ml FGF2 every 2 days. To induce differentiation, NPCs can be cultured in FGF2-depleted N2 medium.

3.2 Neurosphere Formation

Dissect NPCs from E14.5 rat brain cortices as described in Subheading 3.1.1.1. Seed the NPCs into a T25 flask.

3.2.1 Isolation of NPCs

Isolation of NPCs from Brain Tissue

1. At a clean bench, keep the isolated cortices submerged at the bottom of a 15-ml conical tube containing HBSS for 5 min.
2. Aspirate the 1× HBSS solution and add 3 ml N2 medium.
3. Shake the tube lightly, set it down gently, and allow it to sit for 2 min (*see* **Note 2**).
4. Aspirate the N2 medium and add 1 ml fresh medium.
5. Dissociate the cortical cells into a single-cell suspension using N2 medium and only a pipette tip (*see* **Note 3**).
6. Let the cell suspension rest for 1 min to allow other types of cells, such as red blood cells, to settle down to the bottom of the tube.
7. Transfer the supernatant containing NPCs to a new 15-ml conical tube.
8. Count the number of cells.

Formation of Neurospheres from NPCs

1. Seed 2×10^5 NPCs into a T25 flask containing N2 medium supplemented with 10 ng/ml FGF2 and 20 ng/ml EGF (*see* **Note 5**).
2. Culture the NPCs for 4 days. Supplement the N2 medium every day with 10 ng/ml FGF2 and 20 ng/ml EGF.

3.2.2 Passaging the Neurospheres

1. Transfer the neurospheres into a 15-ml conical tube.
2. Keep the tube at the bench for 5 min (*see* **Note 6**).

3. Aspirate the medium and add 1 ml 1× TrypLE to the tube.
4. Incubate the tube for 15 min at 37 °C.
5. Centrifuge for 1 min at 1,000 × *g*.
6. Remove the supernatant and resuspend the cells in 200 μl N2 medium (*see* **Note 7**).
7. Add an additional 5 ml N2 medium and count the number of cells.
8. For monolayer culture, plate the cells on poly-L-ornithine/fibronectin-coated dishes. To maintain neurospheres, seed 2×10^5 NPCs into a T25 flask.

3.2.3 Cryopreservation of Neurospheres

1. Transfer the neurospheres to a 15-ml conical tube.
2. Centrifuge for 5 min at 200 × *g*.
3. Prepare freezing medium with 15 % dimethyl sulfoxide in N2 medium without FGF2 or EGF.
4. Aspirate the supernatant and resuspend the cells in 1 ml freezing medium (*see* **Note 8**).
5. Transfer the neurospheres in freezing medium to polypropylene cryovials.
6. Transfer the cryovials to a freezing container and store at −80 °C for 2 days.
7. After 2 days, move the vials to a liquid nitrogen tank (*see* **Note 9**).

3.2.4 Thawing and Reestablishing the Neurospheres

1. Thaw the vial at 37 °C for 1 min.
2. Add 1 ml N2 medium and transfer the neurospheres to a 15-ml conical tube.
3. Centrifuge for 5 min at 200 × *g*.
4. Aspirate the supernatant and resuspend the neurospheres gently in 1 ml N2 medium.
5. Transfer the neurospheres into a T25 flask filled with 9 ml N2 medium supplemented with 10 ng/ml FGF2 and 20 ng/ml EGF.
6. Culture at 37 °C in 5 % CO_2.

4 Notes

1. The D(+)-glucose solution is dissolved in DMEM/F12 medium and filtered with a 0.22-μm syringe filter.
2. Let the tube rest for 2 min at the bench to settle the cortices, and then aspirate the solution, leaving the pure cortex.

3. This process is a mechanical dissociation process that requires multiple rounds of pipetting. However, to avoid damage to the NPCs, do not pipette more than 40 times.
4. When seeding the NPCs onto a dish, cells should be seeded in a circular swirling motion starting from the center outwards. Let the dish rest for 1 min for attachment of cells. Cells are seeded on the dishes at a confluence of approximately 80 %. The cells should be close to 100 % confluent after 4 days.
5. NPCs must be seeded at a low density (1×10^3 NPCs/well in a 96-well plate) for the formation of neurospheres and their characterization. However, to obtain a high number of neurospheres, NPCs must be seeded at a high density (2×10^5 NPCs/T25 flask). Neurospheres form rapidly at higher cell densities.
6. Centrifugation can induce mechanical stress in the NPCs, and it can affect characteristics and the quality of NPCs. Therefore, mild centrifugation such as $200 \times g$ for 5 min should be utilized.
7. TrypLE remaining in solution with cells can be toxic to NPCs and thus must be removed completely from the cells. Do not pipette more than 40–60 times to resuspend the NPCs because this can induce mechanical stress in the NPCs.
8. Repeated suspension in freezing medium can induce mechanical stress to neurospheres and can destroy the neurospheres. Suspending the neurospheres two to three times in the freezing medium is sufficient.
9. Neurospheres can be stored for 1 week at −80 °C without causing damage to the NPCs.

References

1. Reynolds BA, Tetzlaff W, Weiss S (1992) A multipotent EGF-responsive striatal embryonic progenitor cell produces neurons and astrocytes. J Neurosci 12:4565–4574
2. Gotz M, Huttner WB (2005) The cell biology of neurogenesis. Nat Rev Mol Cell Biol 6:777–788
3. Temple S (2001) The development of neural stem cells. Nature 414:112–117
4. Kriegstein A, Alvarez-Buylla A (2009) The glial nature of embryonic and adult neural stem cells. Annu Rev Neurosci 32:149–184
5. Shimazaki T (2003) Biology and clinical application of neural stem cells. Horm Res 60(Suppl 3):1–9
6. Liste I, Garcia-Garcia E, Martinez-Serrano A (2004) The generation of dopaminergic neurons by human neural stem cells is enhanced by Bcl-XL, both in vitro and in vivo. J Neurosci 24:10786–10795
7. Sher F et al (2008) Differentiation of neural stem cells into oligodendrocytes: involvement of the polycomb group protein Ezh2. Stem Cells 26:2875–2883
8. Tate MC et al (2002) Fibronectin promotes survival and migration of primary neural stem cells transplanted into the traumatically injured mouse brain. Cell Transplant 11:283–295
9. Tate CC et al (2009) Laminin and fibronectin scaffolds enhance neural stem cell transplantation into the injured brain. J Tissue Eng Regen Med 3:208–217
10. Bez A et al (2003) Neurosphere and neurosphere-forming cells: morphological and ultrastructural characterization. Brain Res 993:18–29

Chapter 2

Neural Induction with a Dopaminergic Phenotype from Human Pluripotent Stem Cells Through a Feeder-Free Floating Aggregation Culture

Asuka Morizane, Daisuke Doi, and Jun Takahashi

Abstract

Pluripotent stem cells are promising potential sources for cell replacement therapy and are useful research tools for exploring disease mechanisms. Neural cells are one of the cell types that have been most efficiently differentiated through several established protocols. This chapter describes the feeder-free floating aggregation culture system for the induction of dopaminergic neurons. This method is simple and highly efficient for the production of dopaminergic neurons. It has several advantages for application in clinical usage in comparison to the other protocols using either feeder cells or Matrigel.

Key words Small molecules, Feeder-free, Dopamine neuron, Induced pluripotent stem cells, Embryonic stem cells

1 Introduction

The induction of the proper type of the cells from pluripotent stem cells, such as embryonic stem cells (ESCs) and induced pluripotent stem cells (iPSCs), requires control of both the direction of cell lineages and the stage of the differentiation. Research on neural cell induction from pluripotent stem cells has progressed in comparison to inducing other type of the cells. Various protocols induce pluripotent stem cells to differentiate into neurons through stages that mimic embryonic development (*see* **Note 1**). The pluripotent stem cells (inner cell mass of the embryo as a counterpart) follow the fate of the primitive ectoderm during the first stage of the neural differentiation. Both TGF beta/Activin/Nodal and BMP signal cascades inhibit conversion from the pluripotent state to a neuroectodermal linage at this stage. The inhibition of both cascades, the so-called "Dual-SMAD inhibition protocol," effectively promotes neural induction [1; *see* also **Note 2**]. This chapter describes the feeder-free floating aggregation culture system for neural induction (SFEBq) previously reported by Eiraku et al. [2] with

Renping Zhou and Lin Mei (eds.), *Neural Development: Methods and Protocols*, Methods in Molecular Biology, vol. 1018, DOI 10.1007/978-1-62703-444-9_2,

some modification for generating the dopaminergic phenotype. This system dissociates the human pluripotent stem cells to single cells that quickly re-aggregate in 96-well low cell-adhesion plates (*see* **Note 3**).

Several growth factors, such as SHH and FGF8, have been used for induction of mesencephalic dopaminergic neurons. We have modified the original SFEBq method to yield dopaminergic neurons more stably and effectively by adding Purmorphamine, FGF8, and FGF20.

Several supporting factors, such as Glial cell line-derived neurotrophic factor (GDNF), Brain-derived neurotrophic factor (BDNF), Ascorbic acid and dibutyryl cAMP (dbcAMP) are also needed to maintain those linage-specified neurons and continue maturation to mesencephalic dopaminergic neurons in vitro (*see* **Note 4**).

2 Materials

2.1 Equipment

1. Cell culture disposables: Cell culture dishes, 96-well U-bottom low-attachment plates (Lipidure Coat Plate A-U96; Thermo/NOF corporation), centrifuge tubes, pipettes, multichannel pipettes, pipette tips, chambers for multichannel pipettes, sterile filter units, 8-well chamber slide (BD FALCON).
2. Glass Pasteur pipettes, sterilized by autoclaving and dry heat monitored with indicator tape.
3. Benchtop laminar flow hood with an HEPA filter.
4. CO_2 incubator with CO_2, humidity and temperature control.
5. Cell culture centrifuge.
6. Glass hemocytometer.
7. Cryostat.

2.2 Human Pluripotent Cell Culture

1. Human embryonic stem cells (hESCs) or induced pluripotent stem cells (hiPSCs).
2. SNL feeders (mitomycin-C treated) see details in ref. 3.
3. SNL media: Dulbecco's modified Eagle medium (DMEM) with high glucose (Wako), 10 % fetal bovine serum (CCB).
4. Recombinant FGF-2 (Invitrogen) dissolved in 0.1 % BSA–PBS (−) to 25 μg/mL.
5. hESCs media: DMEM/F12 medium (Invitrogen) with 20 % Knockout serum replacement (KSR; Invitrogen), 1 mM L-Glutamine (SIGMA), 0.1 mM MEM nonessential amino acids (MEM NEAA; Invitrogen), and 100 μM 2-mercaptoethanol (Wako). Supplement the media with FGF-2 to a final concentration of 5 ng/mL at each time of changing media.

6. Sterile 1× PBS (−).
7. Gelatin-coated dishes with 0.1 % gelatin in PBS (−). Allow gelatin to incubate for at least 15 min prior to plating feeder cells.
8. CTK (Collagenase, Trypsin, and KSR solution; *see* the recipe below).
   (a) 5 mL 1 mg/mL Collagenase IV (Invitrogen).
   (b) 0.5 mL 0.1 M $CaCl_2$.
   (c) 5 mL 2.5 % (w/v) trypsin.
   (d) 10 mL Knockout Serum Replacement (KSR).
   (e) 30 mL distilled water.
   (f) Store up to 1 month at −20 °C.
   (g) Do not repeat freeze/thaw cycles.

2.3 Neural Differentiation

1. CTK.
2. Accumax (Innovative cell technologies).
3. GMEM media: Glasgow-MEM (Invitrogen) with 8 % KSR, 0.1 mM MEM NEAA, 100 μM Pyruvate (Sigma), and 100 μM 2-mercaptoethanol.
4. Neurobasal B27 media: Neurobasal (Invitrogen) with B27 supplement (Invitrogen) and 2 mM L-glutamine (SIGMA).
5. Small molecules and growth factors (*see* Table 1).

Table 1
Small molecules and growth factors used in this protocol

Small molecules/ growth factors	Company	Product #	Stock solution	Final concentration
A-83-01	Wako	018-22521	500 μM	500 nM
Ascorbic acid	SIGMA	A4034	200 mM	200 μM
BDNF	R&D	248-BD-025	100 μg/mL	20 ng/mL
dbcAMP	SIGMA	D0260	200 mM	400 μM
FGF20	Peprotech	100-41	1 μg/mL	1 ng/mL
FGF8	Peprotech	100-25	100 μg/mL	100 ng/mL
FGF2	Invitrogen	13256-029	25 μg/mL	20 ng/mL
GDNF	R&D	212-GD	10 μg/mL	10 ng/mL
LDN-193189	STEMGENT	04-0019	100 μM	100 nM
Purmorphamine	Calbiochem	540220	10 mM	2 μM
Y27632	Wako	253-00513	5 mM	10–30 μM

Reconstitute A-83-01, LDN-193189, and Purmorphamine with Dimethyl sulfoxide (DMSO). Reconstitute Ascorbic Acid, Dibutyryl cAMP (dbcAMP), and Y27632 with sterile water. Reconstitute BDNF, FGF20, FGF8, and GDNF with 0.1 % BSA–PBS (−).

2.4 Estimation of Differentiated Cells

1. Accutase (Innovative Cell Technologies).
2. Poly-L-ornithine (Sigma).
3. Laminin (R&D Systems).
4. 4 % Paraformaldehyde (PFA).
5. 30 % Sucrose.
6. O.C.T. compound (Sakura Finetek).

3 Methods

3.1 Preparation of Human Pluripotent Stem Cells

Maintain human pluripotent stem cells (hESCs and hiPSCs) on SNL feeders. Passage the cells every 4–6 days enzymatically with CTK. One of the advantages of using SNL feeders instead of MEF feeder is that the feeder cells are detached earlier than the pluripotent cells by CTK treatment. Incubate the culture with CTK for 2 min, and only the feeder cells are detached and the pluripotent cells remain attached to the dish. Rinsing and aspirating the floating feeder cells at that time point minimize the carryover of the old feeder cells. Add 10 μM of Y27632 to the daily maintenance media overnight 1 day before starting neural induction, to reduce the anoikis of pluripotent stem cells due to the single cell dissociation the following day.

3.2 Initiation of Neural Induction (Day 0)

1. Single cell dissociation. Rinse the culture with PBS (−) twice. Incubate the culture with CTK for 2 min at 37 °C to detach the feeders. Rinse and aspirate the feeders. Add Accumax (e.g., 1 mL for a 10 cm dish) for 5–10 min in 37 °C. Tap the dish, pipette for ten times with a 1,000 μL pipette to dissociate into single cells, and add 2 mL of the differentiation media (*see* below) to stop the enzymatic reaction. Centrifuge the tube for 3 min at $280 \times g$ at 4 °C. Resuspend the pellet with 1–2 mL of the differentiation media, and count the cells with a hemocytometer.
2. Media preparation. Supplement the GMEM media (the details are shown above) with 500 nM A-83-01 (Activin/Nodal inhibitor), 100 nM LDN-193189 (BMP inhibitor) and 30 μM Y27632 (ROCK inhibitor).
3. Plating cells. Suspend the dissociated cells at the concentration of 9,000 cells/150 μL/well (6×10^4 cells/mL). Transfer the cell suspension to a chamber for a multichannel pipette. Distribute the suspension to 96-well U-bottom low-attachment plates with a multichannel pipette. Put the plates into a CO_2 incubator at 37 °C.

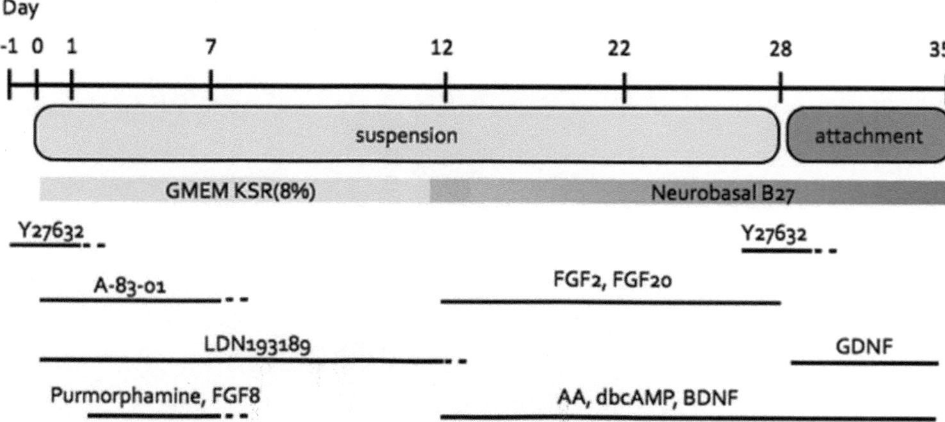

Fig. 1 The time course for generating dopaminergic neurons by feeder free floating-aggregation culture (SFEBq)

3.3 Media Change with the 96-Well Plate

A small aggregate is usually identified with a microscope in each well in a few days. Perform the first media change at day 1. Aspirate 65 μL from 150 μL per well and add 75 μL of new media, then about half of the media is replaced. Add a double concentration of the factors to the new media preparation (for example Purmorphamine at day 1 needs 4 μM for preparing media to yield 2 μM concentration in final) to account for the dilution during media change.

3.4 Inducing Differentiation of Dopaminergic Neurons

Add 2 μM Purmorphamine as a ventralizing factor and 100 ng/mL FGF8 for specification of the rostral–caudal axis from day 1 as indicated in Fig. 1 (*see* also **Note 5**). Switch the basal media from GMEM media to Neurobasal B27 media on day12 to promote neural differentiation. Dissociate the cell aggregation into small clumps or single cells on day 28 for immunocytochemical analyses.

3.5 Making Sliced Sections of Cell Aggregates for Immunocytochemical Analyses of the Differentiated Neurons

Sliced sections of the cell aggregates can be prepared after day 7 (Fig. 2 left panel)

1. Collect cell aggregates from 96-well plates with 1 mL pipette.
2. Rinse with PBS (−) twice.
3. Immerse the aggregates with 4 % PFA for 1 h at 4 °C.
4. Rinse the aggregates twice with PBS (−).
5. Aspirate the PBS (−).
6. Collect the semi-dried aggregates and freeze them with O.C.T. compound.
7. Slice the aggregates into 10 μM sections.
8. Put the sections onto glass slides and proceed to immunostaining.

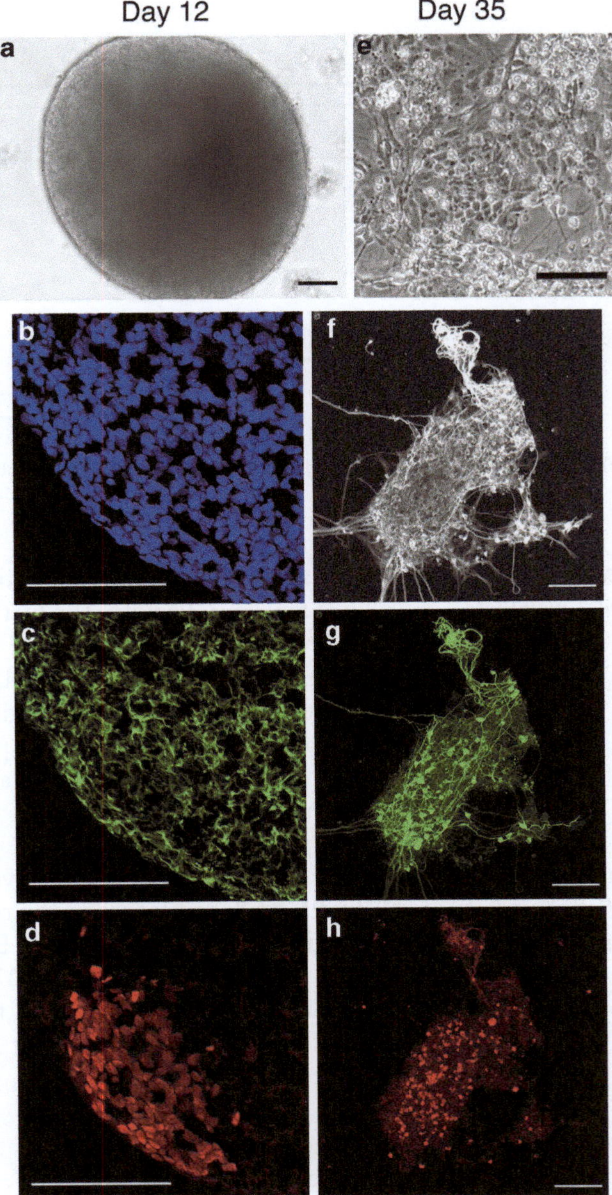

Fig. 2 Phase contrast images (**a** and **e**) and immunocytochemical analyses of the differentiated neurons. The aggregates grow by day 12 (**a**) and immunostaining of the sliced aggregates shows the neural identity of the cells; DAPI (**b**), Nestin (**c**), and Pax6 (**d**) as neuroectodermal markers. The aggregates were dissociated on day 28 and the attachment culture started. The mature neurons showed a dopaminergic neural phenotype on day 35; phase contrast (**e**), Tubβ III (**f**), Tyrosine hydroxylase (**g**), Foxa2 (**h**). Scale bars: 100 μM

3.6 Attachment Culture of the Dissociated Cells for Immunocytochemical Analyses of the Differentiated Neurons at the Late Phase (Fig. 2 Right Panel)

1. Incubate the aggregates with media containing 10 μM Y27632 overnight.
2. Collect the aggregates in a conical tube and rinse with PBS (−).
3. Incubate the aggregates with 1 mL of Accutase for 10 min at 37 °C.
4. Dissociate the aggregates with 1 mL pipette.
5. Spin down and re-suspend the pellet in media containing 10 μM Y27632.
6. Plate the cells at the density of 200,000 cells/cm^2 in an eight-well chamber slide pre-coated with poly-L-ornithine (50 μg/mL) and Laminin (5 μg/mL).
7. Perform half media change every 3–4 days.
8. After culturing for 1 week fix the slide culture with 4 % PFA and proceeds to immunostaining.

4 Notes

1. Pluripotent stem cells have often been cultured on stromal cell lines such as PA6 or MS5 to generate neurons [4, 5]. The protocol starting with formation of embryoid bodies (EBs) has been another option [6]. Such methods were first reported with mouse embryonic stem cells. Those protocols have been modified since the establishment of human ESCs and iPSCs [1, 7, 8].
2. The original protocols used Noggin as a BMP inhibitor and small molecule SB431542 (SIGMA) as a TGF beta/Activin/Nodal inhibitor. Small molecules, such as Dorsomorphin (Sigma) or LDN-193189 (Stemgent) can substitute for Noggin, as Noggin is expensive. In addition to the financial advantage, Dorsomorphin or LDN-193189 has a more stable effect than Noggin in the current system. In our hands, a combination of DMEM, SB431542 and Dorsomorphin [8] were used before but that has been changed to the combination of GMEM, A-83-01 and LDN-193189 that yields more stable neural induction and a higher percentage of dopaminergic progenitors.
3. The advantages of the SFEBq system are the efficiency and the high reproducibility. This system has also a clinical advantage because it does not need xeno-derived feeder or matrix. Flow cytometry shows more than 90 % of the cells are PSA-NCAM positive neural cells on differentiation day 14. Usually one aggregate appears in each of the culture wells. The immunostaining of the sliced aggregates shows that there is few difference between the aggregates in their morphology and the cell content.

4. Dopaminergic neurons are found in the forebrain or olfactory bulb in addition to the mesencephalon. Several transcriptional factors such as Lmx1a and Foxa2 are thought to be important for the differentiation of the mesencephalic dopaminergic phenotype in normal development [9, 10]. Such midbrain specific transcriptional factors are also keys to induce authentic dopaminergic neurons of the mesencephalic type from induced pluripotent stem cells. Several recent papers have reported that some small molecules such as CHIR99021 and PD0325901 effectively induce the phenotype of the mesencephalic floor plate during the early stage of neural induction [11, 12]. Those compounds should be evaluated in the SFEBq protocol. These inductive and supportive factors are effective but the cell population in the culture is not 100 % homogeneous. Safe cell therapy and reliable in vitro disease models are two major goals of pluripotent stem cell research. Therefore, additional method for purification will be needed. Cell sorting with antibodies for cell surface markers specific for dopaminergic neurons is one of the promising strategies.
5. The aggregation yields neural cells with mostly the forebrain phenotype without additional factors, though the detailed ratio depends on the propensity of the original pluripotent stem cell lines. Therefore, inducing the proper phenotype of the neurons requires the addition of appropriate factors during the favorable time window.

Acknowledgments

The authors would like to thank Dr. Sasai Y. for technical advice. This work was supported by JSPS KAKENHI 21591837 (Grant-in-Aid for Scientific Research).

References

1. Chambers SM, Fasano CA, Papapetrou EP et al (2009) Highly efficient neural conversion of human ES and iPS cells by dual inhibition of SMAD signaling. Nat Biotechnol 27: 275–280
2. Eiraku M, Watanabe K, Matsuo-Takasaki M et al (2008) Self-organized formation of polarized cortical tissues from ESCs and its active manipulation by extrinsic signals. Cell Stem Cell 3:519–532
3. Ohnuki M, Takahashi K, Yamanaka S (2009) Generation and characterization of human induced pluripotent stem cells. Curr Protoc Stem Cell Biol Chapter 4:Unit 4A.2
4. Kawasaki H, Mizuseki K, Nishikawa S et al (2000) Induction of midbrain dopaminergic neurons from ES cells by stromal cell-derived inducing activity. Neuron 28:31–40
5. Perrier AL, Tabar V, Barberi T et al (2004) Derivation of midbrain dopamine neurons from human embryonic stem cells. Proc Natl Acad Sci USA 101:12543–12548
6. Lee SH, Lumelsky N, Studer L et al (2000) Efficient generation of midbrain and hindbrain neurons from mouse embryonic stem cells. Nat Biotechnol 18:675–679
7. Watanabe K, Kamiya D, Nishiyama A et al (2005) Directed differentiation of telencephalic

precursors from embryonic stem cells. Nat Neurosci 8:288–296
8. Morizane A, Doi D, Kikuchi T et al (2011) Small-molecule inhibitors of bone morphogenic protein and activin/nodal signals promote highly efficient neural induction from human pluripotent stem cells. J Neurosci Res 89:117–126
9. Andersson E, Tryggvason U, Deng Q et al (2006) Identification of intrinsic determinants of midbrain dopamine neurons. Cell 124:393–405
10. Kittappa R, Chang WW, Awatramani RB et al (2007) The foxa2 gene controls the birth and spontaneous degeneration of dopamine neurons in old age. PLoS Biol 5:e325
11. Kriks S, Shim JW, Piao J et al (2011) Dopamine neurons derived from human ES cells efficiently engraft in animal models of Parkinson's disease. Nature 480:547–551
12. Jaeger I, Arber C, Risner-Janiczek JR et al (2011) Temporally controlled modulation of FGF/ERK signaling directs midbrain dopaminergic neural progenitor fate in mouse and human pluripotent stem cells. Development 138:4363–4374

Chapter 3

Nucleoside Analog Labeling of Neural Stem Cells and Their Progeny

Erik Michael DeBoer and Mladen-Roko Rasin

Abstract

Nucleoside analog pulse labeling is an important technique which can assess the birthdate, cell cycle maintenance, or cycling rates of cells during development. This method has evolved over several decades of use and is now applied to a multitude of tissue subtypes and systems. The methodology in this chapter covers the classic uses for analog pulse labeling as well as their use in conjunction with the newly characterized technique of in utero electroporation (IUE).

Key words Birthdating, Cell cycle, BrdU, CldU, IdU, EdU, Thymidine, Uridine, Pulse-label, In utero electroporation

1 Introduction

Some 35 years ago, the introduction of radioactive triturated thymidine (^{3}H-T) as a thymidine nucleoside analog allowed researchers to determine the birthdate of cells [1]. ^{3}H-T can be pulsed into a developing system and will be incorporated into cells which are synthesizing a new copy of DNA during the S phase of the cell cycle at that time. Importantly, the progeny of the ^{3}H-T-labeled cells also carry the isotope and thus can later be examined histologically through radiographic imaging of the resulting tissue.

Later, this technique was widely used in seminal studies investigating the developing central nervous system (CNS), including those which demonstrated the sequential spatiotemporal generation of distinct subpopulations of neocortical projection neurons. The neuronal progeny which take place in deeper neocortical layers (V–VI) are born from earlier divisions of neural stem cells (NSCs), whereas neurons that reside in the superficial layers (II–IV) are later-born neurons [2–6]. These and others' data demonstrated a fundamental mechanism in the developing CNS; the timing of stem

Renping Zhou and Lin Mei (eds.), *Neural Development: Methods and Protocols*, Methods in Molecular Biology, vol. 1018, DOI 10.1007/978-1-62703-444-9_3, © Springer Science+Business Media, LLC 2013

cell divisions dictates the fate and final placement of the postmitotic progeny [7].

More recently, nonradioactive thymidine analogs were introduced, such as the halogenated deoxyuridine nucleoside analogs: bromodeoxyuridine (BrdU), iododeoxyuridine (IdU), and chlorodeoxyuridine (CldU) [8–10]. In addition, and most recently, 5-ethynyl-2′-deoxyuridine (EdU) (also referred to as "Click-it") technology has been developed, using an alkyne group to foster incorporation into DNA [11]. Similar to ^{3}H-T, these small molecules are incorporated into the DNA of dividing cells during the S phase of the cell cycle. BrdU, IdU, CldU, or EdU not only can be pulsed into developing CNS in a similar manner to ^{3}H-T but can also be applied through drinking water for continuous labeling of cell divisions [12]. Importantly, early validation studies demonstrated that incorporation of halogenated deoxyuridines can be used to distinguish proliferating cell populations in a similar manner to the earlier ^{3}H-T studies [5, 8]. Given the unique chemical structure of the halogenated analogs, some can be subsequently labeled using distinct antibodies. Thus, an advantage of halogenated deoxyuridine labeling is that two analogs can be used to track progeny generated at different time points within the same developing tissue/region. A recent example of co-labeling using CldU in conjunction with IdU has allowed researchers to track sequential birth and final placement of different subpopulations of neocortical neurons generated at discrete time points within normal and mutated developing neocortices [9]. Of note, care should be taken when combining halogenated analog studies with EdU, as the chemistry involved in each and the resulting sensitivity of labeling is somewhat different [11].

In addition to their use in determining the birthdate and final position of postmitotic neuronal subpopulations, deoxyuridine nucleoside analogs can be used to evaluate cell cycle dynamics, exit, or reentry of proliferating stem cell populations in vivo [9, 13]. This is performed by pulsing two thymidine analogs closer together in time such that stem cell populations can incorporate both analogs as they undergo subsequent S phases. Even as they have some notable limitations such as dilution with each cycle [10], halogenated deoxyuridines can be employed in this method to uncover the proportion of cells reentering or exiting the cell cycle during distinct neurogenic phases of normal, mutated or lesioned CNS.

1.1 Using Nucleoside Analogs to Track the Differentiation of Cells That Are Modulated Autonomously via In Utero Electroporation

The recently developed technique, in utero electroporation (IUE), allows an investigator to modulate the gene expression of NSCs autonomously in the developing CNS in vivo and subsequently to track the changes in the progeny of these cells throughout development [14–16]. For example, previous birthdating studies of developing neocortices using nucleoside analogs have elucidated the time points at which distinct subpopulations of projection neurons are born [2–6, 8, 17, 18]. Therefore, the functional gene expression levels of each subpopulation of sequentially generated neocortical projection neurons can be manipulated by performing IUE at the time of their birth.

IUE involves first injecting a purified plasmid into the lateral ventricle(s) of a developing mouse pup, then transfecting the stem cells lining the ventricles with a mild electrical pulse. Usually, these plasmids encode an overexpression vector or a short hairpin RNA (shRNA) specific to a gene of interest and separately a control overexpression or a control, nonspecific scrambled or point mutation shRNA as described [15, 16]. Usually, a plasmid encoding a reporter fluorescent protein, such as eGFP, is co-transfected to allow for later visualization of transfected cells upon tissue processing [14–16]; for more details, *see* Chapter 15 of this book. After the IUE is performed, all progeny of the transfected NSCs will express the gene of interest/targeting shRNA and reporter fluorescent protein. The transfected progeny can then be pulse-labeled using halogenated deoxyuridine nucleoside analogs, such as CldU and IdU, at separate time points, as described below. Using this approach, an investigator can evaluate the cell autonomous effect of altered gene expression on the resulting birthdates, differentiation, or cycling rates of cells in the neocortex [9, 19].

2 Materials

2.1 Nucleoside Pulse Injection Components

1. Nucleoside analogs: 5-bromo-2′-deoxyuridine, 5-chloro-2′-deoxyuridine, or 5-iodo-2′-deoxyuridine; store at −20 °C (B5002, C6891, and I7125, respectively; Sigma Chemical Company).
2. Sterile saline (0.85–0.9 % NaCl in H_2O).
3. Sodium hydroxide: 2 N.
4. Sterile filtration device: 0.22 μM.
5. pH measurement meter.

2.2 Tissue Fixation, Embedding, and Processing Components

1. 1× PBS.
2. Fixation buffer: 4 % paraformaldehyde (PFA) in 1× phosphate-buffered saline (PBS), pH 7.4 (*see* **Note 1**).
3. Agarose gel powder.
4. 1× PBS.
5. Vibratome.
6. Superglue.
7. 30 % sucrose in 1× PBS.

2.3 Immunohistochemistry Components

1. Sectioned, nucleoside-labeled tissue sections.
2. 1× PBS.
3. 10× PBS.

4. Fraction 5 bovine serum albumin (BSA).
5. Hydrochloric acid: 2 N.
6. Normal donkey serum (Jackson ImmunoResearch 017-000-121).
7. Lysine (Sigma L5501).
8. Glycine (BDH BDH5156 or Sigma 50046).
9. Blocking solution with +0.4 % Triton X-100 and without added Triton (*see* **Note 2**).
10. Scintillation vials, glass, 20 mL.
11. Biorockers, at room temperature (RT) and 4 °C.
12. Primary antibodies: monoclonal mouse anti-BrdU (labels BrdU or CldU) (Accurate Chemical OBT0030G) and monoclonal rat anti-BrdU (labels BrdU or IdU) (Becton Dickinson 347580) or desired antibodies for co-staining pulse-labeled cells anti-GFP (Aves GFP1020) (*see* **Note 3**).
13. Secondary antibodies: fluorescent donkey anti-mouse, chicken, and rat Cy2, Cy3, or Cy5 conjugated (Jackson ImmunoResearch Laboratories, Inc.).
14. Dapi: diluted in 1× PBS.
15. Glass slides and cover slips.
16. VECTASHIELD (H-1000 Vector Laboratories).
17. Clear nail polish.

2.4 Imaging and Analysis

1. Confocal microscope.
2. ImageJ (NIH), Neurolucida (MicroBrightField, Inc.), or similar image analysis software.

3 Methods

3.1 Setting Up Timed Pregnancy Breeding for Later Pulse Labeling

1. Use a 12 h dark–light cycle for breeding mice (*see* **Note 4**).
2. At the beginning of the dark cycle, place 1–2 female mice of the desired genetic background or experimental condition into the cage of the appropriate male mouse.
3. Remove the female mice within 4 h of the beginning of the light cycle and mark them, making sure that they are placed in a cage without other male mice.
4. Check female mice for the presence of a sperm plug. If a sperm plug is found, the mouse should be considered to be pregnant and to have embryos at the age of embryonic day 0.5 (E0.5). If no sperm plug is found, repeat **steps 2–4** (*see* **Note 5**).

3.2 Nucleoside Solution Preparation

1. Dissolve 5 mg/mL nucleoside analog in sterile saline by weighing 100 mg nucleoside analog and adding 20 mL saline in a conical tube or similar (*see* **Note 6**) [20].
2. Vortex or rock the powder and saline protecting from light as much as possible until powder is in solution (*see* **Note** 7).
3. Sterile-filter the prepared nucleoside solution using a 0.22 μM syringe filter or similar (*see* **Notes** 7 and **8**).

3.3 Injecting Nucleoside Pulses for Birthdating Analysis

1. Weigh the animal for pulse injection and record weight.
2. Prepare a syringe with a 26.5 gauge needle such that 10 μL/(g of animal weight) of the 5 mg/mL nucleoside solution can be injected.
3. Draw up a sufficient volume of nucleoside solution and ensure that no bubbles are present in the syringe.
4. At the appropriate pulse time point, scruff-hold the animal and inject the nucleoside solution intraperitoneally (*see* **Notes 9** and **10**).
5. Gently place the animal back into its home cage.
6. A second pulse can be applied to label a second population of cells. If desired, follow **steps 1–4** at a second desired time point using a distinct thymidine analog (*see* **Note 11**).
7. Allow the pups to develop until the desired experimental time point, then euthanize according to institutional guidelines, and proceed to tissue processing (*see* **Note 12**).

3.4 Injecting Nucleoside Pulses for Cell Cycle Analysis

1. Follow procedure from Subheading 3.3 numbers 1–5.
2. To label cells in the S phase, pulse a distinct, second nucleoside analog at desired time corresponding to the experimental design (e.g., to be sure that neocortical NSCs underwent at least one full cell cycle, provide ldU 23 h after the first analog was pulsed). This time point can be adjusted as per the investigator's experimental question and design, as cell cycling rates vary across developmental time points and throughout tissue types (*see* **Note 13**).
3. 1 h after the second nucleoside pulse-labeling cells currently in S phase (e.g., 24 h after the first pulse), euthanize the pregnant animal according to institutional guidelines (*see* **Note 13**).
4. Dissect the uterus from the pregnant animal and remove the pups, placing them in an appropriate bath (e.g., cold 1× PBS or HBSS) (*see* **Note 14**).

3.5 Processing of Nucleoside-Labeled Tissue

1. In a bath of cold 1× PBS or HBSS supplemented with 0.5 % d-glucose and 25 mM HEPES, carefully dissect the brains from the experimental animals (*see* **Note 15**).
2. Place brains in 4 % paraformaldehyde (PFA) solution in 20 mL scintillation vials (*see* **Note 16**).

3. Rock the tissue at 100 RPM at RT on a biorocker ensuring that the tissue is moving freely within the solution. Change the PFA solution three times, 10 min apart.
4. Place the scintillation vials on a biorocker at 4 °C for up to 24 h (*see* **Note 17**).
5. Remove tissue from PFA and wash in 1× PBS.
6. Prepare 3 % agarose by weighing 3 g agarose and placing in 100 mL of 1× PBS.
7. Heat the agarose/PBS in a glass container in a microwave until completely homogenous (*see* **Note 18**).
8. Pour the heated 3 % agarose into a 10 cm plastic dish.
9. Place a thermometer into the agarose and allow it to cool to between 37 and 42 °C.
10. Scrape the surface of the agarose with a blade or flat tool to move the surface-cooled agarose away and to the side of the dish (*see* **Note 19**). Remove the thermometer.
11. Remove desired brains from 1× PBS and carefully blot excess moisture from them using a Kimwipe.
12. Place the brains into the agarose, ventral side down (*see* **Note 20**).
13. Allow the agarose to cool to room temperature; it will become opaque.
14. Cut the brains out of the agarose using a blade. Trim agarose block around each brain such that as little excess agarose is left around the brain as possible while maintaining the integrity of the agarose block around the brain. Trim the angle of the block near the brainstem such that the brain can be stood up and is straight both dorsal to ventral and medal to lateral. Superglue the agarose-embedded brain brainstem down to the vibratome cutting boat such that the cortex faces away from the blade mount. Allow to dry for several minutes (*see* **Note 21**).
15. Place a new blade in the vibratome mount and clamp securely.
16. Fill the boat with 1× PBS, submerging the agarose-blocked brain.
17. Adjust the vibratome to cut 70 μM sections, and begin cutting.
18. Ensure that the vibratome is cutting smoothly and that the brain is secure in the agarose block as the vibratome moves through the tissue. If the tissue is not cutting correctly, adjust speed and vibration rate or stop the procedure, cut away the block, and repeat the embedding process.

19. Allow the vibratome to continue cutting until all desired sections are obtained. Using a paintbrush or glass hook, gently move the cut sections from the boat and place in a scintillation vial containing 30 % sucrose in 1× PBS and store at 4 °C (*see* **Note 22**).

3.6 Immunohistochemical Detection of Nucleoside Labels

1. Select desired sections and place in a new 20 mL scintillation vial with roughly 10 mL 1× PBS. Wash the tissue on a biorocker at RT. Perform three washes; 5 min each (*see* **Notes 23** and **24**).
2. Remove the final wash of 1× PBS and place sections in 1 mL 2 N HCL in a fume hood with no motion at RT for 20 min for antigen retrieval (*see* **Note 25**). Scintillation vials should stand upright for this incubation.
3. Remove the 2 N HCL and place sections in 1× PBS. Wash three times for 5 min at RT on a biorocker.
4. Remove 1× PBS as before and place 1 mL blocking solution with 0.4 % Triton X-100 added in each vial. Incubate sections in blocking at RT for 1 h on a biorocker. Scintillation vials should again be upright for this incubation (*see* **Note 26**).
5. After the blocking incubation period, remove the blocking solution from the sections and discard. Replace the blocking with 250 μL to 1 mL primary antibody solutions. These are prepared by pipetting the appropriate amount of each primary antibody(s) into blocking solution with 0.4 % Triton X-100 added. Use enough solution so that sections can move freely within the antibody dilution. For sections from IUE, also be sure to dilute anti-GFP antibody to amplify GFP signal, if needed.
6. Place the scintillation vials at 4 °C on a biorocker and incubate for 48 h (*see* **Note 26**).
7. After the primary antibody incubation period, remove the primary solution from the sections and save for future use or discard. Wash the sections four times for 5 min each in 1× PBS at RT on a biorocker so that sections are freely moving.
8. Dilute required secondary antibodies in blocking solution 1:250. When using sections from IUE experiments, use cy2 to label GFP.
9. Replace final wash of 1× PBS with secondary antibody and incubate on a biorocker at RT for 2 h protecting from light (*see* **Note 26**).
10. Wash sections three times for 5 min at RT on a biorocker in 1× PBS protecting from light so that sections are freely moving.
11. Apply Dapi solution in 1 mL 1× PBS. Incubate at RT on a biorocker for 10 min so that sections are freely moving (*see* **Notes 26** and **27**).

12. Remove and discard Dapi solution. Wash the sections three times for 5 min each in 1× PBS at RT on a biorocker so that sections are freely moving.
13. Mount sections on glass slides (*see* **Note 28**).
14. Use a Kimwipe to blot excess 1× PBS from the slide and allow drying in a dark area checking every 10 min until excess PBS has evaporated, but before sections are completely dried out.
15. Drop a conservative amount of Vectashield onto semi-dried sections and carefully cover slip the slides.
16. Blot excess Vectashield from the sides of the slide using a Kimwipe and apply nail polish around the parameter of the slide to seal it. Allow nail polish to dry for 20 min in a dark place.
17. Store slides at 4 °C for up to 3 months.

3.7 Imaging and Analysis of Pulse-Labeled Sections

3.7.1 Birthdating Studies

1. Using a confocal microscope, focus on the tissue region of interest and take several images containing the channels from all nucleoside labels as well as Dapi (*see* **Note 29**).
2. Export images to a data analysis software such as ImageJ or Neurolucida. Ensure that all channels can be visualized simultaneously in order to confirm that nucleoside labels are Dapi positive; this ensures that analyzed labels are cells and not possible background artifacts.
3. Subdivide the imaged region into several equally sized bins. Here, bins will be used for a spatial analysis of the birthdated cells (*see* **Note 30**).
4. For each of the pulsed thymidine analogs, count the positively labeled (fluorescent) cells in all bins (*see* Fig. 1).
5. For each of the pulsed thymidine analogs, now divide the number of positively labeled cells in each bin by the summed total in all bins (*see* **Notes 31** and **32**).
6. Repeat the above steps between experimental conditions and perform appropriate statistical analyses to determine the differences in the proportion of each label in each bin.
7. To determine a change in proportion of total labeled cells at two distinct time points, first sum the total number of labeled cells in all bins from the first pulse (e.g., with CldU). Next, sum the total number of labeled cells in all bins from the second pulse (e.g., with IdU). Add the total number of counted cells from both labels together. Next, divide the total number counted for each label by the total number of cells counted for both labels. This will yield a proportion of all labeled cells generated at each time point. When repeated between two experimental conditions, this analysis can demonstrate a neurogenic shift or a difference in the generation of subpopulations of neurons at a specific time point when compared to another.

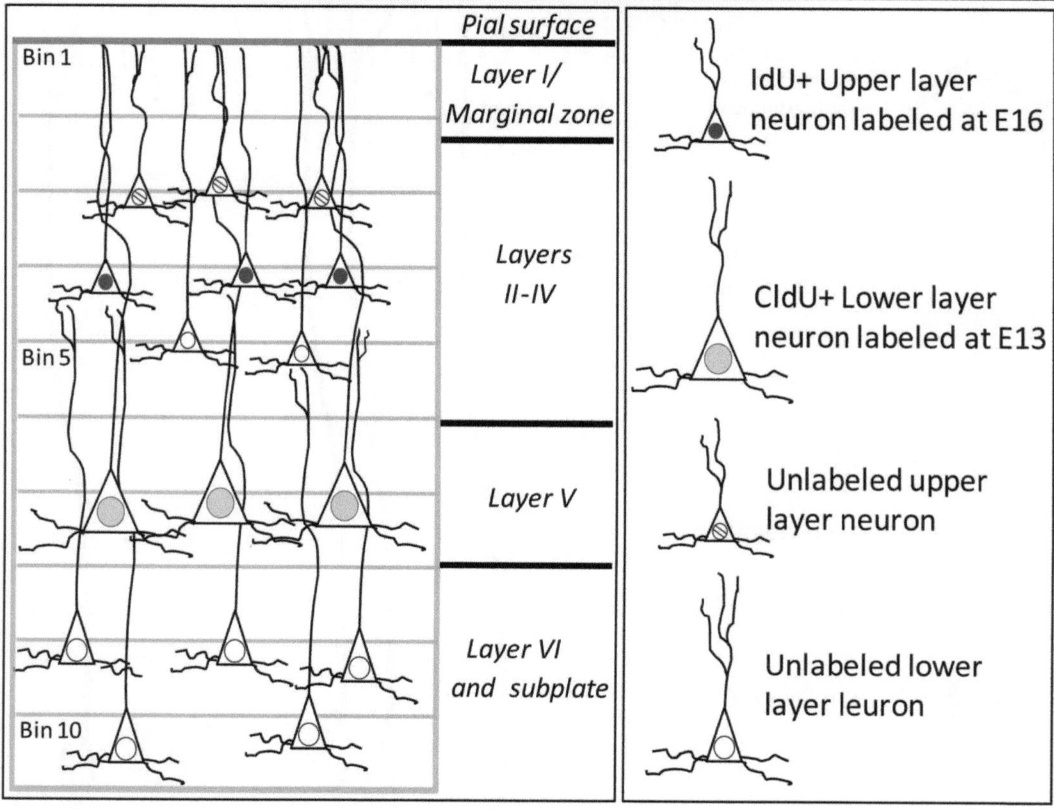

Fig. 1 A schematic example of embryonic birthdating analysis in a postnatal neocortex. Note the presence of postmitotic neuronal progeny in neocortical layers II-VI. CldU pulsed at E13.5 preferentially labels Layer V neurons (light gray nucleus). IdU pulsed at E16.5, however, preferentially labels neurons in layers II-IV (dark gray nucleus). Note that cells which were not born at E13.5 or E16.5 are unlabeled. A grid of 10 bins is overlayed on the cortical plate for distribution analysis as described

3.7.2 Cell Cycle Studies

1. Repeat **steps 1** and **2** from Subheading 3.7.1 (*see* **Note 33**).
2. Count the total number of labeled cells from the first time point and sum (Label 1).
3. Count the total number of labeled cells from the second time point and sum (Label 2).
4. Count the number of co-labeled cells (those that are both Label 1 and Label 2 positive).
5. Divide the number of co-labeled cells by the total number positive for Label 1. This will yield the proportion of cells labeled at the first time point which remained in the cell cycle.
6. Subtract the number of co-labeled cells from the number of cells positive for Label 1. Divide this number by the number of

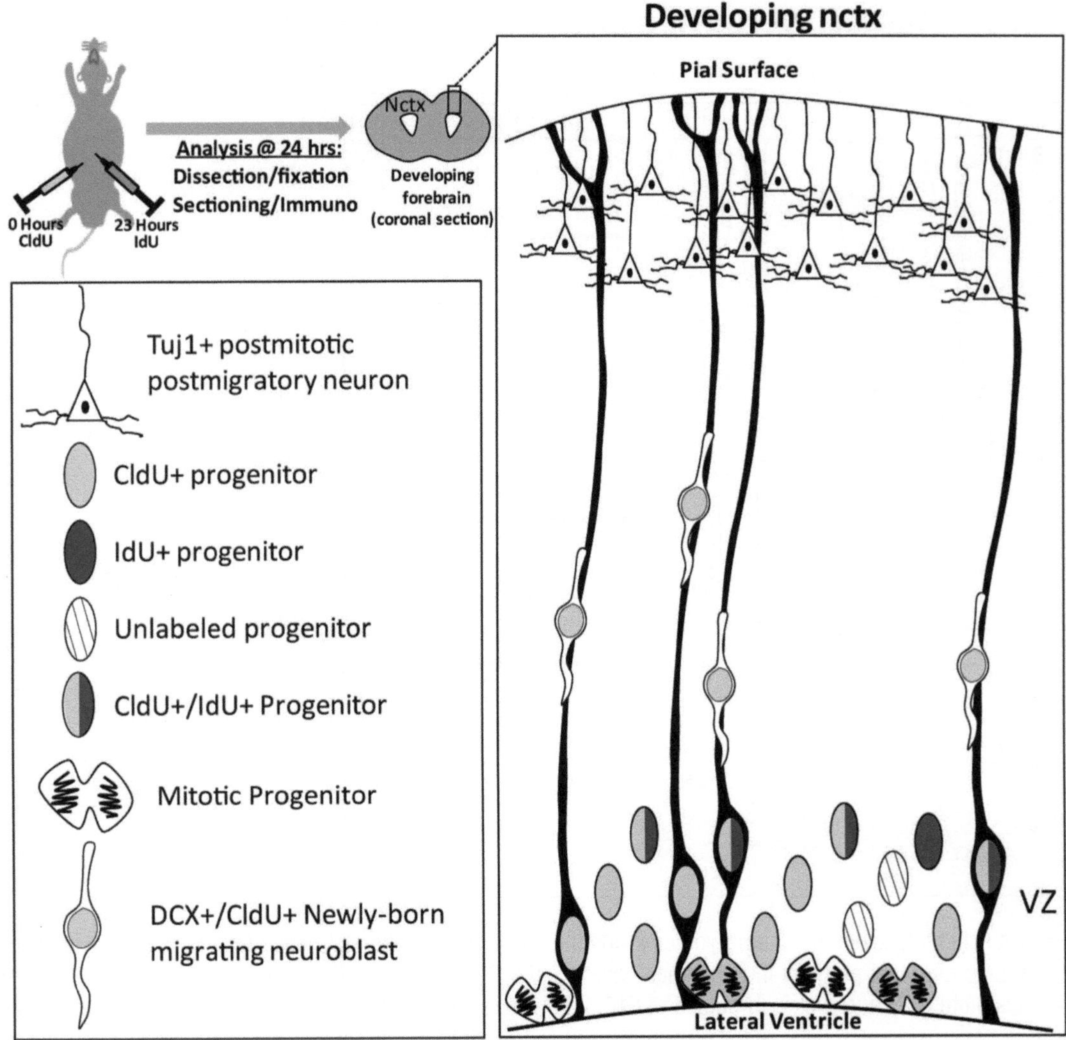

Fig. 2 A schematic for analog pulse injections. A pregnant dam injected with CldU at starting time point 0 (*light gray*), 24 hours before embryo collection. Subsequently the dam is injected with IdU 23 hours after the CldU injection and 1 hour before collection. Processed tissue immunostained for CldU and IdU is imaged with a confocal microscope. This example depicts imaging the neocortex (*box*) for cell cycle analysis. CldU+/IdU-negative (*light gray nucleus*) cells have exited the cell cycle, and are beginning to migrate and differentiate. CldU-negative/IdU+ (*dark gray nucleus*) are cycling at 24 hours but were not in S-phase at start time 0, and are progenitor cells. CldU+/IdU+ (*half light/dark gray nucleus*) cells have re-entered the cell cycle and are progenitor stem cells. Note the presence of some CldU+/DCX+ migrating neuroblasts with leading (thicker) and trailing (thinner) processes. These neuroblasts migrate along the representative black radial glia processes attached at pial surface. Finally, note the presence of mitotic progenitors at the ventricular surface and postmitotic, postmigratory Tuj1+ pyramidal neurons in the neocortex (Nctx). VZ = ventricular zone

Label 1-positive cells. This will yield the proportion of cells labeled at the first time point which have exited the cell cycle (for other possible explanations, *see* **Note 34** and Fig. 2).

3.7.3 Pulse-Label Analysis of In Utero Electroporations

In Utero Electroporation Birthdating Analysis

1. Perform in utero electroporation as described in Chapter 15 of this book.
2. After IUE, inject distinct thymidine analogs at two subsequent time points (e.g., IUE at E13.5, CldU at E14.5, and IdU at E16.5). Follow the above procedures for birthdating.
3. Follow the above procedures for tissue processing, immunohistochemistry, ensuring that an anti-GFP antibody is also used. Follow the above procedures for imaging.
4. Manually count all the GFP+ cells within each bin (*GFP total*).
5. Divide the number of GFP+ cells within each bin by *GFP total*. This will yield the overall distribution of transfected cells.
6. Count the number of GFP+ cells that colocalize with the first analog label within each bin (*Label 1+/GFP+*).
7. Count the number of GFP+ cells that colocalize with the second analog label with each bin (*Label 2+/GFP+*).
8. Divide the number of *Label 1+/GFP+* cells in each bin by the total number of *Label 1+/GFP+* cells counted in all bins. This will yield the distribution of transfected cells within the cortex labeled at the first time point.
9. Divide the number of *Label 2+/GFP+* cells in each bin by the total number of *Label 2+/GFP+* cells counted in all bins. This will yield the distribution of transfected cells within the cortex labeled at the second time point.
10. Repeat the above steps between experimental conditions and perform appropriate statistical analyses to determine the differences in the proportion of each label in each bin.

In Utero Electroporation Cell Cycle Analysis

1. Repeat **steps 1** and **2** from Subheading 3.7.1 (*see* **Note 33**).
2. Count the total number of GFP+ cells that colocalize with the label injected at the first time point (*Label 1+/GFP+*).
3. Count the total number of GFP+ cells that colocalize with the label injected at the second time point (*Label 2+/GFP+*).
4. Count the total number of GFP+ cells that colocalize with both labels (*co-label/GFP+*).
5. Similar to Subheading 3.7.2, begin with the number of cells which are GFP+ as well as positive for *both* labels (*co-label/GFP+*). Divide this number by the *Label 1+/GFP+* total. This will yield the proportion of transfected cells which remained in the cell cycle for the experimental period. Next, subtract the number of *co-label/GFP+* cells from the number of *Label 1+/GFP+*. Divide this number by the total number of *Label 1+/GFP+*. This will yield the proportion of electroporated cells which exited the cell cycle (*see* **Note 34**).

4 Notes

1. Paraformaldehyde is considered to be usable at 4 °C for 1 week, after which it should be disposed of properly and remade for fresh tissue processing.
2. To make blocking:
   - Dissolve 2 g BSA, 0.2 g lysine, and 0.2 g glycine in 100 mL dH_2O.
   - Dissolve normal donkey serum separately in 10 mL dH_2O.
   - Combine both solutions and adjust volume to 180 mL with dH_2O.
   - Add 20 mL 10× PBS and adjust pH to 7.4.
   - Aliquot into 5 or 10 mL and store at −20 °C.
   - Final blocking solution is pH 7.4 in 1× PBS.
3. Care must be taken when selecting anti-BrdU, CldU, and IdU antibodies for co-labeling experiments. Many antibodies will bind to multiple nucleoside analogs. Similarly, and for example, an antibody which labels IdU and not CldU may be sold as "anti-BrdU." Successful use of antibodies for co-labeling CldU and IdU in developing CNS has been recently described [9, 10, 20].
4. Many groups including ours use 6:00 PM to 6:00 AM as the daily 12 h dark cycle.
5. The presence of a sperm plug does not always indicate pregnancy. Pregnancy can be confirmed by tracking weight gain of the female mouse [21]. Stated ages consider that conception occurred 6 h into a 12 h dark cycle. With the light cycle described in **Note 4**, E0.5 would be at 12:00 PM the following day.
6. Smaller volumes can be made at 5 mg/mL, but prepared solutions can be aliquoted into useful volumes and are stable for longer than 1 year at −20 °C with the exception of IdU (*see* **Note 7**).
7. IdU is more difficult to dissolve than CldU or BrdU. We have found it useful to add 500 mL 2N NaOH to 100 mg IdU and vortex to dissolve, then adjust final volume to 20 mL with saline. Once dissolved, the pH of IdU will need to be adjusted to 7.0 in order to be injected (*see* **Note 8**).
8. IdU solution can be stored before being pH adjusted at 4 °C for 3 months. Once pH adjusted, however, IdU precipitates out of solution within several days. Therefore, it is most efficient to pH adjust only the volume you wish to use, then store and label the non-adjusted unused quantity at 4 °C. We have therefore found it useful to store in this way, then pH adjust, and filter before use.

9. Timing is paramount with pulse labeling, and one should label as consistently as possible between animals. For example, a pulse of E13.5 must occur at 12:00 PM on the 13th day after separation from the breeding male.
10. It is advantageous to inject the animal as rostral as possible in the abdominal cavity in order to avoid puncturing the uterus, especially in late pregnancies. However, also be sure not to inject into the thoracic cavity.
11. When temporally separated birthdating of distinct cellular subpopulations is studied, it is useful to always apply analogs in the same order. For example, if CldU and IdU are used often in conjunction for multiple experiments of a study, always pulse at the first time point with CldU and the second time point with IdU.
12. For birthdating analysis, examine the tissue when the cell population of interest is fully developed such that the final position or fate of the cells can be studied.
13. Cell cycle studies must be undertaken while the tissue of interest is in a proliferative phase; therefore, many times embryos must be studied prenatally. A 23 h pulse point can be used as an initial assessment of differences in cell cycle reentry. As there are variations in cell cycling rates across development and between tissue types, the investigator may need to optimize these pulse points to capture the data of interest. For further reading, the following articles are of interest [22–25].
14. If populations of separate genotypes are being studied, be sure to harvest appropriate tissue for genotyping at this step.
15. Unfixed tissue, particularly prenatal tissue, is extremely delicate and can be damaged easily. Take care to gently handle prefixed tissue.
16. If there is no difference in genotype, up to 5 brains per vial can be fixed depending on age. Genotypically different brains must be fixed in individual vials and labeled as such.
17. It is critical to avoid fixing tissue for too long. Overly fixed tissue can add variables to experiments such as immunohistochemistry, and as such, fixation periods greater than 24 h should be avoided.
18. Heated agarose/PBS will eventually boil. Be careful to avoid boiling over the side of the container. This can be accomplished by stopping the microwave periodically and restarting after swirling gently. Do not cover the container to be heated. Be careful to avoid burns by using appropriate personal protection equipment.
19. As the agarose cools to the appropriate temperature, a “skin” forms on the top. This must be moved before a brain is

embedded, or embedding will likely fail and further vibratome processing may destroy the tissue.

20. It is useful to embed the brains around the perimeter of the dish and label the outside with a marker if a group of brains is to be embedded in same dish.
21. This orientation will produce coronal sections. Sagittal, horizontal, or other section planes can also be obtained by maneuvering the block and gluing it appropriately to the boat.
22. Sections can be stored in 30 % sucrose at 4 °C for approximately 3 months.
23. Place the scintillation vials on their sides for a more thorough wash, if needed. Place several vials in a box if desired. It is helpful to label the cap, the side and the bottom of the vials as the caps can easily be switched, and the side labels can often rub off through handling.
24. 1× PBS washes can be changed using a transfer pipette or similar. Take care to avoid aspirating sections into the waste PBS. Often forcing the plastic transfer pipette against the bottom of the scintillation vial is effective to disallow tissue to pass into the pipette as PBS is being aspirated.
25. For tissue younger than E16.5, use a 10 min incubation.
26. Use a cardboard or pipette tip box to stand the scintillation vials upright. This will avoid them being inadvertently knocked over by the motion of the biorocker. Be sure that the sections are floating in solution and are moving freely, and not adhered to the side of the vial, outside the solution volume.
27. We use a stock solution of 1 mg/mL Dapi and dilute 1:1,000 in 1× PBS.
28. Sections can be picked out of the scintillation vial with a paintbrush and rolled onto the slide. Alternatively, a 10 cm dish can be filled partway with 1× PBS, and sections can be floated onto the slide. Lift the slide out of the PBS to attach the section while steadying with a paintbrush.
29. The field of the images must be large enough to capture the entire area of the tissue of interest and of sufficient resolution. In the study of developing neocortical plate, the entire region from the pial surface to the subplate must be imaged in order to determine the relative location of labeled cells. Generally, 10–40× objectives can accomplish this. Always ensure that the confocal laser intensity remains constant for a given channel between images and samples.
30. Many software analysis packages can superimpose a field over an image. Generally, we create a rectangular field over the cortex, from pial surface to subplate. This is then subdivided into 10 equal bins to assess the position of nucleoside-labeled cells.

Table 1
A list of antigens in the cellular subtypes of the developing neocortex

Cortical cell subtype markers	
Antigen	**Population labeled**
Pax6	RG and NSC
Sox2	RG and NSC
BLBP	RG
Glast	RG and astrocytes
Tbr2	IPCs of neocortex
Tbr1, Tle4	SP and layer VI of neocortex
Fezf2, Ctip2	Layers V (high) and VI (low) of neocortex
Er81	Layer Va/b of neocortex
Svet1, Cux1, Cux2	Layers II–IV of neocortex
DCX	Migrating neuroblasts
βIII tubulin	Mature, postmigratory neurons

With this method, changes in the position of birthdated cells can be assessed between experimental conditions because the region of analysis is standardized. This technique can also be easily translated to other regions of the brain (*see* Fig. 1).

31. This analysis will reveal if there is a spatial difference in the location of labeled cells with the tissue field. This is particularly instructive when studying cortex, when birthdates correlate very closely to final position [9]. As the bins are equally divided over a region of interest, data analysis can reveal a shift in a bin position between conditions.
32. Further immunohistochemistry analysis can also include fate markers of known regions, such as in Table 1. Additionally, distinct phases of the cell cycle can be discriminated based on the expression of protein markers. For example, phospho-histone 3 (PH3) is only expressed in the M phase of the cell cycle, whereas cyclin E or thymidine analogs preferentially label cells in G1/S phase. Markers of interest are outlined in Table 2.
33. For cell cycle analysis, select a region of interest where cells are known to be cycling. Take several images to ensure that all areas of the proliferative region of interest are included in analysis.

Table 2
Cell cycle phase-specific markers

Cell cycle phase markers	
Marker	**Phase labeled**
Ki67	Expressed in all actively cycling cells and specifically localized to chromatin in M phase
PCNA	Expressed increasingly in G1 and peaks in S phase
BrdU, CldU, IdU, EdU	Incorporated during S phase
PH3	Expressed in M phase
Phospho-vimentin	Expressed in M-phase of cycling RG
Cyclins	Cyclin D (G1), cyclin E (G1–S), cyclin A (S–G2), cyclin B (late G2–M)
Tis21	Expressed in RG during neurogenic division [26]

34. A difference in double-labeled vs. single-labeled cells can also point to other changes. Cell populations that preferentially reenter or fail to reenter cell cycle may be stem cells or may have differentiated, respectively. These possibilities can be further studied through further immunohistochemistry analysis of stem cell and differentiation markers as outlined in Table 1. Last, proper experimentation may include the assessment of an apoptotic marker, such as activated caspase 3 or a TUNEL assay.

Acknowledgements

We extend our sincere gratitude to Dr. Li Cai of Rutgers University for his assistance in critiquing this chapter.

References

1. Rakic P (1968) S.R., Subcommissural organ and adjacent ependyma: autoradiographic study of their origin in the mouse brain. Am J Anat 122(2):317–335
2. Brand S, Rakic P (1979) Genesis of the primate neostriatum: [3H]thymidine autoradiographic analysis of the time of neuron origin in the rhesus monkey. Neuroscience 4(6):767–778
3. Nowakowski RS, La Vail JH, Rakic P (1975) The correlation of the time of origin of neurons with their axonal projection: the combined use of [3H]thymidine autoradiography and horseradish peroxidase histochemistry. Brain Res 99(2):343–348
4. Rakic P (1974) Neurons in rhesus monkey visual cortex: systematic relation between time of origin and eventual disposition. Science 183(4123):425–427
5. Crandall JE, Herrup K (1990) Patterns of cell lineage in the cerebral cortex reveal evidence

for developmental boundaries. Exp Neurol 109(1):131–139

6. Polleux F, Dehay C, Kennedy H (1997) The timetable of laminar neurogenesis contributes to the specification of cortical areas in mouse isocortex. J Comp Neurol 385(1):95–116
7. Dehay C, Kennedy H (2007) Cell-cycle control and cortical development. Nat Rev Neurosci 8(6):438–450
8. Miller MW, Nowakowski RS (1988) Use of bromodeoxyuridine-immunohistochemistry to examine the proliferation, migration and time of origin of cells in the central nervous system. Brain Res 457(1):44–52
9. Rash BG et al (2011) FGF signaling expands embryonic cortical surface area by regulating notch-dependent neurogenesis. J Neurosci 31(43):15604–15617
10. Breunig JJ et al (2007) Everything that glitters isn't gold: a critical review of postnatal neural precursor analyses. Cell Stem Cell 1(6): 612–627
11. Salic A, Mitchison TJ (2008) A chemical method for fast and sensitive detection of DNA synthesis in vivo. Proc Natl Acad Sci 105(7): 2415–2420
12. Zhao M et al (2003) Evidence for neurogenesis in the adult mammalian substantia nigra. Proc Natl Acad Sci 100(13):7925–7930
13. Hansen DV et al (2010) Neurogenic radial glia in the outer subventricular zone of human neocortex. Nature 464(7288):554–561
14. Bai J et al (2003) RNAi reveals doublecortin is required for radial migration in rat neocortex. Nat Neurosci 6(12):1277–1283
15. Rasin M-R et al (2007) Numb and Numbl are required for maintenance of cadherin-based adhesion and polarity of neural progenitors. Nat Neurosci 10(7):819–827
16. Saito T (2006) In vivo electroporation in the embryonic mouse central nervous system. Nat Protocols 1(3):1552–1558
17. Caviness VS Jr, Sidman RL (1973) Time of origin or corresponding cell classes in the cerebral cortex of normal and reeler mutant mice: an autoradiographic analysis. J Comp Neurol 148(2):141–151
18. Polleux F, Dehay C, Kennedy H (1998) Neurogenesis and commitment of corticospinal neurons in reeler. J Neurosci 18(23): 9910–9923
19. Kosodo Y et al (2011) Regulation of interkinetic nuclear migration by cell cycle-coupled active and passive mechanisms in the developing brain. EMBO J 30(9):1690–1704
20. Tuttle AH et al (2010) Immunofluorescent detection of Two thymidine analogues (CldU and IdU) in primary tissue. J Vis Exp 46:e2166
21. Fekete E (1954) Gain in weight of pregnant mice. J Hered 45(2):88–98
22. Nowakowski RS, Caviness VJ, Takahashi T, Hayes NL (2002) Population dynamics during cell proliferation and neuronogenesis in the developing murine neocortex. Results Probl Cell Differ 39:1–25
23. Caviness VS et al (2003) Cell output, cell cycle duration and neuronal specification: a model of integrated mechanisms of the neocortical proliferative process. Cereb Cortex 13(6): 592–598
24. Cai L, Hayes NL, Nowakowski RS (1997) Local homogeneity of cell cycle length in developing mouse cortex. J Neurosci 17(6):2079–2087
25. Takahashi T, Nowakowski R, Caviness V (1995) The cell cycle of the pseudostratified ventricular epithelium of the embryonic murine cerebral wall. J Neurosci 15(9):6046–6057
26. Attardo A et al (2010) Tis21 expression marks not only populations of neurogenic precursor cells but also new postmitotic neurons in adult hippocampal neurogenesis. Cereb Cortex 20(2):304–314

Chapter 4

Culture of Dissociated Hippocampal Neurons

Yun Peng, Wen-Cheng Xiong, and Lin Mei

Abstract

Hippocampal neurons consist mainly of pyramidal neuron and granule cell, and dissociated hippocampal neurons are a good tool to investigate the molecular and cellular mechanism of neuronal development and neuronal degenerative disease in the central neuronal system (CNS). Here, we describe a general procedure of dissociated hippocampal neuron culture.

Key words Culture, Hippocampus, Dissociated neuron, Coating, Dissection, Digestion, Plating, Feeding, Glial inhibition

1 Introduction

Dissociated neuron culture system has been established decades years ago [1–3]. Neuron development in culture has been well-characterized [3]; the dissociated neurons can develop morphologically and physiologically over time to resemble their in situ counterparts [4]. They can form normally polarity, develop extensive axonal and dendritic arbors, and form numerous, functional synaptic connections with one another [5].

It has been used widely to investigate cellular and molecular mechanisms underlying neuron development, such as neuron polarity formation [6, 7], dendrite outgrowth [8], axon guidance [9], and synaptogenesis [5]. Compared to in vivo experiment, this system can be manipulated in a much more controlled way. The morphological development can be visualized clearly using microscopy techniques. Electrophysiological recording and stimulation can take place globally or occur at a specific location [10]. Moreover, chemical analysis of the neurons and their environment is more easily accomplished than in an in vivo setting. In the case of knockout animals that die at birth from systemic defects, cell culture can be used to study those aspects of neural development that occur postnatally [10].

Renping Zhou and Lin Mei (eds.), *Neural Development: Methods and Protocols*, Methods in Molecular Biology, vol. 1018, DOI 10.1007/978-1-62703-444-9_4,

Neuron populations in the hippocampus are mainly composed of pyramidal neuron and granule cell, and pyramidal neuron is the principal cell type in CA1 and CA3 regions of hippocampus, except that they dominate in number and, morphologically, it exhibits pyramidal-like cell body, and a primary dendrite and several basal dendrites, which can be easily distinguished from other kinds of neuron in hippocampus. In addition, granule cell predominately localizes in the dentate gyrus (DG) of hippocampus. After DG is removed, we can get a mix culture which mainly composes of pyramidal neurons and glia.

Culture of dissociated hippocampal neuron is a relatively complex procedure. It can be divided into several following processes: coverslip preparation, coverslip coating, dissection, digestion, cell plating, and feeding. Each step can be manipulated in different strategies, here, just introduce some common way. This protocol will guide you through the process for generating isolated hippocampal neuron cultures from the brain of late rat embryo. Following these processes, the dissociated hippocampal neurons would polarize appropriately, develop extensive axonal and dendritic arbors, and generate normally synapse with days. These neurons can be used for a variety of applications including immunocytochemistry, biochemistry, electrophysiology [11], calcium and sodium imaging [12], and protein and/or RNA isolation.

2 Materials

2.1 Material for Preparing Coverslips

Instruments: coverslips, 1 L flask, shaker.

Reagents: 200 mL nitric acid.

2.2 Material for Coating

Instruments: 24-well plate, tweezers, burner, and Kimwipes.

Reagents

1. PBS (8 g NaCl, 0.2 g of KCl, 1.44 g Na_2HPO_4, 0.24 g KH_2PO_4). Adjust pH to 7.4 volume to 1 L with additional distilled H_2O.
2. Matrigel solution (the Matrigel must be added slowly to Neurobasal Medium while still frozen, store in 4 °C, and prevent from light; the solution should be used in a month. Matrigel should be prepared at least one day before use to ensure that it has completely dissolved into solution).

2.3 Materials for Dissection

Instruments

1. Anatomy microscope.
2. 2 operating scissors.
3. 2 dissecting tweezers.

4. 2 microdissecting tweezers.
5. 1 microdissecting scissors.
6. 1 spatula (rounded tip, sharp tip).
7. 1 microdissecting tweezers with 45° angled tip.
8. 3 100 mm × 15 mm dishes.
9. A bucket of ice.

Reagents

1. Diethyl ether.
2. Dissection solution (22.50 mL Na_2SO_4 (1 M), 30.00 mL K_2SO_4 (0.25 M), 1.45 mL $MgCl_2$ (1 M), 6.20 mL $CaCl_2$ (0.1 M), 0.50 mL sodium HEPES (500 mM)). Add fill to 250 mL.

2.4 Materials for Digestion

Reagents

1. 2.5 % Neurobasal Medium solution (500 mL Neurobasal Medium, 10 mL B27, 12.5 FBS, 5 mL glutamine).
2. Enzyme solution (2 mg L-cysteine, solution, 100 mL CaCl2 (100 mM), 100 mL EDTA(50 mM), 15 mL NaOH (1 N), 100 mL DNase I), in 10 mL hippocampal dissection. DNase is added immediately before the solution be filtered.
3. Inactivating solution (25 mg bovine albumin, 25 mg trypsin inhibitor, 100 μL DNase I, in 10 mL serum media).

2.5 Materials for Plating

Equipments: hemocytometer and tally counter.

Reagents: 0.4 % trypan blue.

2.6 Materials for Feeding

Reagents

1. Neurobasal Medium (500 mL Neurobasal Medium, 10 mL B27, 5 mL glutamine).
2. FUDR (F-0503 100 mg SIGMA). Add FUDR to 10 mL double-distilled water.

All components and solutions are added in culture hood under sterile conditions, and all tools for dissection must be sterilized by 70 % alcohol treatment.

3 Methods

3.1 Coverslip Preparation

The coverslips used for isolated culture should be clean and sterilized, and all the impurity, such as lipid, ion, and bacteria, must be removed completely. The following processes would ensure that the coverslip is clean enough for neuron culture:

1. Carry carefully 200 mL concentrated nitric acid in a clean 1 L glass flask by a disposable serological pipet (*see* **Note 1**).

Concentrated nitric acid is volatile and corrosive, operator must wear mask and gloves, and this step should be done in ventilated equipment.

2. Place coverslips in the glass flask containing nitric acid, swirl gently, and ensure that the surface of every coverslip can immerse into nitric acid completely.
3. Cap the glass flask with tinfoil and place it on a shaker to shake softly overnight (*see* **Note 2**).
4. The following day, empty nitric acid out of flask, add about 200 mL ddH_2O in the flask containing coverslips, and place the flask in the shaker to shake for 15 min; empty the water. Repeat these processes for about 10 times and ensure that the coverslips are clean and residual nitric acid has been removed completely (*see* **Note 3**).
5. At last, keep these coverslips clean in a 50 mL tube containing 100 % alcohol for storage (*see* **Note 4**).

3.2 Coverslip Coating

The day before, make dissection, and coverslip coating is required. This step is indispensable for neuron culture; the coverslips with substrate coating enable neuron and glia cell attach and survive [13, 14].

1. Prepare a flame and some Kimwipes on bench and keep them at a discreet distance to avoid fire accident (*see* **Note 5**).
2. Grab a coverslip with sterilized microdissecting tweezers from storage tube and tap gently onto Kimwipes to remove excess alcohol.
3. Run it quickly through flame to get rid of alcohol completely (*see* **Note 6**) and place dried coverslip into each well of cell plate.
4. Transfer the cell culture plate containing coverslips into hood and turn on ultraviolet lamp. After illuminated by UV for 30 min, these coverslips are ready to be coated.
5. Add 50 μL Matrigel solution into coverslip of each well and then tap softly with hands to ensure Matrigel covering the whole surface of coverslip, and place the coverslip in incubator overnight (*see* **Note 7**).
6. The next day before plating step, remove the excess Matrigel solution from each coverslip by pipet and add 1 mL 2.5 % FBS Neurobasal Medium solution in each well.

3.3 Dissection

A sterile environment is necessary in the process of dissection. So, all equipment utilized during this procedure must be sterilized in a beaker filled with 70 % alcohol, and the dissection area must be sprayed with 70 % alcohol before using it. Operator should wear mask and gloves during the whole process.

1. Prepare solution. Except for Dissection solution, which is ice cold during dissection process, others solutions have to be warmed up in 37 °C incubator before use.
2. Place pregnant rat (E16–18) (*see* **Note 8**) in a plastic dish and pipet 5 mL diethyl ether in it; after 5 min, the rat fails to move spontaneously and no response to stimulus, such as touch and shaking (*see* **Note 9**).
3. Transfer it on a clean bench. Spray 70 % alcohol at the rat's abdomen, cut the skin along the midline of abdomen, then change a new sterile dissecting scissors to incise the muscle, and expose the uterus.
4. Lift uterus with dissecting tweezers, snip its connection with body by a sterile scissors, and put the uterus in a dish with ice-cold dissection solution.
5. Cut open the uterus with dissecting tweezers to expose embryos. Grasp the body of embryo with tweezers, cut the head of embryo off, and place the head in a new dish containing ice-cold PBS solution.
6. Hold mouse head with microdissecting tweezers, and cut along the scalp with microdissecting scissors to expose the whole brain.
7. Shove the brain carefully out of the skull with the round end of a spatula and place in another new dish containing ice-cold PBS solution.
8. With microdissecting tweezers positioned at the two lobes of the cerebral cortices to hold brain. With another microdissecting tweezers, separate each cortex by carefully nudging it away from the midline of the brain.
9. Hold the cortex with microdissecting tweezers and peel away the meninges starting from the end opposite that of the hippocampus and outer side of the hemisphere. Flip the hemisphere over so that the hippocampus is facing up. Gently remove the remaining meninges away from the tissue.
10. With microdissecting tweezers, hold the cortex, use another one softly fold out the hippocampus, and pinch off the hippocampus (*see* **Note 10**).
11. Hold hippocampus down with microdissecting tweezers. Pinch off dentate gyrus with a 45-degree-angled tip microdissecting tweezers and then transfer all hippocampal tissues into a new dish with 37 °C warmed dissection solution (*see* **Note 11**).

3.4 Digestion

Solutions for the enzymatic digestion and inactivation periods must be prepared at least an hour before dissection. This allows for solutions to warm in incubator and equilibrate before introducing hippocampal tissue.

1. Transfer all hippocampal tissues using a serological pipet and then place the tube in incubator for 30 min. During the incubating period, shake the dish to make sure the fraction of hippocampus completely mixed with the digest enzyme.
2. Aspirate off the enzymatic solution softly without removing these tissue, and add 10 mL inactivating solution in the tube, and incubate at 37 °C for 5 min to terminate the digesting process.
3. With fire-polished glass Pasteur pipet, triturate tissue by sucking them up through narrow tip opening and then pipet them out through narrow tip opening. Allow larger tissue to settle to bottom, and then by pipeting solution near the liquid surface, transfer top suspension of cells to a new tube. Return to original tube containing hippocampal tissue and triturate again in the same manner as mentioned above. Repeat trituration cycle until suspension is completely transferred into new tube.
4. Place a 40 μL filter on a 50 mL tube, filtrate the digestion solution through a 40 μm filter column, and centrifuge at 1,000 rpm for 10 min.
5. Remove carefully top layers of suspension and resuspend the precipitate with warm 1.2 % FBS Neurobasal Medium with a 1:1 ratio (1 mL per hippocampus). Transfer the suspension into a 50 mL conical tube.

3.5 Neuron Plating

3.5.1 Cell Counting

1. Take 100 μL of suspension and put it into a small Eppendorf tube; mix with 100 μL trypan blue (*see* **Note 12**).
2. Spray alcohol on the coverslip and hemocytometer and then wipe off; ensure they are clean and moist.
3. Add approximately 10 μL mixture at the edge of the coverslip and allow the mixture penetrate automatically into the whole space between coverslip and hemocytometer.
4. Visualize the hemocytometer grid under microscope with the 10× objective. Adjust focus until a single counting square fills the field.
5. Count the number of colorless, bright cell by tally counter (*see* **Note 13**).
6. Calculate cell concentration (*see* **Note 14**).

3.5.2 Plating Density

Dissociated cultures are usually plated at "low," "medium," and "high" densities. Dilute the cell with plate solution to final count of 5×10^5 per mL. Add 200 μL suspensions to each well of 24-well plate, at medium density. Add 100 μL to each well, at low density. Add 300 μL to each well, at high density (*see* **Note 15**).

Add suspension to each well and tap softly the plate; ensure that cell has distributed the whole well. Place the cell in incubator with 37 °C, 5 % CO_2 for culture (*see* **Notes 16** and **17**).

3.6 Feed and Glial Inhibition

1. The next day after plating, half of the old Neurobasal Media in culture is replaced with new warm Neurobasal Media.
2. Three days later after neuron plating, glia would have covered almost whole coverslip,the glia exhibit gray layer compared to bright background under microscope. Add 50 μL FUDR in each well to inhibit glia proliferating further (*see* **Note 18**).
3. 6 to 8 days after plating and plates up to 3 weeks old, draw out old Neurobasal Media from each well and replace with 500 μL new Neurobasal Media (*see* **Note 19**).

4 Notes

1. Concentrated nitric acid is volatile and corrosive, operator must wear mask and gloves, and this step should be done in ventilated equipment.
2. Put the flask in a second container and make sure that the flask would not drop from shaker.
3. After washed, residual nitric acid should be neutralized by NaOH firstly; it cannot pour into sink directly.
4. Alcohol would keep these coverslips sterile and protect them from contamination.
5. Kimwipes can also be replaced by other sterile absorbent paper.
6. Coverslips are very fragile; do not hold them too long on flame. Usually less than 1 s.
7. It has been reported at least two substrates, Matrigel and poly-D-lysine, available for coverslip coating. Generally, the neuron plated on Matrigel-coated coverslips would grow relatively faster and, morphologically, more complex compared to that coated by poly-D-lysine. Here, just describe the strategy of Matrigel coating.
8. Hippocampal cultures are often prepared from rats, but the same protocol also works well for mouse cultures; embryos can be replaced by P0 pups.
9. The rat should be satisficed according to LAS policy.
10. Try to stay close to the hippocampus so as not to include extra tissue and do not squeeze hippocampus tissue.
11. It would take 2-3 h for the whole process from hippocampus dissection to neuron plating. If operator spends too longer time on these processes, it would decrease the rate of neuronal viability.
12. After staining with trypan blue, the cells should be counted in 3 min; otherwise, the nonviable cells will begin to take up the dye.

13. Trypan blue can penetrate into dead cell, but exclude from those cells that have intact membrane.
14. The cell number in 1 mL is the number of cells in one large square multiplied by dilution factor. Dilution factor is usually 2 (1:1 dilution with trypan blue).
15. The number of cell used for plating is decided by the area of each well or dish. The surface area of each well is 200 mm^2 in 24-well plate, is 400 mm^2 for 12-well plate, and is 960 mm^2 for 6-well plate.
16. In 4 h after plating, neuron will attach and extend several short neurites. In 48 h, neuron polarity would form, and in 12 days later, neurons start to generate synapse.
17. Compared to in situ neuron, there are two disadvantages of cultured neuron. One is that cultured neuron does not grow in their natural environment, and the treatment to cultured neurons is not a biological way. The other one is, when the neurons are suspended in solution and subsequently dispensed, the connections previously made are disrupted, and a new connection is formed; this kind of connection may not represent biological circuit in the brain.
18. FUDR can be replaced by cytosine arabinoside to inhibit glial overgrowth.
19. If the medium in a well exhibit cloudy and yellowish in appearance that demonstrates the well has been contaminated, aspirate solution of the well completely. Do not feed but by all means avoid these wells.

References

1. Mains RE, Patterson PH (1973) Primary cultures of dissociated sympathetic neurons. I. Establishment of long-term growth in culture and studies of differentiated properties. J Cell Biol 59:329–345
2. Chumasov EN et al (1975) Hippocampus morphology in tissue culture. Tsitologiia 17: 1306–1312
3. Kim SU (1973) Morphological development of neonatal mouse hippocampus cultured in vitro. Exp Neurol 41:150–162
4. Teyler TJ et al (1977) A comparison of long-term potentiation in the in vitro and in vivo hippocampal preparations. Behav Biol 19:24–34
5. Chih B et al (2005) Control of excitatory and inhibitory synapse formation by neuroligins. Science 307:1324–1328
6. Shelly M et al (2010) Local and long-range reciprocal regulation of cAMP and cGMP in axon/dendrite formation. Science 327: 547–552
7. Wang T et al (2011) Lgl1 activation of rab10 promotes axonal membrane trafficking underlying neuronal polarization. Dev Cell 21: 431–444
8. Tan ZJ et al (2010) N-cadherin-dependent neuron-neuron interaction is required for the maintenance of activity-induced dendrite growth. Proc Natl Acad Sci USA 107: 9873–9878
9. Zhu XJ et al (2007) Myosin X regulates netrin receptors and functions in axonal path-finding. Nat Cell Biol 9:184–192
10. Lin YC et al (2002) Development of excitatory synapses in cultured neurons dissociated from the cortices of rat embryos and rat pups at birth. J Neurosci Res 67:484–493
11. Ghezzi D et al (2008) A Micro-Electrode Array device coupled to a laser-based system for the local stimulation of neurons by optical release of glutamate. J Neurosci Methods 175: 70–78

12. Higley MJ, Sabatini BL (2008) Calcium signaling in dendrites and spines: practical and functional considerations. Neuron 59: 902–913
13. Yang H et al (2010) Long-term primary culture of highly-pure rat embryonic hippocampal neurons of low-density. Neurochem Res 35: 1333–1342
14. Liesi P et al (1984) Neurons cultured from developing rat brain attach and spread preferentially to laminin. J Neurosci Res 11: 241–251

Chapter 5

Culturing Mouse Cerebellar Granule Neurons

Tharakeswari Selvakumar and Daniel L. Kilpatrick

Abstract

The cerebellum plays an important role in motor control, motor skill acquisition, memory and learning among other brain functions. In rodents, cerebellar development continues after birth, characterized by the maturation of granule neurons. Cerebellar granule neurons (CGNs) are the most abundant neuronal type in the central nervous system, and they provide an excellent model for investigating molecular, cellular, and physiological mechanisms underlying neuronal development as well as neural circuitry linked to behavior. Here we describe a procedure to isolate and culture CGNs from postnatal day 6 mice. These cultures can be used to examine numerous aspects of CGN differentiation, electrophysiology, and function.

Key words Cerebellum, Cerebellar granule neurons, Differentiation, Neuronal culture, Percoll gradient, Sonic hedgehog

Abbreviations

CGNs	Cerebellar granule neurons
CGNPs	Cerebellar granule neuron progenitors
EGL	External germinal layer
PMZ	Premigratory zone
NB	Neurobasal
PLY	Poly-D-lysine
Shh	Sonic hedgehog
BSA	Bovine serum albumin

1 Introduction

The cerebellum plays a crucial role in several functions coordinating skeletal muscle movement, learning, memory and attention [1–3]. Cerebellar granule neurons (CGNs) are an integral part of cerebellar circuitry [4] and a thorough characterization of their maturation is important for our understanding of cerebellar function. Defects in the development of the cerebellum, and of CGNs per se, are implicated in several nervous system disorders [5]. Further, CGNs are

Renping Zhou and Lin Mei (eds.), *Neural Development: Methods and Protocols*, Methods in Molecular Biology, vol. 1018, DOI 10.1007/978-1-62703-444-9_5, © Springer Science+Business Media, LLC 2013

the most abundant neuronal population in the cerebellum as well as in the CNS as a whole. Thus, ample amounts of purified primary cells can be readily prepared for culture and other applications [6, 7]. They thus provide an excellent model for studying neuronal development and function, both in culture and in vivo.

During the first 1–2 weeks following birth CGN progenitors (CGNPs) proliferate in the external germinal layer (EGL) of the cerebellum. They then exit the cell cycle and take up residence in the pre-migratory zone (PMZ), where immature CGNs form bipolar processes that ultimately become parallel fibers that synapse with Purkinje cell dendrites. During this period, CGNs extend a third, radial process and the cell body then migrates inwardly from the PMZ through the molecular layer and until it finally reaches the internal granule cell layer. In the final phase of the maturation process, CGNs form dendrites and establish synaptic connections with excitatory mossy fibers and inhibitory Golgi type II neuron terminals.

Numerous genes undergo differential gene expression during the various stages of CGN development [8]. Interestingly, much of this differentiation-associated gene expression is recapitulated in primary CGN cultures [7]. Freshly isolated CGN preparations from the postnatal rodent cerebellum consist of CGNPs and immature CGNs derived from the EGL [9, 10]. Upon plating as dissociated cells these cultures become postmitotic and initiate a differentiation program that includes axon extension and subsequent dendrite formation [11, 12]. These events as well as the associated changes in gene expression mirror to a significant extent the sequence and timing of CGN development in vivo [4, 13]. This suggests that CGN cultures express an intrinsically driven differentiation program that reflects many aspects of normal CGN maturation, although clearly extrinsic mechanisms also play an important role [14–16]. Thus, CGN cultures provide a valuable system for exploring intrinsic molecular mechanisms underlying neuronal development. Further, their high abundance provides substantial numbers of cells in high purity that are suitable for numerous molecular and biochemical analyses, including RT-PCR and Northern blotting, Westerns and chromatin immunoprecipitation assays [6, 7]. They also are readily transduced with viral vectors expressing shRNAs or proteins to identify key regulators of their development, morphology, and function [6, 17]. These events can then be confirmed within the intact cerebellum using appropriate in vivo models. In vivo studies are further facilitated by the high abundance of granule neurons within the cerebellum, permitting the analysis of whole tissue extracts in many instances [7, 13]. Lentiviruses also have been used to study gene promoter element function and promoter regulation in CGN cultures by co-transduced transcription factors [7, 13].

This chapter presents a laboratory protocol to isolate a homogenous population of CGNPs and immature CGNs that elaborate an intrinsic program of neuronal maturation in culture.

2 Materials

1. Scalpel (Personna #10).
2. Forceps (Dumont #5, straight and curved).
3. Pasteur pipettes (fire-polished).
4. Petri plates (35 mm, 100 mm).
5. 15 ml and 50 ml conical tubes (Falcon).
6. Centrifuges: Clinical centrifuges (fixed angle and swinging bucket rotors) and Beckman table-top centrifuge (Model GS-6R; Rotor GH 3.8 (refrigerated)).
7. Dissection microscope.
8. 37 °C CO_2 incubator.
9. Water bath at room temperature (RT).
10. Hemocytometer.
11. Phase-contrast microscope.
12. Tissue culture plates (various sizes; Corning).

2.1 Reagents

1. Trypsin (Sigma Cat# T-4799).
2. DNase I (Roche Cat# 10104159001).
3. Soybean trypsin inhibitor (Gibco Cat# 17075-029).
4. Percoll (Sigma Cat# P1644).
5. Bovine Serum Albumin (Sigma Cat# A3294).
6. Neurobasal medium (Gibco Cat# 21103) + supplements (see below).
7. Poly-D-lysine (PLY) (Sigma Cat# P1024).
8. Phenol red.
9. Trypan blue.
10. Mouse laminin (1 mg/ml (100×)) (Invitrogen Cat# 23017-015).
11. 10× Hanks Buffer (Gibco Cat# 14185-052).
12. Sonic hedgehog (Shh) (Shenandoah Cat# 200-55).

2.2 Solutions

1.

Tyrode's–PBS	500 ml
NaCl	4 g
KCl	0.15 g
$NaH_2PO_4 \cdot H_2O$	0.25 g
KH_2PO_4	0.125 g

(continued)

(continued)

$NaHCO_3$	20 mg
Phenol red	0.25 ml
Glucose	1 g

Adjust pH to 7.4 with 1–2 μl 2N NaOH. Filter-sterilize and store at 4 °C.

2.

1 % trypsin–DNase I	20 ml
Trypsin	200 mg
DNase I	20 mg
$MgSO_4.7H_2O$	4.5 mg
Tyrode's–PBS	19.88 ml

Filter-sterilize. Adjust the pH to 7.4 with 1–2 μl 2N NaOH, as indicated by a change in the pH dye indicator. Store at −20 °C in 1 ml aliquots in sterile 1.5 ml conical tubes. Use each aliquot only once after thawing.

3.

Trypsin inhibitor–DNase I solution	20 ml
DNase I	20 mg
Soybean trypsin inhibitor	8 mg
Tyrode's–PBS	Add to make 20 ml final volume

Filter-sterilize. Adjust pH to 7.4 with 1–2 μl 2N NaOH if necessary. Store at −20 °C in 1 ml aliquots in sterile 1.5 ml conical tubes. Use each aliquot only once after thawing.

4.

Percoll (35 %)	50 ml	Percoll (60 %)	50 ml
Percoll	17.5 ml	Percoll	30 ml
10× Hank's	5 ml	10× Hank's	5 ml
Pyruvic acid	5.5 mg	Pyruvic acid	5.5 mg
1 M Hepes pH 7.4	0.5 ml	1 M Hepes pH 7.4	0.5 ml
0.5 M EDTA	0.25 ml	0.5 M EDTA	0.25 ml
Water	27.5 ml	Water	15 ml
Phenol red	150 μl	Trypan blue	100 μl

Adjust pH if necessary with ~10 μl 2N NaOH. Store Percoll solutions at 4 °C and discard after 6 months of storage.

5. Poly-D-lysine (PLY): Dissolve 100 mg PLY in 20 ml sterile water in a tissue culture hood to give a 5 mg/ml solution.

Store as 1 ml aliquots at −20 °C. For each experiment, dilute 1 ml of the 5 mg/ml PLY solution with 49 ml sterile water to give a final concentration of 0.1 mg/ml.

6. Complete Neurobasal (NB) medium (with supplements):

NB medium	96 ml
B27 supplement 50× (Gibco Cat# 17504-044)	2 ml
100× Glutamine (Gibco Cat# 25030)	1 ml
100× Pen/Strep (Gibco Cat# 15140)	1 ml
30 % Glucose	1.5 ml

Filter-sterilize. Store at 4 °C.

7. 4 % BSA in Hank's Buffer (for enriched CGN preparations): Dissolve 2 g BSA in 50 ml Hank's Buffer. Adjust pH to 7.4 with ~100 μl 2N NaOH.

3 Methods

3.1 Steps to Perform Prior to Starting Cell Preparation

1. Typically the evening before the cell preparation, culture plates and chamber slides are coated with PLY by adding diluted PLY solution (0.1 mg/ml) sufficient to cover the surface, and then are left overnight at 37 °C. Approximately 1 h prior to plating cells, remove the PLY from the culture plates or slides and wash with water containing 1× Pen/Strep (three washes) and allow to air-dry completely (*see* **Note 1**).
2. If cells are being cultured on laminin (*see* **Note 1**), add 1× laminin in PBS and incubate at 37 °C for 45–90 min. Following this, the laminin is removed and add 1× sterile PBS to cover the wells. The PBS is removed just before addition of cells to the culture plates.
3. Prepare three Pasteur pipettes with tips of progressively reduced diameters (normal, ~75 and ~50 % that of the first pipette). To reduce the diameter, flame the pipette tip for varying lengths of time (*see* **Note 2**). These can be prepared well in advance.
4. For pre-plating, on the morning of the cell preparation treat 10 cm petri dishes (*do not use tissue culture plates*) with PLY by adding ~11 ml of diluted PLY solution (0.1 mg/ml) and incubate at 37 °C for 60 min. Wash three times with water containing 1× Pen/Strep. Allow the dishes to air-dry completely before use.
5. Complete NB medium is prepared the evening before or the day of the cell preparation (*see* **Note 3**). Frozen stock solutions such as Trypsin–DNase I and Trypsin Inhibitor–DNase I solutions are thawed just prior to use and are left on ice (*see* **Note 4**).

3.2 Purified CGN Cultures

1. Perform this step one pup at a time. Euthanize postnatal day 6 (P6) mouse pups with CO_2 asphyxiation. Decapitate the pups and carefully cut through the skull to remove the brain. Place it on a 35 mm petri dish containing 1 ml Tyrode's positioned on a cold freezer block. Before excising the cerebellum from the rest of the brain tissue, carefully remove the meninges using fine forceps (Dumont #5). The meninges can be identified by the presence of many blood vessels and is readily peeled away from the rest of the brain.
2. After excising the cerebella from five pups, proceed to the next step (*see* **Note 5**).
3. Move the cerebella to a 35 mm petri dish without buffer and place on ice. Mince the tissue with two scalpels using a scissor-like motion. Mince 35 times in one direction and then rotate the plate 90° and repeat the mincing 35 times. Add 1 ml Tyrode's and transfer the minced cerebella to a 15 ml conical tube. Allow the tissue to settle and discard the supernatant.
4. Add 1 ml Trypsin–DNase to the minced tissue and incubate at RT for 3.5 min. Avoid trypsin over-digestion to prevent excessive cell lysis. Some lysis will normally occur resulting in release of nucleic acids and clumping of intact cells. Inclusion of DNase minimizes cell clumping.
5. Centrifuge the digest at $1,000 \times g$ in a clinical swinging bucket rotor for 10 s to pellet the cell clumps. Do not over-spin. Carefully remove the supernatant by aspiration.
6. Add 1 ml Tyrode's–DNase to the pellet. Slowly triturate with a 1 ml pipettor 20 times to wash, suspend and break up cell clumps. Avoid generating bubbles.
7. Centrifuge the cell suspension in a clinical centrifuge at $1,000 \times g$ for 5 min. Discard the supernatant. In the meantime, prepare Percoll gradient as described below.
8. Add 8 ml 35 % Percoll to a 50 ml conical tube. Carefully underlay this with 8 ml of 60 % Percoll from a 10 ml syringe using an 18 gauge needle. Keep the tube on ice while making the gradient (*see* **Note 6**).
9. Suspend the cell pellet in 1 ml trypsin inhibitor–DNase by trituration (15 times) with a 1 ml pipettor. Repeat the trituration 15 times each with three Pasteur pipettes bearing tips of progressively reduced diameter. Trituration must be done slowly to avoid the generation of cell clumps (*see* **Note 7**).

   From here on, all steps should be carried out on ice or at 4 °C.
10. After trituration, add 1 ml Tyrode's to completely suspend cells and to eliminate clumps using a 1 ml pipettor. Add 6 ml Tyrode's buffer to cells and mix well by gentle inversion.
11. *Gently* overlay the cells onto the previously prepared Percoll gradient (*see* **Note 8**).

12. Carefully balance the tubes and centrifuge at 2,500 × *g* in a swinging bucket rotor for 10 min (*see* **Note 9**).
13. After centrifugation, CGNs equilibrate at the interface between the 35 and 60 % Percoll layers. Remove the upper Tyrode's layer and approximately half of the 35 % Percoll layer (*see* **Note 10**).
14. Using a 1 ml pipettor, carefully remove the CGNs by transferring the remaining 35 % Percoll (pink colored layer) as well as *the interface* with the 60 % Percoll layer to a new 50 ml conical tube (*see* **Note 11**).
15. Dilute the Percoll by adding 43 ml Tyrode's. Spin in a clinical centrifuge (swinging bucket rotor) for 5 min at 1,000 × *g*. Discard the supernatant.
16. Suspend the pellet thoroughly in 1 ml complete NB medium by pipetting 15 times with a 1 ml pipettor, taking care to disrupt cell clumps. Add an additional 5 ml complete NB and mix by gentle inversion.
17. Centrifuge at 1,000 × *g* in a clinical centrifuge (swinging bucket rotor) for 5 min. Remove and discard the supernatant.
18. Gently suspend the pellet in 1 ml complete NB medium (15 times). Add 6.5 ml complete NB medium and mix the cells by inversion.
19. *Pre-plating*: Add the cell suspension in complete NB medium to a 10 cm PLY-coated petri dish, making sure that they are evenly dispersed on the plate surface. Incubate at 37 °C for 30 min (*see* **Note 12**).
20. Gently wash off the CGNs from the plate by adding and removing 5 ml complete NB medium. Repeat 6–7 times (*see* **Note 13**).
21. Repeat pre-plating step a second time if further enrichment is desired.
22. Transfer the CGN fraction to a 15 ml conical tube and add complete NB medium to a final volume of 15 ml. Remove a small aliquot, dilute ~10× with PBS, and determine cell yield using a hemocytometer. A typical yield is ~9–12 × 10^6 cells from five P6 mouse cerebella.
23. Plate cells on coated culture dishes (*see* **Notes 14** and **15**).
24. During culturing, replace 50 % of culture medium every 2–3 days (*see* **Note 16**).

3.3 Enriched CGN Cultures

While CGNs prepared using Percoll gradients and pre-plating are highly purified (~95 %), enriched CGNs (80–85 % pure) can be obtained using a protocol that omits these steps. In addition to CGNs, these cultures also contain lesser amounts of glial cells, Purkinje cells, as well as Golgi, basket, and stellate neurons [18]. Enriched preparations may be useful for studying cellular interactions in culture. Due to their considerably higher yields, they are also useful for studies in which larger numbers of CGNs are required.

1. Follow the cell preparation from the previous section up through **step 7**.
2. After trituration with glass pipettes, suspend the cells in 1 ml Tyrode's medium. Suspend the cells completely without clumping using a 1 ml pipettor. Add an additional 5 ml of Tyrode's to the cell suspension and gently mix by inverting. Bring the volume to 6 ml with Tyrode's and transfer cells to a 15 ml conical tube. To this add a 2 ml underlay of 4 % BSA in Hank's buffer (*see* **Note 17**).
3. Centrifuge cells at 1,000 × *g* for 5 min in a clinical centrifuge (swinging bucket rotor) at 4 °C.
4. Aspirate most of the supernatant (~6 ml), then carefully remove the remaining volume with a 1 ml pipettor to minimize cell loss.
5. Suspend the pellet well in 1 ml complete NB medium with a 1 ml pipettor to disperse cell clumps. Add an additional 9.5 ml complete NB and mix well by gentle inversion. Allow the cell suspension to sit at RT for 1–2 min. After the clumps settle, transfer the remaining cell suspension to a new 15 ml tube. An aliquot is taken for cell counting of a 10× dilution in PBS. A typical yield is ~40–50 × 10^6 cells from five P6 mouse cerebella.
6. Plate on PLY-coated culture dishes or chamber slides and culture as outlined for purified CGNs (*see* **Note 18**).

3.4 Proliferating CGNP Cultures

CGNPs can be maintained in the proliferative state by culturing with the mitogen sonic hedgehog (Shh) [13, 19]. This paradigm is valuable for studying cell cycle mechanisms or when it is important to induce specific changes in gene or protein expression (e.g., using lentiviral vectors) prior to onset of growth arrest and differentiation [7, 13]. We typically use a concentration of 2–3 μg/ml Shh (*see* **Note 19**).

1. Add Shh to gradient-purified CGNs at the time of initial plating to stimulate cell proliferation.
2. Remove Shh at the desired time point and replace with fresh complete NB medium to initiate onset of cell cycle exit and differentiation, as appropriate.

4 Notes

1. Discard PLY solutions stored for >3 months, since cells adhere less well to dishes coated with older solutions. PLY is typically sufficient to enable cell preparations of good quality to adhere to the substrate [18]. However, laminin is frequently used as an

additional substrate. For example, laminin is typically used for CGN cultures transduced with viruses [7, 13] since it enhances cell attachment under these conditions. It also can be used to examine substrate effects on CGN neurite outgrowth [20].

2. The relative tip diameters can be checked under a dissection scope. Alternatively, the pipettes can be tested for relative water flow rates.
3. Complete NB medium should be used for no more than ~1–2 weeks following the addition of supplements. It is generally advisable to combine supplements and NB medium the day before the cell preparation.
4. Prepare and thaw all reagents and solutions prior to starting the cell preparation to avoid unnecessary delays. For example, allowing cells to sit for more than ~15 min on ice can lead to clumping. Any leftover solutions should be discarded and not re-frozen. Also discard enzyme solutions stored for >3 months since their use may result in lower cell yields.
5. We typically receive 10 pups per litter. Cerebella from 5 pups are optimal for one CGN preparation, and a single litter is used for 2 cell preps on the same day. Increasing the number of pups does not proportionally increase cell yield, and using fewer pups reduces cell numbers. The age of the pups is also an important factor: using pups older than P7 or younger than P6 reduces yields. This likely reflects the presence of maximal numbers of CGNPs within the murine EGL during this period.
6. Take care to ensure that a distinct interface exists between the two Percoll (pink and blue) layers. Also avoid using Percoll solutions (35 and 60 %) more than 3 months old, which can form precipitates.
7. Proper trituration is important to ensure that isolated cells consist mainly of CGNPs. Excessive trituration or bubbles or frothing can diminish cell yield and/or adversely affect the quality of the cell preparation. Insufficient trituration can result in larger cell aggregates and low yield.
8. Overlay cells onto the Percoll gradient very slowly, at the rate of approximately one drop every 2 s, down the inner side of the tube wall with the tube tilted at an angle and on ice. This minimizes mixing of the cells with the Percoll layer, which can compromise the quality and yield of the CGN preparation. Also avoid disturbing the gradient during its handling.
9. It is crucial to centrifuge the gradient with the brake turned off to maintain distinct gradient interfaces. Poor gradient boundaries can lead to a reduced cell yield and contamination with glial cells.
10. Glial cells sediment within the upper interface between the Tyrode's and 35 % Percoll layers. These are normally discarded;

however, if cerebellar glial cultures are desired, this interface can be collected separately for further washing and plating.

11. Avoid removing significant amounts of the 60 % Percoll (blue colored) layer since higher concentrations of Percoll can precipitate during subsequent steps and also can be toxic to cells.

12. Pre-plating removes remaining glial cells from the cell preparation due to their differential adherence to PLY-coated plastic [18]. CGNs are collected and enriched by repeated washing from the plates.

13. Take care to also wash off cells from the plate edges. A balance between adequate but not excessive washing results in good CGN yield without significant glial cell contamination.

14. CGNs are typically plated onto PLY-coated culture dishes. However, additional substrates may be used depending on experimental circumstances. For example, a laminin coating in addition to the PLY is frequently employed (*see* **Note 1**).

15. The CGN plating density is important for good cell survival, with low cell densities resulting in increased cell death [16]. These conditions are ideal for studying cell survival/death mechanisms [21], but complicate analyses related to neuronal differentiation and mature neuronal function. An optimal plating density for promoting cell survival is in the range of ~3×10^5 cells/cm^2. Note also that culture conditions for optimal CGN survival can vary depending on the rodent species or mouse strain used [22]. Suggested plating densities of purified CGNs that promote survival for different culture dish formats are as follows:

Plate type	Area (cm^2)	Plating cell density
6-well	9.5	3×10^6
12-well	3.8	1–1.3×10^6
24-well	1.9	0.6×10^6

16. Typically replace 50 % of culture medium approximately every 2–3 days. This replenishes nutrients while retaining secreted factors that may promote neuronal function and survival. Complete replacement of medium may be required for certain experiments (e.g., to change culture conditions) and is not deleterious.

17. This gradient step serves to remove cellular debris generated during the tissue disruption steps.

18. Enriched CGNs are typically plated on PLY-coated dishes. These cells also tend to be "heartier" than gradient-purified preparations.

19. Adjust the recommended cell density based on the length of time cells will be treated with Shh. The doubling time of CGNPs in the presence of Shh is ~48 h.

References

1. Timmann D, Daum I (2007) Cerebellar contributions to cognitive functions: a progress report after two decades of research. Cerebellum 6(3):159–162
2. Schmahmann JD, Caplan D (2006) Cognition, emotion and the cerebellum. Brain 129(Pt 2): 290–292
3. Ito M (2006) Cerebellar circuitry as a neuronal machine. Prog Neurobiol 78(3–5):272–303
4. Kilpatrick DL, Wang W, Gronostajski R, Litwack ED (2010) Nuclear factor I and cerebellar granule neuron development: an intrinsic-extrinsic interplay. Cerebellum. doi: 10.1007/s12311-010-0227-0
5. Millen KJ, Gleeson JG (2008) Cerebellar development and disease. Curr Opin Neurobiol 18(1):12–19
6. Wang W, Crandall JE, Litwack ED, Gronostajski RM, Kilpatrick DL (2010) Targets of the nuclear factor I regulon involved in early and late development of postmitotic cerebellar granule neurons. J Neurosci Res 88(2):258–265
7. Wang W, Stock RE, Gronostajski RM, Wong YW, Schachner M, Kilpatrick DL (2004) A role for nuclear factor I in the intrinsic control of cerebellar granule neuron gene expression. J Biol Chem 279(51):53491–53497
8. Kilpatrick DL, Wang W, Gronostajski R, Litwack ED (2012) Nuclear factor I and cerebellar granule neuron development: an intrinsic-extrinsic interplay. Cerebellum 11(1):41–49
9. Goldowitz D, Hamre K (1998) The cells and molecules that make a cerebellum. Trends Neurosci 21(9):375–382
10. Hatten ME, Alder J, Zimmerman K, Heintz N (1997) Genes involved in cerebellar cell specification and differentiation. Curr Opin Neurobiol 7(1):40–47
11. Altman J (1972) Postnatal development of the cerebellar cortex in the rat. 3. Maturation of the components of the granular layer. J Comp Neurol 145(4):465–513
12. Hamori J, Somogyi J (1983) Differentiation of cerebellar mossy fiber synapses in the rat: a quantitative electron microscope study. J Comp Neurol 220(4):365–377
13. Wang W, Shin Y, Shi M, Kilpatrick DL (2011) Temporal control of a dendritogenesis-linked gene via REST-dependent regulation of nuclear factor I occupancy. Mol Biol Cell 22(6): 868–879
14. Lin X, Bulleit RF (1996) Cell intrinsic mechanisms regulate mouse cerebellar granule neuron differentiation. Neurosci Lett 220(2): 81–84
15. Powell SK, Rivas RJ, Rodriguez-Boulan E, Hatten ME (1997) Development of polarity in cerebellar granule neurons. J Neurobiol 32(2): 223–236
16. Gao WO, Heintz N, Hatten ME (1991) Cerebellar granule cell neurogenesis is regulated by cell-cell interactions in vitro. Neuron 6(5):705–715
17. Wang W, Qu Q, Smith FI, Kilpatrick DL (2005) Self-inactivating lentiviruses: versatile vectors for quantitative transduction of cerebellar granule neurons and their progenitors. J Neurosci Methods 149(2):144–153
18. Hatten ME (1985) Neuronal regulation of astroglial morphology and proliferation in vitro. J Cell Biol 100(2):384–396
19. Wechsler-Reya RJ, Scott MP (1999) Control of neuronal precursor proliferation in the cerebellum by Sonic Hedgehog. Neuron 22(1): 103–114
20. Nagata I, Nakatsuji N (1990) Granule cell behavior on laminin in cerebellar microexplant cultures. Brain Res Dev Brain Res 52(1–2): 63–73
21. Gallo V, Kingsbury A, Balazs R, Jorgensen OS (1987) The role of depolarization in the survival and differentiation of cerebellar granule cells in culture. J Neurosci 7(7):2203–2213
22. Fujikawa N, Tominaga-Yoshino K, Okabe M, Ogura A (2000) Depolarization-dependent survival of cultured mouse cerebellar granule neurons is strain-restrained. Eur J Neurosci 12(5):1838–1842

Chapter 6

Preparation of Primary Cultured Dopaminergic Neurons from Mouse Brain

Won-Seok Choi, Hyung-Wook Kim, and Zhengui Xia

Abstract

Dopaminergic neurons are involved in a variety of normal brain functions; degenerations of these neurons cause diseases in human. Investigation of how dopaminergic neurons respond to extracellular signals and molecular mechanisms regulating dopaminergic neuron survival and death often requires reliable, reproducible, and high-quality primary cultures of dopaminergic neurons. Here, we described methods to dissect and culture these neurons from embryonic mesencephalon of mouse brain. We utilize coverslips made from Aclar film to maximize the number of surviving dopaminergic neuron in the culture and immunocytochemistry of tyrosine hydroxylase (TH) to identify dopaminergic neuron.

Key words Dopaminergic neuron, Primary culture, Immunocytochemistry

1 Introduction

Midbrain dopaminergic neurons project axons to many forebrain regions, including striatum, amygdala, septum, olfactory tubercle, and frontal cortex [1, 2]. These neurons are involved in various brain functions including control of motor behaviors and mediation of reward behaviors [3, 4]. Loss of dopaminergic neurons in the substantia nigra pars compacta (SNpc) causes Parkinson's disease [5]. Dopaminergic neurons in the midbrain are also one of the targets of major classes of addictive drugs [6]. Overactivity of dopaminergic neurons may be a factor in psychotic behavior including schizophrenia [6, 7]. Consequently, primary cultured midbrain dopaminergic neurons provide a very useful model system to investigate signaling events of dopaminergic neurons and their regulation at a cellular level, including their mis-regulation in various disease states. We have developed a method to culture primary dopaminergic neurons from embryonic (E) day 14 mouse or rat ventral mesencephalon (midbrain) [10]. This method is reproducible and cost-effective and yields about 1–5 % TH-positive,

Renping Zhou and Lin Mei (eds.), *Neural Development: Methods and Protocols*, Methods in Molecular Biology, vol. 1018, DOI 10.1007/978-1-62703-444-9_6, © Springer Science+Business Media, LLC 2013

dopaminergic neurons, consistent with other reports in the literature [8]. We also describe in detail the immunocytochemistry procedure for identifying and quantifying TH-positive, dopaminergic neurons.

2 Materials

Prepare all solutions using distilled and deionized or similar quality water. All reagents are prepared and stored at room temperature or 4 °C unless indicated otherwise.

2.1 Plate or Aclar Film Coating

1. Cell culture plates (24, 48, or 96 wells; *see* **Note 1**).
2. Aclar film sheets (Electron Microscopy Sciences, Hatfield, PA; *see* **Note 2**).
3. 70 % ethanol.
4. Poly-D-lysine and laminin (BD Bioscience, San Jose, CA), stored at −80 °C as small aliquots for single use (*see* **Note 3**).

2.2 Primary E14 Ventral Mesencephalic Culture

1. Embryonic day 14 (E14) mouse or rat embryos from pregnant dam (Harlan, Indianapolis, IN; *see* **Note 4**).
2. Dissection tools: large scissors, medium to small scissors, and fine forceps (Fine Science Tools, Foster City, CA; *see* **Note 5**).
3. Dissection scope (Nikon SMZ-10A, Melville, NY).
4. Phosphate-buffered saline (PBS), pH 7.2 (Gibco, Carlsbad, CA; *see* **Note 6**).
5. Primary culture media: Dulbecco's Modified Eagle's Medium (DMEM) (Sigma-Aldrich, Inc., St. Louis, MO) supplemented with 10 % fetal bovine serum (FBS) (Gibco), 100 U/mL penicillin and 100 μg/mL streptomycin, 25 mM HEPES, 4 mM glutamine, and 30 mM glucose (Sigma). Supplements are stored at −20 °C as small aliquots. Media are made fresh for each culture (*see* **Note 7**).
6. Filtered 0.4 % trypan blue solution in PBS.
7. Serum-free media: DMEM-F12 (Gibco) supplemented with 1 % N2 supplement (Invitrogen, Carlsbad, CA) and 10 μg/mL bovine serum albumin (BSA, Sigma). Supplements are stored at −20 °C as a small aliquot. Media are freshly made before each culture (*see* **Note 8**).

2.3 Immunocytochemistry and Quantification of Dopaminergic Neurons

1. Fixing solution: 4 % paraformaldehyde (Fisher Scientific, Pittsburgh, PA)/4 % sucrose (Mallinckrodt Baker, Phillipsburg, NJ) in PBS, pH 7.5. Heat 700 mL water to 60 °C in fume hood. Add 40 g paraformaldehyde and stir. To help dissolve, add 5N sodium hydroxide drop-wise until the solution is clear. Add

40 g sucrose and stir until dissolved (*see* **Note 9**). Add 100 mL of 10× PBS. Adjust pH to 7.5 and bring to a final volume of 1,000 mL. Solution can be stored as aliquots at −20 °C.

2. Postfix washing solution: 10 mM glycine in PBS (*see* **Note 10**).
3. Permeabilization solution: 0.5 % IPEGAL (Sigma) in PBS.
4. 0.1 % Triton X-100 (Sigma) in PBS (PBST).
5. Blocking solution: 2.5 % BSA (Equitech-Bio, Inc., Kerrville, TX), 5 % horse serum, and 5 % goat serum (Gibco) in PBST. Aliquot serum and store at −20 °C.
6. Humidified container: 10 cm Petri dish with damp filter paper or paper towel in it.
7. Primary antibodies: rabbit polyclonal antibody against TH (1:5,000; Pel-Freez, Rogers, AR) or mouse monoclonal antibody against tyrosine hydroxylase (TH; 1:500; Sigma) (*see* **Note 11**).
8. Secondary antibodies: Alexa Fluor 488, 568, or 660 goat anti-mouse IgG (1:1,000; Molecular Probes, Carlsbad, CA) and Alexa Fluor 488, 568, or 660 goat anti-rabbit IgG.
9. Nuclear staining solution: 2.5 μg/μL Hoechst 33258 (Molecular Probes) in PBST. Solution can be stable at 4 °C for several months.
10. Aqua-Poly/Mount (Polysciences, Inc., Warrington, PA).
11. Microscope equipped with fluorescence and digital camera (Axiovert 200 M, Zeiss, Thornwood, NY).

3 Methods

All procedures with animals should be done following the regulation of Institutional Animal Care and Use Committee.

3.1 Plate or Aclar Coating

1. Punch out Aclar coverslips, using disc punch, from the film sheet to discs of 1 cm in diameter (for 24-well plates) or other sizes appropriate for the sizes of the wells, at least 2 days prior to culture. Collect coverslips in a 50 mL conical tube and sterilize the coverslips with 70 % ethanol. Vortex several times and soak the coverslips overnight. Coverslips can be kept in 70 % ethanol until the experiment (*see* **Note 12**).
2. Aspirate 70 % ethanol (*see* **Note 13**).
3. Add autoclaved water to completely cover coverslips, vortex, and aspirate off water. Repeat five times to wash coverslips.
4. Place one Aclar coverslip per well in 24-well plate and coat coverslip with 50 μg/mL poly-D-lysine and 5 μg/mL laminin in water, 600 μL per well. Make sure that the entire coverslip is fully covered in coating solution (Fig. 1a, b).

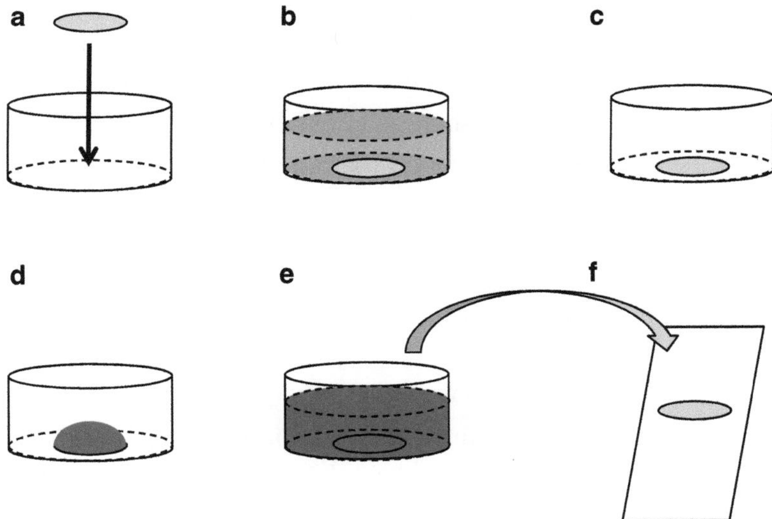

Fig. 1 Aclar coating, primary culture, and mounting of mesencephalic neurons. (**a**) Place Aclar coverslips in the wells of a 24-well plate. (**b**) Coat coverslips with poly-D-lysine and laminin. (**c**) Wash coverslips and dry completely. (**d**) Plate cells onto the coverslip as a microisland form. (**e**) Add medium, incubate, and treat as needed. (**f**) Mount coverslips onto glass slides

5. Place the 24-well plate with coverslips in coating solution in a 37 °C tissue culture incubator overnight (*see* **Note 14**).
6. On day of culture, wash plates twice with double-distilled water, making sure that coverslips are completely immersed in water during each wash (*see* **Note 15**). Let plate dry completely in tissue culture hood with lid off (Fig. 1c).

3.2 Primary E14 Ventral Mesencephalic Culture

1. All dissection tools should be sterilized with 70 % ethanol.
2. Euthanize pregnant dam using the procedure appropriate for fresh tissue harvesting that is approved by Institutional Animal Care and Use Committee, such as CO_2 euthanasia followed by decapitation.
3. Spray 70 % ethanol on abdomen and extract embryos by making a longitudinal cut from the sternum to the tail and two lateral cuts across the lower abdomen using large scissors. Excise through the muscle wall to expose the uterine horn. Eliminate both sides of the uterine horn and remove carefully the embryonic sacs from each embryo using small or medium scissors. Place all the embryos in a dish with PBS on ice (*see* **Note 16**). Decapitate the embryos using small scissors and place the heads in PBS solution on ice.
4. Dissect ventral mesencephalon (VM) in cold PBS pH 7.2 on ice using fine forceps under a dissecting microscope. To excise VM region, make two vertical cuts on the brain, one to remove the diencephalon from the midbrain and the other at the isthmus to remove the hindbrain from the midbrain.

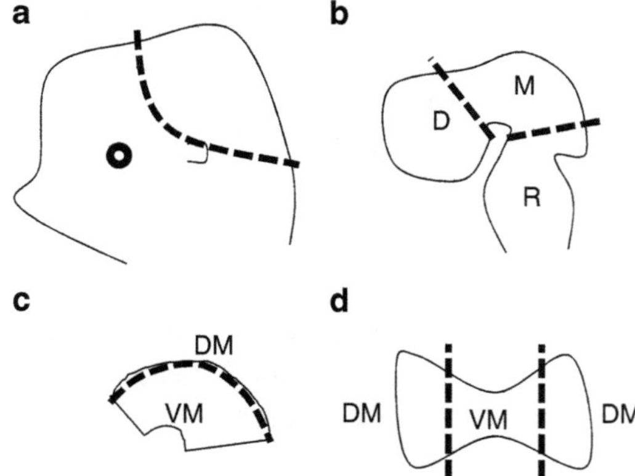

Fig. 2 Excision of ventral mesencephalon (VM). (**a**) Remove the skin and the scull to expose brain. (**b**) Make two vertical cuts, as indicated by *dashed lines*, one separating the diencephalon (D) from the midbrain (M) and the other at the isthmus separating the midbrain from the hindbrain (rhombencephalon, R). (**c**) Make one longitudinal cut along the top of the neural tube separating the dorsal mesencephalon (DM) of the midbrain into two spreading "wings." (**d**) Make two horizontal cuts to remove the dorsal mesencephalon from the ventral mesencephalon

Then make a longitudinal cut along the top of the neural tube separating the dorsal mesencephalon (DM) of the midbrain into two spreading "wings." Make two horizontal cuts to remove the dorsal mesencephalon from the ventral portion. See diagram in Fig. 2 for details.

5. Cut tissue blocks mechanically with fine forceps into smaller pieces to help further dissociation (*see* **Note 17**).
6. Transfer tissue pieces to a 15 mL conical tube containing DMEM. Let tissue pieces settle and remove the medium (*see* **Note 18**).
7. Add 5 mL of DMEM and incubate for 10 min in a 37 °C water bath.
8. Pre-warm primary culture media at 37 °C (DMEM supplemented with 10 % FBS, 25 mM HEPES, 30 mM glucose, 100 U/mL penicillin, and 100 μg/mL streptomycin).
9. After incubation at **step 7**, remove medium and add 5 mL of primary culture media.
10. Dissociate the tissue pieces by pipetting five to seven times using a 1 mL pipette tip (*see* **Note 19**).
11. Take 10 μL of the cell suspension and dilute tenfold with trypan blue solution.
12. Count the number of cells using hemocytometer.
13. Seed remaining cells at **step 11** at a density of 1.0×10^6 cells/mL by adding one drop (100 μL) per coverslip to form micro-island. Handle the plate very carefully to ensure that the liquid

drop stays on Aclar coverslips but does not spread outside of the coverslip (Fig. 1d; *see* **Note 20**).

14. Incubate cells overnight at 37 °C in a 7 % CO_2 incubator (*see* **Note 21**).
15. On next day (day in vitro 1, DIV1), add 400 μL pre-warmed primary culture media. Add media carefully along the side of the tissue culture well to avoid disturbing cells (*see* **Note 22**).
16. On DIV3, change half of the media by removing 250 μL of the conditioned media and adding 250 μL of fresh pre-warmed (37 °C) media.
17. On DIV5, replace half of the conditioned media with fresh DMEM-F12 media containing 1 % N2 supplement and 10 μg/mL BSA (*see* **Note 8**).
18. On DIV7, change half of the conditioned media with fresh DMEM-F12 media supplemented with 1 % N2 supplement and 10 μg/mL BSA with appropriate drug concentration for any desired drug treatment (*see* **Note 23**).

3.3 Immunocytochemistry and Quantification of Dopaminergic Neurons

1. Remove the culture media. Fix cells with fixing solution (4 % paraformaldehyde/4 % sucrose in PBS, pH 7.5) at room temperature for 15 min (*see* **Note 24**). If cells are poorly attached, do not remove the culture media. Instead, add equal volume of 8 % paraformaldehyde/8 % sucrose in PBS, pH 7.5 directly to the media to fix cells.
2. Wash cells three times with 10 mM glycine in PBS for 5 min each. Cells can be kept in the last wash solution and stored at 4 °C until staining.
3. Permeabilize cells with 0.5 % IPEGAL in PBS at room temperature for 30 min (*see* **Note 25**).
4. Wash once with PBST.
5. Incubate with blocking buffer (PBST 2.5 % BSA, 5 % horse serum, 5 % goat serum) for 1–4 h at room temperature or overnight at 4 °C (*see* **Note 26**).
6. Remove coverslips from each well of the tissue culture plate and lay on a piece of parafilm placed at the bottom of a humidified container. Add 40 μL of primary antibody diluted in blocking solution onto each coverslip. Incubate for 2 h at room temperature or overnight at 4 °C (*see* **Note 27**).
7. Wash each coverslip three times for 10 min with 500 μL of PBST.
8. Remove PBST and immediately add 40 μL of secondary antibody diluted in blocking solution on each coverslip. Incubate at room temperature for 1 h (*see* **Note 28**).
9. Wash each coverslip for 10 min with 500 μL of PBST.

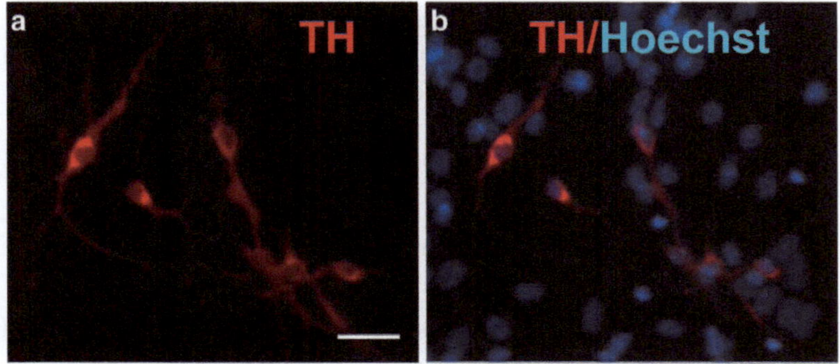

Fig. 3 Staining of (**a**) TH (rabbit, 1:5,000) alone and (**b**) together with Hoechst to visualize nuclei. Scale bar = 25 μM

10. Incubate cells with 2.5 μg/μL Hoechst 33258 in PBST for 10 min to stain nuclei.
11. Wash each coverslip one to two times for 10 min with 500 μL of PBST.
12. Add one drop of Aqua-Poly/Mount onto a glass slide and mount coverslip onto it (Fig. 1f). Air-dry the mounted coverslip in the dark.
13. Immunostained dopaminergic neurons are detected using fluorescence microscopy. *See* Fig. 3 and Refs. 9–11 for an example.
14. Count the number of all TH-positive cells per Aclar coverslip [9–11].

4 Notes

1. Cell culture plates should be used.
2. In our experience, Aclar film is better than glass coverslips for cells to attach and grow.
3. Poly-D-lysine and laminin could be stored as mixed concentrated stock solution at −80 °C.
4. E12–15 mouse or rat embryos can be used for the culture. However, embryos earlier than E14 are more difficult to dissect because of their small sizes, while the survival rate of primary TH+ neurons cultured from later stages (E16 or older) tends to be lower.
5. Select the size of the forceps to fit to the hands best.
6. PBS should be sterilized before use. After sterilization, the lid should be open in the culture hood.
7. Depending on the existing concentration of glucose, add proper amount of additional glucose to make the final

concentration approximately 30 mM. After mixing, the media should be filter-sterilized, aliquoted, and stored at 4 °C to minimize contamination.

8. BSA is optional and can be excluded especially if the chemical used for treatment of the cells in one's experiment binds to BSA tightly, which lowers the effective concentration of the drug.
9. When weighing paraformaldehyde in the chemical hood, wear a mask and gloves. Dissolve the paraformaldehyde completely before taking the solution out of the hood. The final solution could be filtered to remove any undissolved residuals.
10. Glycine solution should be freshly made, kept in the refrigerator, and used within several hours.
11. Sheep or chicken anti-TH antibodies are also available.
12. Make sure that coverslips are completely covered in 70 % ethanol.
13. In Subheading 3.1, **steps 1–6** should be done in a tissue culture hood.
14. Incubation for 4–6 h could also work. Avoid incubation longer than 24 h.
15. Push the coverslip down to the bottom using 1 mL pipette tip.
16. Change the dish with fresh PBS to remove blood from the embryos.
17. Cut the tissue no more than three to four times to minimize cell loss.
18. Do not use vacuum to aspirate medium from this point. Use a glass pipette or 1 mL pipette to remove medium to avoid tissue loss. DMEM washing could be skipped and the tissue in the DMEM could be directly incubated at 37 °C and proceed to **step 9**.
19. Avoid over-dissociation. Pipette up and down no more than seven times to dissociate tissue. Care must be taken to pipette gently and avoid foaming or bubbling which lowers the survival rate of neurons.
20. The microisland could be made with 70–100 μL cell suspension.
21. Move the plate with extreme care not to shake it to avoid breaking down of the microislands. Place the plate in the most stable place like along the wall inside the incubator and mark it clearly so other investigators do not touch or move the plates by accident.
22. Always pre-warm media to 37 °C in a water bath before adding to cells. Add media very slowly and gently especially on DIV1 to minimize the impact on the cells growing in the microisland.

23. Make sure to prepare double concentration of drug in the serum-free media because it will be diluted twofold after being added to the remaining same volume of media.
24. Fix cells no longer than 30 min to avoid over-fixation that could increase nonspecific staining.
25. Alternatively, cells could be permeabilized by incubating in ice cold methanol at −20 °C for 10 min and washing five times with PBS.
26. For faster staining, blocking could be done in 30 min. Shaking is not recommended because neurons are easily detached.
27. Be careful not to dry the cells from this step onward. Drying can result in massive nonspecific staining.
28. Secondary antibody could be incubated up to 2 h to increase staining intensity.

Acknowledgments

This work was supported by the National Institutes of Health (ES013696 and ES 012215 to Z. X.).

References

1. Fallon JH, Loughlin SE (1985) Substantia nigra. In: Paxinos GP (ed) The rat nervous system. Academic Press, New York, pp 353–369
2. Swanson LW (1982) The projections of the ventral tegmental area and adjacent regions: a combined fluorescent retrograde tracer and immunofluorescence study in the rat. Brain Res Bull 9:321–353
3. Björklund A, Lindvall O (1975) Dopamine in dendrites of substantia nigra neurons: suggestions for a role in dendritic terminals. Brain Res 83:531–537
4. Le Moal M, Simon H (1991) Mesocorticolimbic dopaminergic network: functional and regulatory roles. Physiol Rev 71:155–234
5. Moore DJ, West AB, Dawson VL, Dawson TM (2005) Molecular pathophysiology of Parkinson's disease. Annu Rev Neurosci 28: 57–87
6. Fibiger HC, Phillips AG (1986) Reward, motivation, cognition, psychobiology of mesotelencephalic dopamine systems. In: Bloom FE (ed) Handbook of physiology: the nervous system, vol IV, 4th edn. Waverly Press, Baltimore, pp 647–675
7. Barch DM, Ceaser A (2012) Cognition in schizophrenia: core psychological and neural mechanisms. Trends Cogn Sci 16(1):27–34, Epub 2011 Dec 12
8. Shimoda K, Sauve Y, Marini A, Schwartz JP, Commissiong JW (1992) A high percentage yield of tyrosine hydroxylase-positive cells from rat E14 mesencephalic cell culture. Brain Res 586:319–331
9. Klintworth H, Newhouse K, Li T, Choi WS, Faigle R, Xia Z (2007) Activation of c-Jun N-terminal protein kinase is a common mechanism underlying paraquat- and rotenone-induced dopaminergic cell apoptosis. Toxicol Sci 97:149–162
10. Choi WS, Kruse SE, Palmiter RD, Xia Z (2008) Mitochondrial complex I inhibition is not required for dopaminergic neuron death induced by rotenone, MPP+, or paraquat. Proc Natl Acad Sci USA 105:15136–15141
11. Choi WS, Abel G, Klintworth H, Flavell RA, Xia Z (2010) JNK3 Mediates Paraquat- and Rotenone-Induced Dopaminergic Neuron Death. J Neuropathol Exp Neurol 69: 511–520

Chapter 7

Culturing Astrocytes from Postnatal Rats

Yangyang Huang and Cheryl F. Dreyfus

Abstract

The use of cultures has informed us of functions and regulation of astrocytes that were previously unknown. This chapter details the methods that result in such cultures.

Key words Astrocyte, Cell culture, Rat, Postnatal, Glial cultures

1 Introduction

Astrocytes are a prominent glial cell type in the central nervous system (CNS) that is being found to exhibit multiple important functions (reviewed in refs. 1–3). For example, astrocytes regulate the ionic milieu in the intercellular space, provide trophic support to neurons, regulate levels of neurotransmitters, participate in formation of the blood–brain barrier, and participate in scar formation following CNS injury. However, the biochemical and physiological properties of astrocytes are difficult to study in the intact brain and spinal cord due to the complexity of these regions. To address this problem, multiple laboratories have adopted astrocyte cultures as an approach to generate a relatively well-defined cell population with which to study biochemical and physiological functions of astrocytes at a single-cell or molecular level. Many different culture methods have been utilized, including brain slice cultures and mixed glial cocultures, as well as highly enriched astrocyte cultures. The slice cultures or mixed cell culture methods attempt to create accessible systems that can examine function in a mixed cell environment. The isolated astrocyte cultures attempt to examine function of astrocytes divorced from other cell types.

One of the earliest methods to obtain relatively pure astrocyte cultures was developed by McCarthy and de Vellis [4]. This method takes advantage of the observation that when mixed brain cultures are grown, for 7–12 days, astrocytes settle and attach to a polylysine substrate, while oligodendrocytes settle on top. If these mixed

Renping Zhou and Lin Mei (eds.), *Neural Development: Methods and Protocols*, Methods in Molecular Biology, vol. 1018, DOI 10.1007/978-1-62703-444-9_7, © Springer Science+Business Media, LLC 2013

cultures are derived from postnatal animals, neurons die in a rich serum medium. Selective detachment of oligodendrocytes in the top layer is achieved by shaking culture flasks on an orbital shaker for 15–18 h at 37 °C and then removing these loosely adherent cells resulting in a relatively enriched astrocyte preparation.

Here we present our current method [5] for culturing isolated astrocytes that is derived from the McCarthy and de Vellis approach. Before adopting this approach, it must be appreciated that the cells cultured from postnatal animals may be representative of reactive astrocytes *in vivo* [6]. Moreover, cells grown in isolation may be distinct from cells interacting with other populations in the intact central nervous system. Nevertheless, the use of these isolated astrocyte cultures continues to inform us about astrocyte function and is leading to further assessment of astrocyte function in vivo, recently aided by transgenic mice models [7] in which manipulations can be directed specifically to astrocyte populations.

It is to be noted that materials and methods for growing astrocytes and oligodendrocytes as isolated cultures are similar initially. They become distinct when one or the other of these glial populations are the cells of choice (see accompanying Chapter 8). Differences in the two techniques are indicated by ****.

2 Materials

2.1 Equipment and Lab Supplies

1. Horizontal laminar flow hood.
2. Time-mated Sprague–Dawley rats (Taconic) to generate P0–P2 rat pups (*see* **Note 1**).
3. Dissection kit (*see* **Note 2**). This kit contains 1 pair of serrated 12″ thumb dressing forceps, 4 pairs of curved #7 Dumont forceps, 4 scalpel handles #3 4″, sterile stainless steel #11 surgical blades, 1 pair of straight Bonn scissors, 3.5″, and 4 pairs of straight #5 Dumont forceps.
4. Dissecting microscope.
5. Bench centrifuge.
6. Humidified incubator set to 37 °C with 5 % CO_2/95 % air (*see* **Note 3**).
7. 37 °C water bath.
8. Orbital shaking incubator: must be capable of shaking at 400 or 250 rpm while maintaining a temperature of 37 °C for 16–18 h (*see* **Note 4**).
9. Hemocytometer (VWR Scientific Counting Chamber, Stock #15170-173).
10. Bunsen burner.
11. 15 and 50 ml conical tubes.

12. Tissue culture-treated 75 cm^2 flasks (Nunc, 156472) (*see* **Note 5**).
13. 100 mm plastic petri dishes (Becton, Dickinson and Company, 100 mm Cell Cultureware, Catalog #353003).
14. 35 mm plastic petri dishes (Becton, Dickinson and Company, 35 mm Cell Cultureware, Catalog #354456).
15. Sterile 10 ml pipettes.
16. 100 mm glass petri dishes for dissection (*see* **Note 6**).
17. 35 mm glass petri dishes for dissociation of tissue (*see* **Note 6**).
18. Autoclaved Pasteur pipettes.
19. Sterile filters (Millipore, Steriflip Vacuum Filtration System, Catalog #SCGP00525).
20. Stericup filter units (Millipore Catalog #SCGVU11RE).
21. Two buckets with ice.
22. Two glass beakers (100 ml).
23. Black garbage bags.

2.2 Reagents

1. Poly-D-lysine (Sigma P0899-100MG).
2. Eagle's MEM with Earle's salts and L-glutamine (MediaTech, MT-10-010CV).
3. Fetal bovine serum (FBS) (Serum Source International, Heat Inactivated Premium Quality FBS, Catalog # FB02-100HI) (*see* **Note 7**).
4. Glucose (Sigma, D-(+)-Glucose, Catalog #G8270).
5. Penicillin/streptomycin (10,000 U/ml penicillin, 10,000 μg/ml streptomycin) (Gibco, Pen/Strep, Catalog #15140-148).
6. 80 % ethanol.
7. Phosphate-buffered saline (PBS) power, pH 7.4 (Sigma-Aldrich, #P3813).
8. Sodium metasilicate nonahydrate (Sigma, S4392).
9. Hydrochloric acid (36.5–38 % HCl, J.T.Baker, 9535–04).

****10. Cytosine β-D-arabinofuranoside hydrochloride (Ara-C) (Sigma, C6645-100 mg).

****11. Trypsin/EDTA (10×) (Gibco #15400-054-100 ml).

2.3 Solutions

All solutions are prepared using ultrapure water (prepared by purifying deionized water to attain a sensitivity of 18 MΩ cm at 25 °C) and analytical grade reagents. All solutions are filtered through 0.22 μm pore filter system (Millipore). Solutions are stored +4 °C, −20 °C, or −80 °C as noted by each specific reagent.

1. Sterile ultrapure water: add 500 ml of ultrapure water into the filter unit system and apply vacuum. Store at 4 °C.
2. 1× poly-D-lysine (0.1 mg/ml): dissolve 100 mg poly-D-lysine powder in 100 ml ultrapure water to make 10× solution and filter (*see* **Note 8**). Aliquot to 10 ml/conical tube and store the solution in −20 °C for up to 6 months. Dilute 10× to 1× solution in sterile ultrapure water prior to use.
3. 0.01 M phosphate-buffered saline (PBS), pH7.4. Add 1 pouch of PBS powder to 1,000 ml ultrapure water, dissolve, and filter. Store at 4 °C.
4. NM-15 contains the following: Eagle's MEM with Earle's salts and 2 mM L-glutamine, 15 % heat-inactivated fetal bovine serum, 6 mg/ml glucose, 0.5 U/ml penicillin, and 0.5 μg/ml streptomycin. To make NM-15 (200 ml), add 160 ml MEM into a sterile flask. In a 50 ml falcon tube, add 10 ml MEM, 1.2 g glucose, and 10 μl Pen/Strep. Mix and dissolve the powders well. Filter the 10 ml into the sterile flask containing MEM. Add 30 ml sterile FBS to the flask. Mix well and keep in +4 °C.
5. 1 % sodium metasilicate: dissolve 500 g sodium metasilicate in 2.5 l ultrapure water to make 20 % stock solution. Dilute 20 times in ultrapure water prior to use (*see* **Note 6**). Keep in a chemical hood at room temperature.
6. 1 % HCl: dilute 230 ml 36.5–38 % HCl to a final volume of 9 l. Keep in a chemical hood at room temperature.

****7. Ara-C (0.01 mM): weigh 0.045 g Ara-C and dissolve in 15 ml ultrapure water. Filter it and aliquot into 500 μl; store in −20 °C.

****8. 1× trypsin/EDTA: dilute the 10× trypsin/EDTA to 1× in sterile ultrapure water. Keep in 4 °C.

3 Methods

Carry out all procedures in a horizontal laminar flow hood until otherwise mentioned. The hood should be exposed to UV light for at least 30 min before use.

3.1 Preparation for Dissection

1. Before dissection, spray the laminar flow hood with 80 % ethanol to fully sterilize the surface.
2. Place dissecting tools on a sterile surface in the hood (*see* **Note 9**).
3. Retrieve two buckets with ice and place in hood. Fill two 100 ml beakers with 80 % ethanol. Label 1 and 2, and then place in the first ice bucket.

4. Coat 75 cm^2 culture flask with 8 ml 1× poly-D-lysine. Leave the poly-D-lysine on the flasks for more than 30 min or overnight in 37 °C incubator (*see* **Note 10**). Remove poly-D-lysine. Wash the flasks twice with 10 ml sterile water twice. Add 9 ml NM-15 to each flask and keep in the incubator until ready for use.
5. Place paper towel on the second bucket in the hood. Prepare three 100 mm sterile petri dishes (1 glass, 2 plastic). Add 15 ml sterile PBS to the glass petri dish and one plastic dish (*see* **Note 11**). Place on ice with paper towel. Leave the other empty plastic petri dish in the hood. Prepare one 35 mm small glass petri dish, add 2 ml NM-15 to it, and place on ice.
6. Set up the dissecting microscope (that has been wiped down with ethanol) in the hood.
7. Set up a body bag in garbage container.

3.2 Pup Dissection

1. Place postnatal 0–2-day-old rat pups on paper towel on ice until fully anesthetized (*see* **Note 12**). Decapitate the pups. Discard the bodies into the body bag.
2. With large dressing forceps, dip rat pup head quickly into beaker #1 of ice-cold 80 % ethanol, followed by beaker #2 of ice-cold 80 % ethanol.
3. Transfer sterilized head to empty, 100 mm petri dish.
4. Hold rat pup head in place with large dressing forceps. With curved #7 Dumont forceps, grasp loose skin on the animal's neck and pull forward towards the eyes so that the skull is fully exposed.
5. Holding head steady with large dressing forceps, carefully insert tip of small Bonn scissors into foramen magnum at the base of the skull. Cut the skull along midline fissure with the small scissors. Being careful not to damage underlying cortex, cut the skull along the midline towards the animal's nose (*see* **Note 13**).
6. With curved #7 Dumont forceps, peel back flaps of skull to expose underlying cerebral cortex.
7. Using curved #7 Dumont forceps, scoop underneath brain to sever cranial nerves. Then, lift brain from skull and place in the first 100 mm plastic petri dish with PBS on ice.
8. Place head in body bag (*see* **Note 14**).
9. Repeat **steps 3–8** for remaining pups.
10. Carefully attach sterile stainless steel #11 scalpel blades to blade handles.
11. Transfer 3–4 rat brains to 100 mm glass petri dish with curved #7 Dumont forceps.

12. Carefully isolate one brain under the microscope. Remove all meninges. Dissect out region of interest under the microscope. Place region in a 35 mm glass petri dish with 2 ml NM-15.
13. Repeat 11–12 until the remaining brains are all dissected. Discard the remaining tissue in the body bag.

3.3 Plating the Cells

1. Replace used blades with new, sterile stainless steel #11 blades. Carefully dice the tissue into small pieces.
2. Using the Bunsen burner, flame the tips of the glass Pasteur pipettes to make them smooth. Increase the length of flaming to make a series of pipettes with tips of reduced pore size.
3. With a glass Pasteur pipette, transfer all tissue and medium into a 15 ml conical tube.
4. Triturate with a series of glass pipettes, usually 2. Dissociate the tissue until a cloudy appearance is seen (*see* **Note 15**).
5. Make a 1–5 dilution by adding one drop of cell suspension to 4 drops of NM-15 using polished glass pipette. Mix well. Add 10 μl diluted cell suspension each to the top and bottom chambers of the hemocytometer.
6. Count cells using the hemocytometer (*see* **Note 16**).
7. Dilute the cells with NM-15 to a concentration of 1.5×10^7 cells/ml. Add 1 ml of the cells to 9 ml NM-15 in each 75 cm^2 flask. Gently shake each flask on horizontal plane to distribute cells evenly over poly-D-lysine-coated surface. The final plating density will be about 2×10^5 cells/cm^2.
8. Make sure that caps are placed loosely on each flask and place in 37 °C incubator (*see* **Note 17**).
9. The culture will reach confluence between 7 and 12 days.

3.4 Feeding the Cells

1. Feed the cells with fresh NM-15 every 3–4 days.
2. Aspirate approximately one-half of the medium. Replace one-half of the medium until confluent monolayers are established. After that time all the medium can be replaced. Add 37 °C prewarmed fresh NM-15.

3.5 Isolating and Culturing Astrocytes****

1. After 7–12 days in culture, secure flasks with caps tightened on orbital shaker set to 37 °C (*see* **Note 18**). Shake flasks twice at 400 rpm for 20 min each.
2. Aspirate 10 ml NM-15 and add 10 ml of prewarmed NM-15 into the flasks.
3. Incubate the flasks for at least 2 h.
4. Secure the flasks on orbital shaker set to 37 °C. Shake flasks at 250 rpm for 16–18 h at 37 °C.
5. Aspirate 10 ml NM-15 and add 10 ml of prewarmed NM-15 into the flasks. Supplement each flask with 100 μl of Ara-C

stock (0.01 mM) (*see* **Note 19**). Gently shake the flasks on a horizontal surface to mix the medium and Ara-C well. Leave the flasks in Ara-C for 3–5 days before replating cells for subculture and cell treatment.

6. Place petri dishes (35 mm or 100 mm) in laminar flow hood (*see* **Note 20**). Add enough 1× poly-D-lysine to coat bottom of each petri dish and incubate at 37 °C for 30 min. Aspirate 1× poly-D-lysine from dishes and rinse thoroughly with sterile ultrapure water twice. Add 0.5 ml NM-15 to 35 mm dishes or 6 ml NM-15 to 100 mm dishes. Place culture dishes in incubator until ready for use.
7. Aspirate medium in the flasks. Rinse with 10 ml PBS twice. Add 3 ml 1× trypsin/EDTA to each flask to cover the whole cell surface. Incubate at 37 °C for 5 min (*see* **Note 21**).
8. Tap flasks firmly on hood countertop to dislodge cells. Add 5 ml NM-15 to each flask to stop enzymatic reaction. Rinse the cells off flask surface with glass Pasteur pipette. Transfer cells to 50 ml conical tube.
9. Rinse each flask with 8 ml NM-15 and transfer it to the same 50 ml conical tube. Combine different flasks of cells in one conical tube.
10. Centrifuge the 50 ml conical tube at 2,000 rpm for 10 min. Aspirate off the supernatant carefully.
11. Using a polished glass pipette, carefully resuspend the cells in 2 ml NM-15.
12. Dissociate cells about 15 times with flame polished fine glass pipette.
13. Dilute 1:5 (1 drop of cells +4 drops of NM-15) and mix well. Place on hemocytometer.
14. Count cells (*see* **Note 16**).
15. Dilute the cell suspension to desired concentration using warm NM-15 and plate onto poly-D-lysine-coated dishes. Add 1 ml suspension (containing 1.8×10^5 cells) in NM-15 to 35 mm dishes and 2 ml suspension (containing 1.5×10^6 cells) in NM-15 to 100 mm dishes.
16. Mix well on a horizontal surface and place in the incubator.
17. Cells can be treated after 1–5 days depending on the experimental design.

4 Notes

1. Average litter size ranges from 10 to 20 pups.
2. Except for the sterile stainless steel #11 surgical blades, all dissecting tools are boiled in soapy water for 15 min after each

dissection. A bar of Ivory soap must be used that is cut into small pieces and added to ultrapure water. All tools are thoroughly rinsed in ultrapure water. Tools are dried and kept in test tubes cushioned with glass fiber. All dissection tools are then baked at 180 °C for 5 h for sterilization. It is best to autoclave the dissecting kit the day before this protocol is begun.

3. To avoid contamination, it is best to maintain a dedicated incubator for the culture of mixed glial cells. Avoid culturing bacteria and yeast in dedicated tissue culture incubator if possible.
4. A variety of shaking incubators exist and can be used for this protocol. This incubator must be able to hold 75 cm^2 flasks securely. Furthermore, it is again recommended to avoid using this incubator to shake bacterial cultures.
5. It is extremely important that tissue culture-treated, non-vent/close flasks be used for the culture of mixed glia. Culture flasks with filter caps are not compatible with this protocol, as the media will come into contact with the membrane during the shaking step and become contaminated.
6. Prior to use all glass dishes are boiled in a 1 % sodium metasilicate solution for 1 h in a chemical hood, soaked in a bucket of 1 % HCl in chemical hood overnight, followed by a bucket of ultrapure water overnight. Plates are rinsed thoroughly under ultrapure water, dried, and baked at 180 °C for 5 h for sterilization.
7. FBS must be heat inactivated to destroy complement proteins in serum which negatively affect cell survival. Heating at 56 °C for 30 min is recommended. Extended length of time or higher temperature may damage other growth factors in serum.
8. Poly-D-lysine is viscous and hard to dissolve. Add 10 ml ultrapure water to poly-D-lysine bottle. Pipette solution up and down many times with a 10 ml pipette to make sure all the powder dissolves completely. Transfer 10 ml poly-D-lysine to 90 ml water and filter it through 22 μm filter system. Because of the viscosity of poly-D-lysine solution, it is normal that the filtering process takes longer than other solutions.
9. Dissection tools are placed on a sterile surface that can be a plastic tissue culture dish or sterile gauze.
10. Prepare enough 1× poly-D-lysine-coated 75 cm^2 flasks to accommodate the number of cells to be cultured. The concentration of cells should be 1.5×10^7 cells/flask.
11. The plastic plate is used to keep the brains on ice and slow down tissue degradation. The glass plate is used to dissect under the microscope because it avoids damage due to scalpel blades. We usually transfer 3–4 brains each time from plastic plate to the glass plate for dissection. This preserves tissue freshness and minimizes contamination.

12. To determine that animals are anesthetized, pinch limbs to check responses before dissection.
13. The brain is extremely soft and is easily injured during this step. To avoid cutting into the cortex, it is best to work slowly and take care to keep the scissors pointed away from the brain. Alternatively, the skull of the animal is very soft (particularly at P0 and P1). It is sometimes possible to cut the skull using the pointed edge of the curved #7 Dumont forceps. If this is possible, simply make a cut with the forceps along the midline and then peel back the flaps of skull to expose the underlying cerebral hemispheres.
14. If at any point in the procedure a dissecting tool touches a non-sterile surface (such as the dissecting microscope), immediately begin to use one of the extra tools that were autoclaved with the dissecting kit.
15. Triturate about 20 times each with 2 glass pipettes until a homogenous cloudy solution appears. The objective is to triturate sufficiently to avoid tissue clumps and avoid mechanical damage.
16. The hemocytometer that we use has 25 small grid boxes in the center of each chamber, and its volume is 1 mm (wide) × 1 mm (long) × 0.1 mm deep (or 10^{-4} ml). We generally count bright cell bodies in four corner grid boxes and one center box both in the top and bottom chambers and add up numbers in 10 grid boxes. Total cell number in 2 ml suspension = (counted cell number in 10 small grid boxes/10) × 5 (dilution factor) × 2 (initial cell volume) × 0.25×10^6.
17. Flask caps are kept loose in the incubator for adequate gas exchanges.
18. Do not move flasks during the first 2 days of culture to allow better cell attachment. According to our initial plating density, astrocytes normally reach confluence at day 7. Shake cell only when astrocytes are confluent.
19. Mitotic inhibitor Ara-C is added to culture to kill any remaining proliferating cells. Since astrocytes are confluent at this stage, proliferation is contact inhibited and therefore astrocytes are not killed by Ara-C.
20. It is important to consider the nature of the experiment you wish to perform before plating the astrocytes. 35 mm^2 dishes are particularly amendable to immunocytochemical experiments, while 100 mm dishes are amenable to Western Blot analyses. We do not second passage astrocytes for experiments.
21. Trypsin/EDTA is prewarmed at 37 °C water bath for best reaction. Allow only 5 min for enough cell detachment.

Acknowledgment

The development of this protocol was supported by NIH HD 23315.

References

1. Ransom BR, Ransom CB (2012) Astrocytes: multitalented stars of the central nervous system. In: Milner R (ed) Methods in molecular biology 814. Human Press, New York, NY, pp 3–7
2. Oberheim NA, Goldman SA, Nedergaard M (2012) Heterogeneity of astrocytic form and function. In: Milner R (ed) Methods in molecular biology 814. Human Press, New York, NY, pp 23–45
3. Barres BA (2008) The mystery and magic of glia: a perspective on their roles in health and disease. Neuron 60:430–440
4. McCarthy KD, de Vellis J (1980) Preparation of separate astroglial and oligodendroglial cell cultures from rat cerebral tissue. J Cell Biol 85:890–902
5. Jean YY, Lercher LD, Dreyfus CF (2008) Glutamate elicits release of BDNF from basal forebrain astrocytes in a process dependent on metabotropic receptors and the PLC pathway. Neuron Glia Biol 4:35–42
6. Zamanian JL, Xu L, Foo LC et al (2012) Genomic analysis of reactive astrogliosis. The J of Neuroscie 32(18):6391–6410
7. Casper KB, Jones K, McCarthy KD (2007) Characterization of astrocyte-specific conditional knockouts. Genesis 45:292–299

Chapter 8

Culturing Oligodendrocyte Lineage Cells from Neonatal Rats

Clifton G. Fulmer and Cheryl F. Dreyfus

Abstract

The use of enriched oligodendrocyte lineage cell cultures has yielded insight into functions of these cells and regulatory mechanisms. This chapter details methods that result in such cultures.

Key words Oligodendrocytes, Myelin, Glia, Cell culture, Rat

1 Introduction

Mature oligodendrocytes are the myelinating cells of the central nervous system (CNS). In addition to producing myelin sheaths that facilitate salutatory conduction along axonal tracks, oligodendrocyte lineage cells also produce trophic factors and other molecules that can support the development and survival of neighboring neurons [1–4]. Thus, oligodendrocytes are critical to normal CNS function and loss of these cells or their impaired function may have clinical implications. Examination of oligodendrocyte biology in the normal case and after demyelinating insults and analysis of factors that influence these processes have become the focus of numerous studies. To evaluate these events one approach has been to isolate oligodendrocyte lineage cells in a simplified system and then apply information learned in this way to the more complex CNS using newly developed transgenic mice that permit the assessment of oligodendrocyte lineage cells in vivo.

A variety of methods have been developed to facilitate the initial process of growing relatively pure oligodendrocyte lineage cells in culture. In particular, many investigators have expanded upon a protocol first developed by Ken McCarthy and Jean de Vellis that takes advantage of the fact that viable neurons are not easily cultured from postnatal rat cerebral cortex, facilitating the isolation of pure mixed glial cultures with no neuronal contamination [5, 6]. When cultured from 7 to 12 days, these mixed rat glial cultures undergo stratification, with the astrocytic population adhering

Renping Zhou and Lin Mei (eds.), *Neural Development: Methods and Protocols*, Methods in Molecular Biology, vol. 1018,
DOI 10.1007/978-1-62703-444-9_8, © Springer Science+Business Media, LLC 2013

strongly to the tissue culture-treated substrate and the oligodendrocyte progenitors settling on this astrocyte monolayer. The loosely adherent oligodendrocyte progenitor cells can be selectively detached from the astrocyte monolayer with the sheer forces generated by an orbital shaker. After the collection and filtration of the progenitor-rich supernatant, it is possible to grow enriched oligodendrocyte cultures [6]. This basic approach has been modified further to eliminate microglial contamination by briefly plating these cells onto tissue culture-treated dishes [7]. The protocol results in highly enriched oligodendrocyte lineage cell cultures that are >95 % pure. Although immunopanning methods have been developed to enhance the purity of such cultures, they are limited because the yield of cells is low largely limiting the analysis of such cultures to morphological and physiological study. The modified McCarthy and de Vellis techniques remain valuable for biochemical analyses where the number of cells cultured is enhanced. Here, we present our current method for the culture of neonatal rat oligodendrocytes isolated via the shaking technique [8].

It is to be noted that Materials and Methods for growing astrocytes and oligodendrocytes as isolated cultures are similar initially. They become distinct when one or the other of these glial populations is the cell of choice (see accompanying Chapter 7). Differences in the two techniques are indicated by ****.

2 Materials

2.1 Equipment and Lab Supplies

1. Horizontal laminar flow hood.
2. Time-mated Sprague–Dawley rats (Taconic) to generate P0–P2 rat pups (*see* **Note 1**).
3. Dissecting kit (*see* **Note 2**): The kit contains 1 pair of serrated 12″ Thumb dressing forceps, 4 pairs of curved #7 Dumont forceps, 4 scalpel handles #3 4″, stainless steel #11 surgical blades, 1 pair of straight Bonn Scissors, 3.5″, and 4 pairs of straight #5 Dumont forceps.
4. Dissecting microscope.
5. Bench centrifuge.
6. Humidified tissue culture incubator set to 37 °C with 5 % CO_2 (*see* **Note 3**).
7. 37 °C water bath.
8. Orbital shaking incubator: Must be capable of shaking at 250 rpm while maintaining a temperature of 37 °C for 16–18 h (*see* **Note 4**).
9. Hemocytometer (VWR Scientific Counting Chamber, Stock #15170-173).
10. Bunsen burner.
11. 15 and 50 ml conical tubes.

12. Tissue culture-treated 75 cm^2 flasks (Nunc, 156472) (*see* **Note 5**).
13. 100 mm plastic petri dishes (Becton, Dickinson and Company, 100 mm Cell Cultureware, Catalog #353003).
14. 35 mm plastic petri dishes (Becton, Dickinson and Company, 35 mm Cell Cultureware, Catalog #354456).
15. Sterile 10 ml pipettes.
16. 100 mm glass petri dishes for dissection (*see* **Note 6**).
17. 35 mm glass petri dishes for dissociation of tissue (*see* **Note 6**).
18. Autoclaved Pasteur pipettes.
19. Sterile filters (Millipore, Steriflip Vacuum Filtration System, Catalog #SCGP00525).
20. Stericup filter units (Millipore Catalog #SCGVU11RE).
21. 2 buckets with ice.
22. 2 glass beakers (100 ml).
23. Black garbage bags.

****24. Non-tissue culture-treated (150 × 25 mM) culture dishes (Fisher Scientific, Sterile 150 × 15 mm Polystyrene Petri Dish, Catalog #0875714).

2.2 Reagents

1. Poly-D-lysine (Sigma, P-0899-100MG).
2. Eagle's MEM with Earle's salts and L-glutamine (MediaTech, MT-10-010CV).
3. Fetal bovine serum (Serum Source International, Heat Inactivated Premium Quality FBS, Catalog # FB02-100HI) (*see* **Note 7**).
4. Glucose (Sigma, D-(+)-Glucose, Catalog #G8270)
5. Penicillin/streptomycin (10,000 U/ml penicillin, 10,000 μg/ml streptomycin) (Gibco, Pen/Strep, Catalog #15140-148).
6. 80 % Ethanol.
7. Phosphate buffered saline (PBS) (Sigma-Aldrich, #P3813).
8. Sodium metasilicate nonahydrate (Sigma, S4392).
9. Hydrochloric acid (36.5–38 % HCl, J.T. Baker, 9535–04).

****10. Ham's F-12 Nutrient Mix (Gibco, Catalog #11765-054).

****11. Basal Medium Eagle (BME) (Gibco, Catalog #21010-046).

****12. Apo-transferrin (Sigma, Catalog #T2252).

****13. Insulin (Sigma, Catalog #I5500).

****14. 0.01 N hydrochloric acid.

****15. Progesterone (Sigma, Catalog #P-6149).

****16. Putrescine (Sigma, Catalog #P-7505).

****17. Selenium (Sigma, Catalog #S-5261).

****18. L-glutamine (200 mM) (Gibco, Catalog #25030-149).

****19. 3,3′,5-Triiodo-L-thyronine (T3, Sigma, Catalog #2752).

****20. L-thyroxine sodium salt pentahydrate (T4, Sigma, Catalog #2501).

****21. 0.1 M NaOH.

2.3 Solutions

All solutions are prepared using ultrapure water (prepared by purifying deionized water to attain a sensitivity of 18 MΩ cm at 25 °C) and analytical grade reagents. All solutions are filtered through a 0.22 μm pore filter system (Millipore). Solutions are stored at +4 °C, −20 °C, or −80 °C as noted by each specific reagent.

1. Sterile ultrapure water: Add 500 ml of ultrapure water into the filter unit system and apply vacuum. Store at 4 °C.
2. 1× Poly-D-lysine (0.1 mg/ml): Dissolve 100 mg poly-D-lysine powder in 100 ml ultrapure water to make 10× solution and filter (*see* **Note 8**). Aliquot to 10 ml/conical tube and store the solution in −20 °C for up to 6 months. Dilute 10× to 1× solution in sterile ultrapure water prior to use.
3. 0.01 M phosphate buffered saline (PBS), pH 7.4: Add 1 pouch of PBS powder (Sigma-Aldrich, #P3813) to 1,000 ml ultrapure water, dissolve, and filter. Store at 4 °C.
4. NM-15 contains: Eagle's MEM with Earle's salts and 2 mM L-glutamine, 15 % heat-inactivated fetal bovine serum, 6 mg/ml glucose, 0.5 U/ml penicillin, and 0.5 mg/ml streptomycin. To make NM-15 (200 ml), add 160 ml MEM into a sterile flask. In a 50 ml falcon tube, add 10 ml MEM, 1.2 g glucose, and 10 ml pen/strep. Mix and dissolve the powders well. Filter the 10 ml into the sterile flask containing MEM. Add 30 ml sterile FBS to the flask. Mix well and keep in +4 °C.
5. 1 % sodium metasilicate: Dissolve 500 g sodium metasilicate in 2.5 l ultrapure water to make 20 % stock solution. Dilute 20 times in ultrapure water prior to use. Keep in chemical hood at room temperature (*see* **Note 6**).
6. 1 % HCl: Dilute 230 ml 36.5–38 % HCl to a final volume of 9 l. Keep in chemical hood at room temperature.

****7. Progesterone (21 mg/ml): Add 1 ml of 100 % ethanol to 1 mg of progesterone. This can be done directly in the stock bottle of progesterone if 1 mg size was purchased. Add this 1 mg/ml solution of progesterone to 46.6 ml of sterile ultrapure or distilled water to generate stock solution. Aliquot solution and store at −20 °C.

****8. Putrescine (10 mg/100 ml): Dissolve 100 mg of putrescine in 1 ml of sterile ultrapure or distilled water to generate stock solution. Aliquot solution and store at −20 °C.

****9. Selenium (1 mg/ml): Dissolve 1 mg of selenium in 1 ml of sterile ultrapure or distilled water to generate working dilution. Aliquot solution and store at −20 °C.

****10. 3,3′,5-Triiodo-L-thyronine (T3) (0.08 mg/ml): Dissolve 2 mg of T3 in 1.5 ml of 0.1 M NaOH. Add 23.5 ml of sterile ultrapure or distilled water and sterile filter. Aliquot solution and store at −80 °C.

****11. L-thyroxine sodium salt pentahydrate (T4) (0.786 mg/ml): Dissolve 0.011 g of T4 in 1.5 ml of 0.1 M NaOH. Add 11 ml of sterile ultrapure or distilled water and sterile filter. Aliquot solution and store at −80 °C.

****12. Oligodendrocyte medium (OM) (500 ml): Oligodendrocyte medium contains a 1:1 mixture of Ham's F-12 and Basal Medium Eagle, 6 mg/ml glucose, 100 μg/ml transferrin, 25 μg/ml insulin, 20 nM progesterone, 60 μM putrescine, 30 nM selenium, 6.6 μM glutamine, 0.5 units/ml penicillin, 0.5 μg/ml streptomycin, 0.08 μg/ml T3, and 0.5 μM T4. To make this medium retrieve one 500 ml bottle of Ham's F-12 Nutrient Mix and one 500 ml bottle of Basal Medium Eagle (BME) from 4 °C. Remove 15 ml of F-12 medium from stock bottle and place in a 50 ml conical tube. Weigh out 3 g of glucose and dissolve in previously mentioned 15 ml aliquot of F-12. Add 50 mg of apo-transferrin to solution of glucose and Ham's F-12 and vortex to dissolve. In a separate 15 ml conical tube, dissolve 12.5 mg of insulin in 1.5 ml of 0.01 N HCl. Once insulin is dissolved, add to original 15 ml aliquot of F-12 containing glucose and transferrin. To this solution, add 150 μl of 21 μg/ml stock solution of progesterone, 50 μl of 10 mg/100 μl stock solutionof putrescine, and 2.6 μl of 1 mg/ml stock solution of selenium. Add 16.5 ml of 200 mM L-glutamine and 25 μl of pen/strep to solution. Transfer this solution to a 500 ml vacuum filtration unit. Add 235 ml of Ham's F-12 and 250 ml of BME and filter sterilize. Add 500 μl of 0.08 mg/ml T3 and 250 μl of 0.786 mg/ml T4 to the sterile medium.

3 Methods

Carry out all procedures in a horizontal laminar flow hood until otherwise mentioned. The hood should be exposed to UV light for at least 30 min before use.

3.1 Preparation for Dissection

1. Before dissection, spray the laminar flow hood with 80 % ethanol to fully sterilize the surface.
2. Place dissecting tools on a sterile surface in the hood (*see* **Note 9**).
3. Retrieve two buckets with ice and place in hood. Fill two 100 ml beakers with 80 % ethanol. Label 1 and 2, and then place in first ice bucket.
4. Coat 75 cm^2 culture flask with 8 ml 1× poly-D-lysine. Leave the poly-D-lysine on the flasks for more than 30 min or overnight in 37 °C incubator (*see* **Note 10**). Remove poly-D-lysine. Wash the flasks twice with 10 ml sterile water twice. Add 9 ml NM-15 to each flask and keep in the incubator until ready for use.
5. Place paper towel on the second ice bucket in the hood. Prepare three 100 mm sterile petri dishes (1 glass, 2 plastic). Add 15 ml sterile PBS to the glass petri dish and one plastic dish (*see* **Note 11**). Place on ice with paper towel. Leave the other empty plastic petri dish in the hood. Prepare one 35 mm small glass petri dish, add 2 ml NM-15 to it, and place on ice.
6. Set up the dissecting microscope (that has been wiped down with ethanol) in the hood.
7. Set up a body bag in garbage container.

3.2 Pup Dissection

1. Place postnatal 0–2 day-old rat pups on paper towel on ice until fully anesthetized (*see* **Note 12**). Decapitate the pups. Discard the bodies into the body bag.
2. With large dressing forceps, dip rat pup heads quickly into the first beaker of ice-cold 80 % ethanol, followed by the second beaker of ice-cold 80 % ethanol.
3. Transfer sterilized head to empty, 100 mm petri dish.
4. Hold rat pup head in place with large dressing forceps. With curved #7 Dumont forceps, grasp loose skin on the animal's neck and pull forward towards the eyes so that the skull is exposed.
5. Holding head steady with large dressing forceps, carefully insert tip of small Bonn scissors into foramen magnum at the base of the skull. Being careful not to damage underlying cortex, cut the skull along the midline towards the animal's nose (*see* **Note 13**).
6. With curved #7 Dumont forceps, peel back flaps of skull to expose underlying cerebral cortex.
7. With curved #7 Dumont forceps, scoop underneath of the brain to sever cranial nerves. Then, lift brain from skull and place in first 100 mm petri dish with PBS on ice.
8. Place head in body bag (*see* **Note 14**).
9. Repeat **steps 3–8** for remaining rat pups.

10. Carefully attach sterile stainless steel #11 scalpel blades to blade handles.
11. Transfer 3–4 rat brains to 100 mm glass petri dish with curved #7 Dumont forceps.
12. Carefully isolate one brain under the microscope. Remove all meninges. Dissect out region of interest under the microscope. Place region in a 35 mm glass petri dish with NM-15.
13. Repeat steps 11–12 until the remaining brains are all dissected. Discard the remaining tissue in the body bag.

3.3 Plating the Cells

1. Replace used blades with new, sterile stainless steel #11 blades. Carefully dice the tissue into small pieces.
2. Using the Bunsen burner, flame the tips of the glass pipettes to make them smooth. Increase the length of flaming to make a series of pipettes with tips of reduced pore size.
3. With a glass Pasteur pipette transfer all tissue and medium into a 15 ml conical tube.
4. Triturate with a series of glass pipettes, usually 2. Dissociate the tissue until a cloudy appearance is seen (*see* **Note 15**).
5. Make a 1–5 dilution by adding one drop of cell suspension to 4 drops of NM-15 using polished glass pipet. Mix well. Add 10 μl diluted cell suspension each to the top and bottom chambers of the hemocytometer.
6. Count cells using the hemocytometer (*see* **Note 16**).
7. Dilute the cells with NM-15 to a concentration of 1.5×10^7 cells/ml. Add 1 ml of the cells to 9 ml NM-15 in each 75 cm^2 flask. Gently shake each flask on horizontal plane to distribute cells evenly over poly-D-lysine-coated surface. The final plating density will be approximately 2×10^5 cells/cm^2.
8. Make sure that caps are placed loosely on each flask and place in 37 °C incubator (*see* **Note 17**).
9. The culture will reach confluence between 7 and 12 days.

3.4 Feeding the Cells

1. Feed the cells with fresh NM-15 every 3–4 days.
2. Aspirate approximately one-half of the medium. Replace one-half of the medium until confluent monolayers are established. After that time all the medium can be replaced. Add 37 °C pre-warmed fresh NM-15.

3.5 Isolating and Culturing Oligodendrocyte Progenitor Cells****

1. Grow mixed glial cultures until they are confluent.
2. Remove NM-15 media from 4 °C and warm to 37 °C in water bath.
3. Place petri dishes (35 mm or 100 mm) in laminar flow hood (*see* **Note 18**). Add enough 1× poly-D-lysine to coat the bottom of each petri dish and incubate at 37 °C for 30 min.

Aspirate 1× poly-D-lysine from flask and rinse thoroughly with sterile ultrapure water twice. Add .5 ml NM-15 to 35 mm dishes or 6 ml NM-15 to 100 mm dishes. Incubate culture dishes for 30–60 min at 37 °C.

4. Remove culture flasks from the incubator. Aspirate used media and replace with fresh NM-15.
5. Return flasks to incubator for 2 h.
6. Remove culture flasks from the incubator, and tighten the caps completely (*see* **Note 19**).
7. Secure flasks to the orbital shaker set to 37 °C. Shake cultures at 250 rpm for 16–18 h.
8. Remove flasks from shaking incubator.
9. With a sterile pipette, transfer media from flasks to 150 mm non-tissue culture-treated culture dishes (*see* **Note 20**).
10. To increase yield of oligodendrocyte progenitors, gently rinse each flask with approximately 5 ml of warm NM-15 to recapture oligodendrocytes that remain attached to astrocyte substrate. With a sterile pipette, transfer this media to the 150 mm non-tissue culture-treated culture dishes. Flasks containing astrocytes can now be discarded.
11. Collect media from culture dishes in 50 ml conical tubes. Contaminating microglia will remain attached to the culture dish.
12. Centrifuge cell suspension for 5 min at 100 × *g*.
13. Carefully aspirate the supernatant and discard (*see* **Note 21**).
14. Dissociate pellet in approximately 1 ml of warm NM-15 by carefully pipetting up and down with a polished glass pipette.
15. Remove 10 μl from the cell suspension and dilute in 40 μl of warm NM-15. Mix well by pipetting up and down, then count cells using a hemocytometer.
16. Dilute the cell suspension to desired concentration using warm NM-15 and plate onto poly-D-lysine-coated dishes. Add 1 ml of cells in NM-15 to 35 mm dishes and 2 ml of cells in NM-15 to 100 mm dishes.
17. After 24 h of growth in NM-15, remove OM media from 4 °C and warm to 37 °C in water bath.
18. Rinse oligodendrocyte progenitor cultures with warmed OM twice. Then, add OM or treatment media to each dish.

4 Notes

1. Average litter size ranges from 10 to 20 pups.
2. Except for the sterile stainless steel #11 surgical blades, all dissecting tools are boiled in soapy water for 15 min after each

dissection. A bar of Ivory soap must be used that is cut into small pieces and added to ultrapure water. All tools are thoroughly rinsed in ultrapure water. Tools are dried and kept in test tubes cushioned with glass fiber. All dissection tools are then baked at 180 °C for 5 h for sterilization. It is best to autoclave the dissecting kit the day before this protocol is begun.

3. To avoid contamination, it is best to maintain a dedicated incubator for the culture of mixed glial cells. Avoid culturing bacteria and yeast in dedicated tissue culture incubator if possible.
4. A variety of shaking incubators exist and can be used for this protocol. This incubator should be able to hold 75 cm^2 flasks securely. Furthermore, it is again recommended to avoid using this incubator to shake bacterial cultures.
5. It is extremely important that tissue culture-treated, non-vent/close flasks be used for the culture of mixed glia. Culture flasks with filter caps are not compatible with this protocol, as the media will come into contact with the membrane during the shaking step and become contaminated.
6. Prior to use all glass dishes are boiled in a 1 % sodium metasilicate solution for 1 h, soaked in a bucket of 1 % HCl in chemical hood overnight, followed by a bucket of ultrapure water overnight. Plates are rinsed thoroughly under ultrapure water, dried, and baked at 180 °C for 5 h for sterilization.
7. FBS must be heat inactivated to destroy complement proteins in serum which negatively affect cell survival. Heating at 56 °C for 30 min is recommended. Extended length of time or higher temperature may damage other growth factors in serum.
8. Poly-D-lysine is viscous and hard to dissolve. Add 10 ml ultrapure water to poly-D-lysine bottle. Pipette solution up and down many times with a 10 ml pipette to make sure all the powder dissolves completely. Transfer 10 ml poly-D-lysine to 90 ml water and filter it through 22 μm filter system. Because of the viscosity of poly-D-lysine solution, it is normal that the filtering process takes longer than other solutions.
9. Dissection tools are placed on a sterile surface that can be a plastic tissue culture dish or sterile gauze.
10. Prepare enough 1× poly-D-lysine-coated 75 cm^2 flasks to accommodate the number of cells to be cultured. The concentration of cells should be 15×10^6 cells/flask.
11. The plastic dish is used to keep the brains on ice and slow down tissue degradation. The glass dish is used to dissect under the microscope because it avoids damage due to scalpel blades. We usually transfer 3–4 brains each time from the plastic dish to the glass dish for dissection. This preserves tissue freshness and minimizes contamination.

12. To determine that animals are anesthetized, pinch limbs to check responses before dissection.
13. The brain is extremely soft and is easily injured during this step. To avoid cutting into the cortex, it is best to work slowly and take care to keep the scissors pointed away from the brain. It is sometimes possible to cut the skull using the pointed edge of the curved #7 Dumont forceps. If this is possible, simply make a cut with the forceps along the midline, then peel back the flaps of skull to expose the underlying cerebral hemispheres.
14. If at any point in the procedure a dissecting tool touches a non-sterile surface (such as the dissecting microscope), immediately begin to use one of the extra tools that was autoclaved with the dissecting kit.
15. Triturate about 20 times each with 2 glass pipettes until a homogenous cloudy solution appears. The objective is to triturate sufficiently to avoid tissue clumps but also avoid mechanical damage.
16. The hemocytometer that we use has 25 small grid boxes in the center of each chamber, and its volume is 1 mm (wide) × 1 mm (long) × 0.1 mm deep (or 10^{-4} ml). We generally count bright cell bodies in four corner grid boxes and one center box both in the top and bottom chambers and add up numbers in 10 grid boxes. Total cell number in 2 ml suspension = (counted cell number in 10 small grid boxes/10) × 5 (dilution factor) × 2 (initial cell volume) × 0.25×10^6.
17. Flask caps are kept loose in the incubator for adequate gas exchange.
18. It is important to consider the nature of the experiment you wish to perform before plating the oligodendrocyte progenitors. 35 mm dishes are particularly amendable to immunocytochemical experiments. For these types of studies, typically 250,000 cells/dish are plated. For Western blot experiments, it is better to use coated 100 mm^2 dishes with approximately 1,000,000 cells/dish.
19. It is essential to tighten caps here to prevent media from spilling during the overnight shake. This is the only time during the culture protocol that these non-vented flasks should be completely sealed.
20. After the overnight shake, the media will contain oligodendrocyte progenitor cells and microglia. Transferring this media to non-tissue culture-treated dishes before plating will significantly reduce microglial contamination, as the oligodendrocyte progenitors will not adhere to a non-tissue culture-treated surface that has not been coated with poly-D-lysine [7]. Typically,

media from 3 to 4 flasks can be pooled into each 150 mm culture dish.

21. The oligodendrocyte progenitor cell pellet will be very loosely adherent to the side of the conical tube. Be extremely careful not to disturb the pellet while removing the conical tube from the rotor and while aspirating the supernatant.

Acknowledgments

The development of this protocol was supported by NIH NS036647 and grants from the National Multiple Sclerosis Society.

References

1. Hartline DK, Colman DR (2007) Rapid conduction and the evolution of giant axons and myelinated fibers. Curr Biol 17:R29–35
2. Schafer DP, Rasband MN (2006) Glial regulation of the axonal membrane at nodes of Ranvier. Curr Opin Neurobiol 16:508–514
3. Dai X, Lercher LD, Clinton PM, Du Y, Livingston DL et al (2003) The trophic role of oligodendrocytes in the basal forebrain. J Neurosci 23:5846–5853
4. McTigue DM, Horner PJ, Stokes BT, Gage FH (1998) Neurotrophin-3 and brain-derived neurotrophic factor induce oligodendrocyte proliferation and myelination of regenerating axons in the contused adult rat spinal cord. J Neurosci 18:5354–5365
5. Booher J, Sensenbrenner M (1972) Growth and cultivation of dissociated neurons and glial cells from embryonic chick, rat and human brain in flask cultures. Neurobiology 2:97–105
6. McCarthy KD, de Vellis J (1980) Preparation of separate astroglial and oligodendroglial cell cultures from rat cerebral tissue. J Cell Biol 85: 890–902
7. Gallo V, Zhou JM, McBain CJ, Wright P, Knutson PL et al (1996) Oligodendrocyte progenitor cell proliferation and lineage progression are regulated by glutamate receptor-mediated K+ channel block. J Neurosci 16:2659–2670
8. Du Y, Lercher LD, Zhou R, Dreyfus CF (2006) Mitogen-activated protein kinase pathway mediates effects of brain-derived neurotrophic factor on differentiation of basal forebrain oligodendrocytes. J Neurosci Res 84:1692–1702

Chapter 9

Isolation and Culture of Schwann Cells

Yanmei Tao

Abstract

Primarily cultured Schwann cells are essential for the investigation of molecular mechanisms regulating proliferation, survival, differentiation, and myelination of Schwann cell and for the development of efficient transplantation for regeneration of injured spinal cord or peripheral nervous system. Here we describe a basic protocol for isolation and purification of primary Schwann cell from neonatal rat or mouse and discuss some modifications adapted to the culturing from adult nerves and optional methods for Schwann cell enrichment.

Key words Primary cultures, Schwann cells, Rat, Mouse, Immunopanning

1 Introduction

Axons are wrapped or ensheathed by Schwann cells in the peripheral nervous system. Fully differentiated Schwann cells extend plasma membrane processes to wrap axons and form myelin. Myelin sheath thickness and internodal distance are important determinants of nerve conduction velocity, which is critical for precise control of timing impulse conduction. In contrast to central nerves, peripheral nerves are able to regenerate and restore the function on peripheral innervation, in which Schwann cells play a decisive role. Primary Schwann cells cultured in vitro, which are mostly isolated from sciatic nerves of neonatal rodents, are important for investigation of the molecular mechanisms driving each developmental event of Schwann cell [1–5]. Proliferation, differentiation, survival, and functional maturation of Schwann cells are controlled by different sets of extrinsic and intrinsic factors [6]. The differentiation and maturation of Schwann cells in vitro can be conducted by switches of supplemented nutrition and co-culturing with neurons [7, 8]. Since mature Schwann cells can dedifferentiate and regain the proliferating ability after nerve injury, it is also possible to culture Schwann cells from adult nerves, providing the route for development of autologous transplantation and genetic manipulation of Schwann cells in animals and patients [9–12].

Renping Zhou and Lin Mei (eds.), *Neural Development: Methods and Protocols*, Methods in Molecular Biology, vol. 1018, DOI 10.1007/978-1-62703-444-9_9, © Springer Science+Business Media, LLC 2013

This chapter will provide details for primarily culturing Schwann cells from neonatal rodents based on the literature and personal experience. Briefly, sciatic nerves are dissected out from rats or mice at postnatal day 2 (P2) to P4, digested and triturated to obtain dissociated immature Schwann cell which are usually contaminated with fibroblasts and axonal debris. Dissociated cells are cultured for 2 days before supplementing cytosine arabinoside (Ara-C) for another 2 days to suppress the growth of fibroblast cells. Seven days later, cells are resuspended and processed to immunopanning by antibody against Thy1.1 followed by rabbit complement to eliminate fibroblasts. Immunopanned cells are replated into dishes coated with poly-L-lysine (PLL) and laminin. Schwann cell proliferation is promoted by insulin and forskolin, an activator to adenylate cyclase, at the presence of serum. Two weeks later, Schwann cells grow confluent and can be subjected to characterization, differentiation, and other investigations.

It is much harder to obtain large amount of primary Schwann cells from mouse in vitro than that from rat. There are many varied protocols adopted by different laboratories for mouse Schwann cell culturing [13–16]. For either rat or mouse Schwann cell culturing, modifications are major the age of animal used [from embryonic day 12.5 [13], P2–P4 [14] to adult [17]], dissociating cells before or after tissue culturing [1, 5, 13, 18, 19], extracellular matrix proteins (such as PLL, laminin, or poly- L-ornithine) for dish coating [5, 14, 18, 19], or supplemented growth factor or signaling activator (such as insulin, FGF, neuregulin, forskolin, dibutyryl cAMP, NGF, IGF, neurite extracts, or pituitary extracts) [15, 16, 19–24]. The present protocol has been optimized for both mouse and rat Schwann cell culturing [19].

To study the regulative mechanisms for each cellular event or to develop the therapeutic application using cultured Schwann cells, molecular pathways can be manipulated by addition of growth factors, cytokines, recombinant proteins, or cell signaling agonists or antagonists into culture media, introducing genes or siRNA into cells by transfection agents, virus, or electroporation, or obtaining genetic manipulated Schwann cells from mutant mice [13, 17, 19, 25, 26].

2 Materials

All material should be sterilized or filtered with 0.22 μm filter.

2.1 Materials Stored at 4 °C

Dulbecco's Modified Eagle's Medium (DMEM): Containing high glucose with L-glutamine and pyruvate.

Dulbecco's phosphate-buffered saline without Ca^{2+} and Mg^{2+} (Ca^{2+}/Mg^{2+}-free PBS).

2.2 Materials Stored in Aliquots at –20 °C

Fetal bovine serum (FBS): Thaw the serum at room temperature and heat-inactivate at 56 °C for 30 min. Cool to room temperature and aliquot for freezing.

Trypsin: 2.5 % in PBS containing EDTA, 10×.

Collagenase: 10 mg/ml solved in PBS.

DNase: 1 % solved in PBS.

Poly-L-lysine (PLL): 2.5 mg/ml, 100×.

Penicillin/streptomycin (pen/strep): 5,000 U/ml of penicillin and 5 mg/ml of streptomycin, 100×.

Cytosine arabinoside (Ara-C): 10 mM, 1,000×.

Insulin: 10 mg/ml, 1,000×. 10 mg Insulin solved in 9 ml PBS plus 50 μl 1 N HCl and 1 ml 1 % BSA, filtered through 0.22 μm filter.

Forskolin (7-deacetyl-7-[*O*-(*N*-methylpiperazino)-γ-butyryl]-dihydrochloride): 2 mM, 1,000×, solved in DMSO.

2.3 Materials Stored in Aliquots at –80 °C

Laminin.

Anti-Thy1.1 monoclonal antibody.

Rabbit complement.

2.4 Equipment

Surgery tools: one medium dissecting scissors, one blunt forceps, two microscissors, two serrated forceps, and two No. 5 microforceps.

Sterilized culture dishes: Φ100 mm, for holding animal bodies and dissection.

Pre-coated culture dishes: Φ100 mm dishes containing 3 ml PLL (25 ng/ml) diluted in PBS. Swirl dishes gently to allow PLL solution to cover the dish bottom. Wrap the dishes with the aluminum foil and incubate in a laminar fume hood at room temperature overnight. Suck out PLL solution in the next morning and wash twice with PBS. Store the pre-coated culture dishes in –20 °C. For double coating with laminin, PLL-coated dishes are incubated at room temperature with 10 ng/ml laminin covering the dish bottom for 2 h before washing twice by PBS. Laminin coating is done freshly prior to the culturing.

1 ml sterilized syringe and 18-G, 21-G needles.

Phase-contrast microscope.

1.5 Eppendorf (EP) tubes.

15/50 ml Falcon tubes.

Pasteur plastic pipette.

Pipette aid.

Centrifuge.

3 Methods

3.1 Dissection and Nerve Harvesting

1. Rat or mouse pups at postnatal day 2–4 (P2–P4) are grouped and placed in ice to be anesthetized (*see* **Note 1**). Group 6 pups maximal at a batch (*see* **Note 2**).
2. Prepare one dish with 20 ml 70 % alcohol at room temperature and two dishes with 20 ml sterilized PBS precooled on ice. When animals stop moving and become numb, rinse the animals by 70 % alcohol for 2 s and then rinse in sterilized PBS twice (*see* **Note 3**). Transfer the animals between dishes with sterilized blunt forceps.
3. Decapitate animals using dissecting scissors. Keep the bodies on a precooled sterilized dish on ice.
4. Place one animal back up each time on a precooled sterilized dish lid on ice for sciatic nerve dissection.
5. Open an incision in the skin of lateral mid-thigh using microscissors.
6. Remove the back skin from thigh to waist as well as from thigh to knee by using two serrated forceps (*see* **Note 4**). Can be helped by using scissors.
7. Tear the connective tissue between the vastus lateralis muscle and semitendinosus muscle apart using two serrated forceps and expose the femur underneath.
8. Look for the sciatic nerve next to femur under the semitendinosus muscle. Gently dissociate nerves from connected tissues by tearing the connective tissues using two No. 5 forceps. Do not stretch or directly clamp the nerves. Dissociate tissues from the nerve down to the knee and up to the hip.
9. Find the efferent route for the sciatic nerve from the back of pelvis cavity. Cut sacroiliac joint longitudinally using microscissors to open the pelvis cavity. Take care not to cut too deep otherwise sciatic nerves might be cut off.
10. Track the sciatic nerve to the lumbosacral plexus from the vertebrate (Fig. 1). Cut the sciatic nerve as rostrally as possible and then cut the other end at the knee (*see* **Note 5**).
11. Dissect out the sciatic nerve from the other leg as described above. Pool the sciatic nerves into one 35 mm dish containing 2 ml Ca^{2+}/Mg^{2+}-free PBS (*see* **Note 6**).

3.2 Nerve Digestion and Mixed Cell Culturing

1. Pipette out PBS; add 1 ml DMEM and 300 μl 10 mg/ml collagenase. Shake gently to mix well and incubate at 37 °C for 30 min.
2. Pipette out the DMEM/collagenase slowly and carefully. Be sure not to disturb the nerves (*see* **Note 7**).

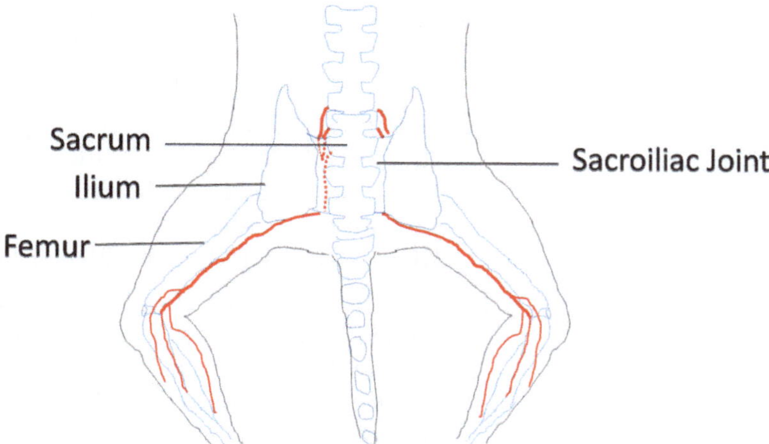

Fig. 1 Illustration of the anatomy of the sciatic nerve in rat or mouse from a back view. *Light blue line* sketches the skeleton and *red line* indicates the sciatic nerves. *Dash red line* indicates the nerve hidden under the Sacrum and Ilium

3. Add 1 ml pre-warmed 0.25 % trypsin diluted in PBS (Ca^{2+}/Mg^{2+}-free). Shake gently to mix well and incubate at 37 °C for 10 min.
4. Add 300 μl 1 % DNase and tilt the dish to gather the tissue solution to one edge (*see* **Note 8**).
5. Triturate the nerves by pipetting up and down using a 1 ml pipette tip gently for 10 times.
6. Using a 1 ml syringe equipped with an 18-G needle, draw up the triturated tissue and expel back into the dish slowly. Pass the tissue through the needle for 15 times.
7. Replace the needle to a 21-G needle, and draw up and pass the tissue through the needle slowly for 15 times.
8. After the last passage, draw the tissue and pass the cell suspension through a 70 μm cell strainer into a 50 ml Falcon tube. Wash the tissue pellet on the mesh by passing 10 ml pre-warmed DMEM+10 % heat-inactivated FBS (*see* **Note 9**).
9. Centrifuge the tube at 500 × *g* for 10 min.
10. Discard the supernatant by vacuum. Resuspend the cell pellet by 10 ml DMEM supplemented with 10 % FBS and 1 % pen/strep (DMEM/10% FBS).
11. Plate the cell suspension into a 100 mm petri dish pre-coated with PLL+laminin (*see* **Notes 10** and **11**). Shake gently to distribute the cells evenly.
12. Culture the cells in 37 °C incubator supplemented with 5 % CO_2 for one day.

13. Change the medium thoroughly to 10 ml fresh DMEM/10 % FBS the next day (*see* **Note 12**). Put culture dishes back into 37 °C incubator supplemented with 5 % CO_2.
14. The third day after plating, change medium thoroughly to 10 ml fresh DMEM/10% FBS supplemented with 10 μm Ara-C (*see* **Note 13**). The fifth day after plating, discard the medium by vacuum and wash the dish once with pre-warmed 10 ml fresh DMEM/10% FBS (*see* **Note 14**).
15. Refill the dish with 10 ml of fresh DMEM/5% FBS supplemented with 10 μg/ml insulin and 2 μm forskolin (*see* **Note 15**). Culture for 1–2 weeks until cells reach 100 % confluence. Replace medium to fresh DMEM/5% FBS supplemented with 10 μg/ml insulin and 2 μm forskolin every 3 days.

3.3 Purification of Schwann Cells by Immunopanning

Until indicated, all pipetting processes are done by 10 ml Pasteur plastic pipette controlled by electric pipette aid.

1. When the cells reach 100 % confluence, wash the dishes once with pre-warmed Ca^{2+}/Mg^{2+}-free PBS and then add 3 ml 0.05 % trypsin/EDTA. Use 1.5 ml of trypsin/EDTA for 60 mm dishes and 1 ml for 35 dishes (*see* **Note 16**).
2. Incubate the dishes with trypsin/EDTA at 37 °C. Observe under the microscope within 1 min. Pipette up and down gently by 10 ml pipette to detach the cells as soon as the cells become partially dissociated from the dish bottom. It may take 2 min to see the detachment. Add 10 ml DMEM/10 % FBS immediately after pipetting to stop trypsinization (*see* **Note 16**).
3. Mix well by 10 ml pipette and transfer the cell suspension into 15 ml Falcon tube.
4. Centrifuge the tube at 500 × *g* for 10 min.
5. Discard the supernatant, and then add 10 ml pre-warmed DMEM/10% FBS and resuspend the cell pellets by pipetting up and down gently.
6. Centrifuge the tube at 500 × *g* for 10 min again. Discard the supernatant. This step is to wash the cells.
7. Prepare 1 ml anti-Thy1.1 antibody (final concentration 30 ng/μl) in DMEM/10% FBS. Precool the antibody solution in ice during the centrifugation of **step 6**.
8. Resuspend the cell pellet by precooled 1 ml Thy1.1 antibody/DMEM/10% FBS solution. Incubate the cells in ice for 1 h to allow antigen/antibody interaction (*see* **Note 17**). Mix the cells gently by tapping every 15 min.
9. Centrifuge the tube 5 min at ~300 × *g* at 4 °C.

10. Freshly mix the precooled 1 ml rabbit complement with 3 ml DMEM during the centrifugation at **step 9**. Keep the solution in ice (*see* **Note 18**).
11. Discard the supernatant. Resuspend the cell pellet by rabbit complement/DMEM solution.
12. Incubate the cell suspension containing rabbit complement at 37 °C for 1 h.
13. Add 10 ml pre-warmed DMEM into tube to dilute the rabbit complement. Centrifuge at 500 × *g* for 10 min.
14. Resuspend the cell pellet in 12 ml of pre-warmed DMEM and centrifuge again to wash the cells.
15. Resuspend the cell pellet in 10 ml pre-warmed DMEM/5% FBS supplemented with 10 μg/ml insulin and 2 μm forskolin. Plate the cells into a new 100 mm petri dish pre-coated with PLL+laminin.
16. Purity of Schwann cells after immunopanning can reach 98–100 % by characterization by immunostaining with antibody against Schwann cell marker S100 (Fig. 2).
17. Change medium thoroughly every the other day to supplement fresh mitogens. Grow cells to 100 % confluence.

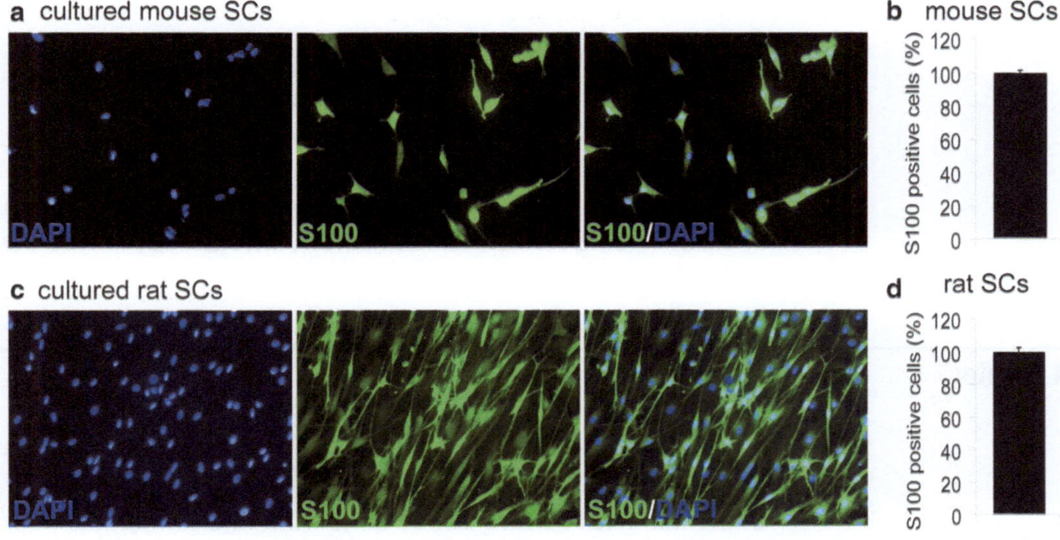

Fig. 2 Purity of primary Schwann cells cultured according to the present protocol reaches almost 100 %. (**a**, **c**) Immunofluorescence results of primarily cultured mouse (**a**) or rat (**c**) Schwann cells using anti-S100 antibody (*green*). DAPI-stained nucleus (*blue*). (**b**, **d**) Quantitative results of the percentage of S100-positive cells in primarily cultured mouse (**b**) or rat (**d**) Schwann cells

3.4 Passaging and Freezing Primary Schwann Cells

1. When Schwann cells reach confluence, cells from one dish can be dissociated by trypsinization and split into five of similar dishes pre-coated with PLL+laminin.
2. Schwann cells can be passaged twice to expand the numbers before freezing or application for further experiments.
3. For freezing of Schwann cells, cells can be trypsinized by 5 ml 0.05 % trypsin/EDTA and dissociated from dishes and pooled into 15 ml Falcon tube. Add 8 ml DMEM/10% FBS to the tube and mix well by pipetting up and down to stop trypsinization.
4. Centrifuge the tube at 500 × *g* for 10 min.
5. Resuspend the cells with freezing medium (DMEM/15% FBS supplemented with 7 % DMSO). Use 1.5 ml freezing medium for cells from one 100 mm petri dish.
6. Aliquot cell suspension into cryotubes. 1 ml each tube.
7. Freeze the sample boxes with cryotubes containing Schwann cell overnight at −80 °C and then transfer into liquid nitrogen the next day. Cells can be frozen in liquid nitrogen infinitely in theory.
8. To recover the frozen Schwann cells, thaw a cell tube from liquid nitrogen at 37 °C water bath quickly with gentle swirling.
9. Transfer the thawed Schwann cells suspension into 15 ml Falcon tube containing 10 ml pre-warmed DMEM/5% FBS. Mix well by pipetting gently.
10. Centrifuge at 500 × *g* for 10 min.
11. Discard the supernatant; resuspend cell pellets with 10 ml pre-warmed DMEM/5% FBS supplemented with 10 μg/ml insulin and 2 μm forskolin.
12. Plate cells into a 100 mm petri dish pre-coated with PLL+laminin. Change medium every the other day thoroughly (*see* **Note 19**).

4 Notes

1. We found cultures from P2 to P4 rodents have better potential to expand Schwann cells than those from younger animals.
2. Schwann cell culture can be done from one to multiple pups. To ensure cell viability, do dissection for maximal 6–7 pups at a batch.
3. 70 % alcohol is used to decontaminate the animal skin. Dip the whole body of animals in alcohol but DO NOT keep the animals in 70 % alcohol for too long. Otherwise alcohol will penetrate into skin and damage tissues.

4. Skin can be removed for both sides at this step. Not necessary to be very clean but removal of skin will eliminate the possibility of bacterial contamination.
5. Do not cut the nerve before isolating the nerves as needed. Otherwise the nerve will retract and becomes hard to see. Wet the tissues by dropping some precooled PBS if dissection takes too long.
6. If culturing Schwann cells from adult nerves, nerve fragments should undergo further microdissection to remove epineurium under stereo microscope to ensure efficient digestion in the next step [17, 27]. It has been reported pre-degenerating nerve stumps for 7–14 days in vivo or in vitro increases the proliferating Schwann cells from adult rat or human nerves in the subsequent cultures in vitro [10, 18, 28].
7. Digested nerves become flat and sticky; thus, they are easily sucked into the pipette if pipetting too hard. Not necessary to pipette the solution very clean.
8. DNase is necessary to shear the linear DNA released from broken cells and free the single cells. However, it is not suggested to incubate the cells with DNase at 37 °C; otherwise intact cells will be affected.
9. The most common procedure for cell cultures is to stop trypsinization before mechanical dissociation of cell clumps. We found it is better to dissociate the cells before stopping trypsinization. The modified procedure reduces the damage of cells from the mechanical shearing.
10. Dishes should be pre-coated with extracellular matrix proteins. Laminin is required for efficient Schwann cell proliferation [29–31]. It might be enough to coat dishes with PLL for rat Schwann cell expansion. However, we suggest to double precoat dishes with PLL and laminin especially for culturing of mouse Schwann cells.
11. Cell density is important for cell viability and proliferation because Schwann cells will autocrine or paracrine growth factors to support each other [15, 16]. Lower density culturing will dilute the released growth factors. To ensure the cell density, nerve cells from 1 to 2 pups can be plated into one 35 mm petri dish, while those from 3 to 4 pups into one 60 mm petri dish.
12. Changing the medium completely in the next day helps removal of debris formed by axonal fragments and dead cells.
13. Cells dissociated from nerves contain mostly the Schwann cells besides about 10 % fibroblasts. Serum-containing medium will promote the growth of fibroblasts. Cytotoxic Ara-C can suppress efficiently the fibroblast proliferation. However, Ara-C

has toxic effect on Schwann cells as well and should not be supplemented before Schwann cells proliferate to a certain amount. It has been suggested that culturing Schwann cells in serum-free medium or low-serum culture medium suppresses the growth of fibroblasts and improves the purity of Schwann cells. Examples for serum-free medium are Eagle's Minimum Essential Medium (MEM) supplemented with 100 μg/ml endothelial mitogen, 1 μg/ml vitamin C, 5 μg/ml insulin, 0.01 μg/ml cholera toxin, 8 μg/ml putrescine, 4.3 ng/ml sodium selenite, 40 μg/ml $MgCl_2.6H_2O$, albumin, fatty acids, amino acids, vitamins and other trace elements [32], or melanocyte growth medium (MGM) supplemented with 10 ng/ml FGF_2, 2 μm forskolin and 5 μg/ml bovine pituitary extracts [17]. Low-serum culture medium is composed by MEM plus 0.5 % horse serum, 100 μM dibutyryl cyclic AMP, 1 mM insulin, and 10 ng/ml β-neuregulin 1 [14]. It is not necessary to deprive serum in present protocol since immunopanning is a highly efficient method to eliminate fibroblasts and purify Schwann cells.

14. Wash the dish to ensure elimination of the cytotoxic Ara-C. Do not use cold medium to wash. Schwann cells are easy to detach from the culture dishes under cold solution jet which has been developed to enrich the pure Schwann cell for subsequent culture [17]. It is reported that culturing cells from predegenerated adult rat sciatic nerves in serum-free medium and passaging cells via cold jet generated by cold medium expelled from 1 ml Gilson tip can yield Schwann cells with 95 % purity [17, 18].
15. Reducing serum helps to limit the growth of remained fibroblasts. Schwann cell proliferation requires the activation of growth factor signaling pathway. Forskolin, the reversible activator of adenylate cyclase, and insulin are used frequently to stimulate Schwann cell proliferation in vitro with serum presence [1, 24, 33]. The most potent mitogen for Schwann cell proliferation is members of neuregulin/heregulin families, such as glial growth factor (GGF), type II neuregulin I [34], and neurite membrane extracts which contain membrane-associated neuregulin 1[25]. Insulin or insulin-like factor 1 promotes efficiently the Schwann cell proliferation in the presence of activators for cAMP pathway [24]. In the study of neuregulin effect on Schwann cell development and myelination, insulin combined with forskolin is an applicable substitute to stimulate Schwann cell yield in vitro [19].
16. Schwann cell purification by immunopanning takes advantage of the fact that the Thy 1.1 antigen expresses specifically on the surface of fibroblast cells. Anti-Thy1.1 antibody recognizes fibroblasts and binds to the surface antigen. Cells bound with

antibody will be killed by cytotoxic factors in rabbit complement through antibody-mediated cell apoptosis pathway [32]. Thus, it is important to preserve the Thy1.1 antigen on the surface of fibroblast cells by limiting the time of cells in trypsin. Limiting the trypsin volume also helps to preserve the surface Thy1.1 antigen.

17. Incubating at 4 °C is critical to prevent the endocytosis of antigen/antibody complex.
18. Complement contains immune factors to induct antibody-mediated cell apoptosis. Rabbit complement should be stored at −80 °C and thawed on ice to preserve the factors' activity.
19. Frozen Schwann cells can be passaged twice maximally. More passages are not recommended since it will change Schwann cells' property. Primary Schwann cells will transform after 20 passages and become growth factor independent and form foci [35]. We found that primary Schwann cells after five passages were more flat and refractory for transfection or hard to develop myelin when co-culturing with neurons in differentiation medium.

References

1. Salzer JL, Bunge RP (1980) Studies of Schwann cell proliferation. I. An analysis in tissue culture of proliferation during development, Wallerian degeneration, and direct injury. J Cell Biol 84:739–752
2. Jessen KR, Brennan A, Morgan L, Mirsky R, Kent A, Hashimoto Y, Gavrilovic J (1994) The Schwann cell precursor and its fate: a study of cell death and differentiation during gliogenesis in rat embryonic nerves. Neuron 12:509–527
3. Brockes JP, Fields KL, Raff MC (1979) Studies on cultured rat Schwann cells. I. Establishment of purified populations from cultures of peripheral nerve. Brain Res 165:105–118
4. Bottenstein JE, Sato GH (1979) Growth of a rat neuroblastoma cell line in serum-free supplemented medium. Proc Natl Acad Sci USA 76:514–517
5. Weinstein DE, Wu R (2001) Isolation and purification of primary Schwann cells. Curr Protoc Neurosci Chapter 3: Unit 3 17
6. Jessen KR, Mirsky R (2005) The origin and development of glial cells in peripheral nerves. Nat Rev Neurosci 6:671–682
7. Eldridge CF, Bunge MB, Bunge RP, Wood PM (1987) Differentiation of axon-related Schwann cells in vitro. I. Ascorbic acid regulates basal lamina assembly and myelin formation. J Cell Biol 105:1023–1034
8. Moya F, Bunge MB, Bunge RP (1980) Schwann cells proliferate but fail to differentiate in defined medium. Proc Natl Acad Sci USA 77:6902–6906
9. Haastert K, Lipokatic E, Fischer M, Timmer M, Grothe C (2006) Differentially promoted peripheral nerve regeneration by grafted Schwann cells over-expressing different FGF-2 isoforms. Neurobiol Dis 21:138–153
10. Haastert K, Mauritz C, Matthies C, Grothe C (2006) Autologous adult human Schwann cells genetically modified to provide alternative cellular transplants in peripheral nerve regeneration. J Neurosurg 104:778–786
11. Takagi T, Ishii K, Shibata S, Yasuda A, Sato M, Nagoshi N, Saito H, Okano HJ, Toyama Y, Okano H, Nakamura M (2011) Schwann-spheres derived from injured peripheral nerves in adult mice–their in vitro characterization and therapeutic potential. PLoS One 6:e21497
12. Scarpini E, Kreider BQ, Lisak RP, Pleasure DE (1988) Establishment of Schwann cell cultures from adult rat peripheral nerves. Exp Neurol 102:167–176
13. Kim HA, Pomeroy SL, Whoriskey W, Pawlitzky I, Benowitz LI, Sicinski P, Stiles CD, Roberts TM (2000) A developmentally regulated switch directs regenerative growth of Schwann cells through cyclin D1. Neuron 26:405–416

14. Stevens B, Tanner S, Fields RD (1998) Control of myelination by specific patterns of neural impulses. J Neurosci 18:9303–9311
15. Seilheimer B, Schachner M (1987) Regulation of neural cell adhesion molecule expression on cultured mouse Schwann cells by nerve growth factor. EMBO J 6:1611–1616
16. Dong Z, Sinanan A, Parkinson D, Parmantier E, Mirsky R, Jessen KR (1999) Schwann cell development in embryonic mouse nerves. J Neurosci Res 56:334–348
17. Haastert K, Mauritz C, Chaturvedi S, Grothe C (2007) Human and rat adult Schwann cell cultures: fast and efficient enrichment and highly effective non-viral transfection protocol. Nat Protoc 2:99–104
18. Mauritz C, Grothe C, Haastert K (2004) Comparative study of cell culture and purification methods to obtain highly enriched cultures of proliferating adult rat Schwann cells. J Neurosci Res 77:453–461
19. Tao Y, Dai P, Liu Y, Marchetto S, Xiong WC, Borg JP, Mei L (2009) Erbin regulates NRG1 signaling and myelination. Proc Natl Acad Sci USA 106:9477–9482
20. Salzer JL, Bunge RP, Glaser L (1980) Studies of Schwann cell proliferation. III. Evidence for the surface localization of the neurite mitogen. J Cell Biol 84:767–778
21. Dong Z, Brennan A, Liu N, Yarden Y, Lefkowitz G, Mirsky R, Jessen KR (1995) Neu differentiation factor is a neuron-glia signal and regulates survival, proliferation, and maturation of rat Schwann cell precursors. Neuron 15:585–596
22. Manthorpe M, Skaper S, Varon S (1980) Purification of mouse Schwann cells using neurite-induced proliferation in serum-free monolayer culture. Brain Res 196:467–482
23. Salzer JL, Williams AK, Glaser L, Bunge RP (1980) Studies of Schwann cell proliferation. II. Characterization of the stimulation and specificity of the response to a neurite membrane fraction. J Cell Biol 84:753–766
24. Schumacher M, Jung-Testas I, Robel P, Baulieu EE (1993) Insulin-like growth factor I: a mitogen for rat Schwann cells in the presence of elevated levels of cyclic AMP. Glia 8:232–240
25. Taveggia C, Zanazzi G, Petrylak A, Yano H, Rosenbluth J, Einheber S, Xu X, Esper RM, Loeb JA, Shrager P, Chao MV, Falls DL, Role L, Salzer JL (2005) Neuregulin-1 type III determines the ensheathment fate of axons. Neuron 47:681–694
26. Haastert K, Semmler N, Wesemann M, Rucker M, Gellrich NC, Grothe C (2006) Establishment of cocultures of osteoblasts, Schwann cells, and neurons towards a tissue-engineered approach for orofacial reconstruction. Cell Transplant 15:733–744
27. Morrissey TK, Kleitman N, Bunge RP (1991) Isolation and functional characterization of Schwann cells derived from adult peripheral nerve. J Neurosci 11:2433–2442
28. Komiyama T, Nakao Y, Toyama Y, Asou H, Vacanti CA, Vacanti MP (2003) A novel technique to isolate adult Schwann cells for an artificial nerve conduit. J Neurosci Methods 122:195–200
29. Chen ZL, Strickland S (2003) Laminin gamma1 is critical for Schwann cell differentiation, axon myelination, and regeneration in the peripheral nerve. J Cell Biol 163: 889–899
30. Yu WM, Feltri ML, Wrabetz L, Strickland S, Chen ZL (2005) Schwann cell-specific ablation of laminin gamma1 causes apoptosis and prevents proliferation. J Neurosci 25:4463–4472
31. Armstrong SJ, Wiberg M, Terenghi G, Kingham PJ (2007) ECM molecules mediate both Schwann cell proliferation and activation to enhance neurite outgrowth. Tissue Eng 13:2863–2870
32. Needham LK, Tennekoon GI, McKhann GM (1987) Selective growth of rat Schwann cells in neuron- and serum-free primary culture. J Neurosci 7:1–9
33. Morgan L, Jessen KR, Mirsky R (1991) The effects of cAMP on differentiation of cultured Schwann cells: progression from an early phenotype (04+) to a myelin phenotype (P0+, GFAP-, N-CAM-, NGF-receptor-) depends on growth inhibition. J Cell Biol 112:457–467
34. Mahanthappa NK, Anton ES, Matthew WD (1996) Glial growth factor 2, a soluble neuregulin, directly increases Schwann cell motility and indirectly promotes neurite outgrowth. J Neurosci 16:4673–4683
35. Haynes LW, Rushton JA, Perrins MF, Dyer JK, Jones R, Howell R (1994) Diploid and hyperdiploid rat Schwann cell strains displaying negative autoregulation of growth in vitro and myelin sheath-formation in vivo. J Neurosci Methods 52:119–127

Part II

Gene Expression and Analysis

Chapter 10

DNA Transfection: Calcium Phosphate Method

Munjin Kwon and Bonnie L. Firestein

Abstract

The calcium phosphate transfection is a widely used method for introducing foreign DNA plasmids into cells. Mechanisms underlying this transfection method are not yet defined; however, DNA–calcium phosphate precipitates are internalized by the cells and DNA is efficiently expressed in almost all cell types. The cost-efficiency and simplicity of this method allows for use in primary neuronal cultures, despite issues of neurotoxicity. Here, we describe an optimized calcium phosphate transfection method for the delivery of DNA plasmid into primary dissociated neuronal cultures.

Key words Transfection, Calcium phosphate, Primary neuronal culture

1 Introduction

The introduction of DNA plasmids into cells is necessary for the study of various aspects of neuronal cell biology, including the investigation of gene and protein function by overexpression or knockdown, tracking of expressed proteins to subcellular compartments, and the study of protein turnover. Since no transfection method is suitable for all applications and cell types, researchers tend to use techniques suitable for specific applications and concerns, such as transfection efficiency, cell survival, expression levels, and applicability to experiments. There are four different categories of transfection methods: electrical transfection, chemical transfection, virus-based transfection, and physical transfection.

Chemical transfection methods, such as calcium phosphate coprecipitation and lipofection, are relatively simple to perform and do not require any specialized equipment. These advantages allow this method to be used widely in many cell types. Calcium phosphate transfection results in DNA crystals that are complexed with the calcium ions in the phosphate buffer. These crystals precipitate onto the cells and are presumably taken up by endocytosis [1].

Renping Zhou and Lin Mei (eds.), *Neural Development: Methods and Protocols*, Methods in Molecular Biology, vol. 1018, DOI 10.1007/978-1-62703-444-9_10, © Springer Science+Business Media, LLC 2013

Although it is very easy and inexpensive, the calcium phosphate transfection method has not been used very often for primary neuronal cultures because it induces neuronal toxicity. However, when optimized, this method results in relatively low toxicity and high transfection efficiencies [2–4]. This neuronal gene delivery can be accomplished using standard eukaryotic expression vectors. Co-transfection of several plasmids [5] and many different gene combinations can be applied. In addition, except at very early stages, neurons at almost all stages of differentiation in culture can be transfected using this method. The time course and levels of protein expression can be easily manipulated by altering the amount of plasmid used [6]. This method can also be used with a number of experimental procedures and treatments.

This method is suitable for experiments requiring a relatively low number of transfected cells, such as those investigating the overexpression or knockdown effect of a gene of interest by imaging neuronal morphology of single cells in a neuronal network, which requires the identification of dendrites and axons from individual neurons of interest [7]. Moreover, subcellular localization, colocalization, and trafficking studies of proteins can be performed.

2 Materials

Prepare all solutions using ultrapure water and analytical grade reagents. All solutions and equipment must be sterile.

1. Plasmid DNA in H_2O (purified by endotoxin-free plasmid prep kit).
2. 2.5 M $CaCl_2$ (sterile-filtered).
3. 2× HEPES-buffered saline (2× HeBS): 274 mM NaCl (3.2 g), 10 mM KCl (142 mg), 1.4 mM $Na_2HPO_4{\cdot}7H_2O$ (76 mg), 15 mM dextrose (540 mg), 42 mM HEPES (2 g) in 180 ml H_2O and adjust pH to 7.03 with 5 N NaOH. Bring to 200 ml with H_2O. Adjust pH to 7.12 with 1 N NaOH (*see* **Note 1**). Sterile filter and aliquot 1 ml in microcentrifuge tubes. Store up to several month at −80 °C (*see* **Note 2**), thawing individual aliquots as needed.
4. Primary neuronal cultures of interest.
5. Transfection medium: Neurobasal medium with 2 % B27 or NS21 (*see* **Note 3**) supplement.
6. Full neurobasal medium: Neurobasal medium with 2 % B27 or NS21 supplement, 1 % Pen/Strep, 1 % Glutamax.
7. 5 % CO_2 incubator, 37 °C.
8. Microcentrifuge tubes (sterile).

3 Methods

Carry out all procedures in a laminar flow hood.

This method is based on transfection of 1 well of a 24 well plate. Scale up or down as necessary.

1. Equilibrate transfection medium in 37 °C, 5 % CO_2 incubator overnight.
2. Thaw and aliquot 15 μl of 2× HeBS per well in a microcentrifuge tube (*see* **Note 4**).
3. Prepare $CaCl_2$–DNA precipitates in microcentrifuge tubes.
   (a) 1.5 μl of 2.5 M $CaCl_2$.
   (b) 3 μg of DNA.
   (c) Add H_2O up to 15 μl.
4. Add $CaCl_2$–DNA dropwise to the tube of 2× HeBS while swirling (*see* **Note 5**).
5. Incubate in the dark for 25 min at room temperature (*see* **Note 6**).
6. While incubating, prepare cells.
   (a) Collect the medium from the well into 50 ml conical tubes (*see* **Note 7**).
   (b) Wash cells with transfection medium four times.
   (c) Add transfection medium into the well and put in incubator.
7. Add 30 μl of transfection mixture dropwise over the culture.
8. Shake the plate or rock back and forth gently and incubate for 40 min.
9. Aspirate the transfection medium and wash two times with fresh transfection medium.
10. Replace with conditioned medium (saved from **step 6**).

4 Notes

1. Making and trying several different pHs (from 7.06 to 7.14) of the HeBS might be necessary because pH meters can vary in their measurements. Precise pH of buffer is important for the formation of precipitates.
2. Storing at −20 °C also works.
3. NS21 is redefined and modified supplement B27 for neuronal cultures [8].
4. Do not reuse the buffer once it was thawed and opened. It might cause changes in pH of the buffer.

5. Drop solution while holding tube with HeBS on a vortex. Vortex strength can be set to shake at levels 1–3. Solutions have to be dropped in a buffer directly, not onto the tube wall.
6. We usually use aluminum foil to protect from light.
7. Add new Neurobasal medium to replace lost volume as needed in final replacement.
8. We have tried this protocol for cortical and hippocampal dissociated cultures from E18 rat embryo. In our experience, this protocol works best between 5 days in vitro (DIV) and 15 DIV.

References

1. Karra D, Dahm R (2010) Transfection techniques for neuronal cells. J Neurosci 30(18):6171–6177. doi:10.1523/JNEUROSCI.0183-10.2010
2. Dudek H, Ghosh A, Greenberg ME (2001) Calcium phosphate transfection of DNA into neurons in primary culture. Curr Protoc Neurosci Chapter 3:Unit 3.11. doi: 10.1002/0471142301.ns0311s03
3. Goetze B, Grunewald B, Baldassa S, Kiebler M (2004) Chemically controlled formation of a DNA/calcium phosphate coprecipitate: application for transfection of mature hippocampal neurons. J Neurobiol 60(4):517–525. doi:10.1002/neu.20073
4. Xia Z, Dudek H, Miranti CK, Greenberg ME (1996) Calcium influx via the NMDA receptor induces immediate early gene transcription by a MAP kinase/ERK-dependent mechanism. J Neurosci 16(17):5425–5436
5. Eguchi M, Yamaguchi S (2007) Double transfection into primary dissociated neurons. J Biosci Bioeng 103(5):497–499. doi:10.1263/jbb.103.497
6. Dahm R, Zeitelhofer M, Gotze B, Kiebler MA, Macchi P (2008) Visualizing mRNA localization and local protein translation in neurons. Methods Cell Biol 85:293–327. doi:10.1016/S0091-679X(08)85013-3
7. Dudek H, Datta SR, Franke TF, Birnbaum MJ, Yao R, Cooper GM, Segal RA, Kaplan DR, Greenberg ME (1997) Regulation of neuronal survival by the serine-threonine protein kinase Akt. Science 275(5300):661–665
8. Chen Y, Stevens B, Chang J, Milbrandt J, Barres BA, Hell JW (2008) NS21: re-defined and modified supplement B27 for neuronal cultures. J Neurosci Methods 171(2):239–247. doi:10.1016/j.jneumeth.2008.03.013

Chapter 11

The Gene-Gun Approach for Transfection and Labeling of Cells in Brain Slices

Anna Dunaevsky

Abstract

Biolistic transfection and diolistic labeling are techniques in which subcellular-sized particles, coated with DNA and lipophilic dyes, respectively, are propelled into cells. The gene-gun approach is particularly applicable for use on ex vivo organized tissue such as brain slices, where cells are not accessible for transfection with methods used in dissociated cell preparations. This simple and rapid method results in targeting of individual cells in a Golgi-like manner, allowing investigating structural and functional aspects of neuronal development.

Key words Gene-gun, Biolistics, Diolistics, Transfection, Lipophilic dyes, Brain slices

1 Introduction

The introduction of genes into postmitotic neurons and glia cells is an important approach that allows to probe gene function in various developmental processes such as dendritic outgrowth and synaptogenesis [1–3]. Biolistic transfection was initially developed as a method of gene transfer into plants [4], as it allows transfer across cell walls, but has been since extensively used in the field of developmental neuroscience. Biolistic transfection using a hand-held gene gun is a mechanical method for introducing genes into cells. Although multiple methods are available for transfection of cells with DNA (i.e., lipofection and calcium phosphate precipitation), these work well for dissociated cells in culture and are generally not suitable for cells in thick preparations. The biolistic approach relies on the high-speed propulsion of micrometer-sized particles coated with DNA that can reach cells deep inside organized tissue such as brain slices [5].

Although the use of recombinant viruses that can be injected into ex vivo and in vivo brain tissue and target postmitotic neurons is an increasingly popular approach [6], it has its limitations as compared to the biolistic approach. Many viral constructs require

Renping Zhou and Lin Mei (eds.), *Neural Development: Methods and Protocols*, Methods in Molecular Biology, vol. 1018, DOI 10.1007/978-1-62703-444-9_11, © Springer Science+Business Media, LLC 2013

the use of special biosafety conditions. There can also be concerns regarding immunogenicity and toxicity of viral constructs. Another significant drawback is that the size of the virally packaged DNA insert can be limiting. Finally, the preparation of viral constructs is time-consuming and costly. The biolistic transfection approach does not suffer from the above-mentioned limitations. Plasmids can be coated directly unto particles without the need of cloning into viral constructs. A single particle can be coated with multiple plasmids thus allowing the simultaneous expression of several genes [7]. Finally, the preparation of coated particles and the delivery into cells is relatively rapid.

The gene-gun approach can be used not only to transfect cells but also to label them with dyes. For diolistics, instead of coating particles with DNA, they can be coated with lipophilic dyes such as DiI and DiO [8]. This approach can be used to label cells in live preparations for live imaging [9] as well as used to investigate neuronal structure and connectivity in fixed tissue [10] such as postmortem brain samples of neurodevelopmental and other disorders.

2 Materials

2.1 Biolistic Transfection Components

1. Organotypic brain slices.
2. Gold microcarriers, 1 μm (Bio-Rad, Hercules, CA, USA) (*see* **Note 1**).
3. Polyvinylpyrrolidone (PVP) (Sigma, Chemical Company, St. Louis, MO, USA): Stock solution (20 mg/ml) in ethanol. Working solution 0.01–0.1 mg/ml.
4. 1 M $CaCl_2$ (Sigma).
5. 0.05 M spermidine (Sigma) (*see* **Note 2**).
6. Plasmid DNA (1 mg/ml).
7. 100 % ethanol (*see* **Note 3**).

2.2 Diolistic Labeling Components

1. Fixed brain sections (100 μm).
2. Tungsten microcarriers, 1.6 μm (Bio-Rad).
3. Polyvinylpyrrolidone (PVP, Sigma): Stock solution (20 mg/ml) in ethanol. Working solution 0.01–0.1 mg/ml.
4. Lipophilic dyes (DiO, DiI, DiD) (Molecular Probes, Invitrogen).
5. Methylene chloride (Sigma).

2.3 Equipment

1. Helios gene gun (Bio-Rad).
2. Tubing preparation station (Bio-Rad).
3. 165-2441 Tefzel tubing (Bio-Rad).

4. Tubing cutter (Bio-Rad).
5. Sonicator.
6. Helium gas cylinder and regulator.
7. Nitrogen gas cylinder and regulator.
8. Vortexer.
9. Microcentrifuge.
10. 10 ml syringe.

3 Methods

Prepare live organotypic brain slices of interest for biolistic transfection [3] and fixed brain sections (100–200 μm) for diolistic labeling.

3.1 Preparation of Tubing

Cut Tefzel tubing to the correct length (allow for 6–8 cm to protrude from the end of the tubing preparation station). Attach a syringe to the Tefzel tubing via adaptor tubing and draw the working PVP solution into the Tefzel tubing (*see* **Note 4**). Allow to stand for 5 min and then eject the PVP solution and detach the tubing from the syringe. Load the Tefzel tubing into the tubing preparation station. Dry the Tefzel tubing while running dry nitrogen through the tube at 0.4 psi for at least 15 min.

3.2 Preparation of Plasmid-Coated Particles

1. Weigh out 12 mg of 1 μm gold microcarriers into an Eppendorf tube. Add 50 μl of 0.05 M spermidine. Vortex the mixture for a few seconds and then sonicate for 3–5 s. Add 24 μg of plasmid DNA (*see* **Notes 5** and **6**). Volume of DNA should not exceed 50 μl. Vortex the DNA/gold/spermidine mix for 5 s. While vortexing, add 50 μl of 1 M $CaCl_2$ dropwise to the mix. Allow the DNA to precipitate onto the beads for 10 min at room temperature.
2. Spin the tube with the gold and DNA in a microfuge for about 15 s to pellet. Remove the supernatant with vacuum (a glass pipette covered with a plastic pipette tip works well). Resuspend the pellet in the remaining supernatant by vortexing briefly.
3. Wash the pellet four times with 1 ml of freshly opened 100 % ethanol each time. Each time vortex, spin 5 s in microfuge and aspirate the ethanol. Change tips each time. Resuspend the gold particles in 1.5 ml of ethanol. Parafilm the tube until ready to use.

3.3 Preparation of DNA-Coated Particles

1. Turn off the nitrogen flow on the tubing preparation station. Remove the Tefzel tubing from the preparation station and attach to a 10 ml syringe using a connector tubing. Vortex the

gold mixture and invert the tube as needed to keep it mixed. *The following steps need to be done very quickly.* Place the end of the Tefzel tubing in the gold mixture and draw the solution to the middle of the tubing using the syringe. Avoid introduction of bubbles. Immediately bring the tubing to a horizontal position and slide it back into the tubing preparation station. Do not push the end into the gasket yet. Leave the syringe attached. Allow the gold to settle for 1–2 min. Carefully draw out the ethanol using the syringe (1–2 cm/s). Detach the tubing from the syringe and twirl tube with your fingers quickly three to four times. Allow the gold to settle "upside down" for 3–4 s. Push the tubing into the gasket, start the rotor, and, after 20–30 s, turn on the nitrogen with a flow of 0.4 psi for 5 min.

2. Cut the tubing at the edge of where gold particles can be seen. Use the tubing cutter to cut the individual bullets. Store loaded bullets in a scintillation vial with a desiccant pellet at 4 °C.
3. Bullets should be used within 2 weeks.

3.4 Preparation of Dye-Coated Particles

1. Weigh out ~8 mg Tungsten particles in glass jar. Add ~2.5 mg of DiI dye and 250 μl of methylene chloride and swirl the mix. Let the mix dry in a chemical hood a few minutes, leaving just a little liquid. Add 1.5 ml distilled water. Sonicate the jar 10 min to emulsify.
2. Immediately transfer emulsion to PVP-coated Tefzel tubing, drawing in with a syringe (specific for DiI dye). Carefully thread tubing into the preparation station. Do not push the end into the gasket yet and leave the syringe attached. Make sure the nitrogen flow is turned off. Allow the particles to settle for 30 min. Very slowly remove the solution using the syringe. Detach the syringe, push the tubing all the way through the gasket, and rotate the tube for 30 s. Turn on the nitrogen at 0.4 psi and dry the tubing for additional 30 min.
3. Cut the tubing at the edge of where the particles can be seen. Use the tubing cutter to cut the individual bullets. Store the bullets in a scintillation vial with a desiccant pellet at 4 °C.
4. Bullets can be used for several months.

3.5 Shooting of Organotypic Slices with DNA-Coated Particles

1. Prepare the gene gun by attaching the sterilized nosepiece with the diffuser screen in place. Insert cartridge holder and lock in place (Fig. 1). Attach the gene gun to the helium gas and adjust the pressure to 200 psi. Discharge the gun first with no bullets to make sure the pressure is correct. Insert bullets into the cartridge holder.
2. Remove the 6-well dish with the organotypic slices from the incubator just before shooting. Hold the gun upright with the

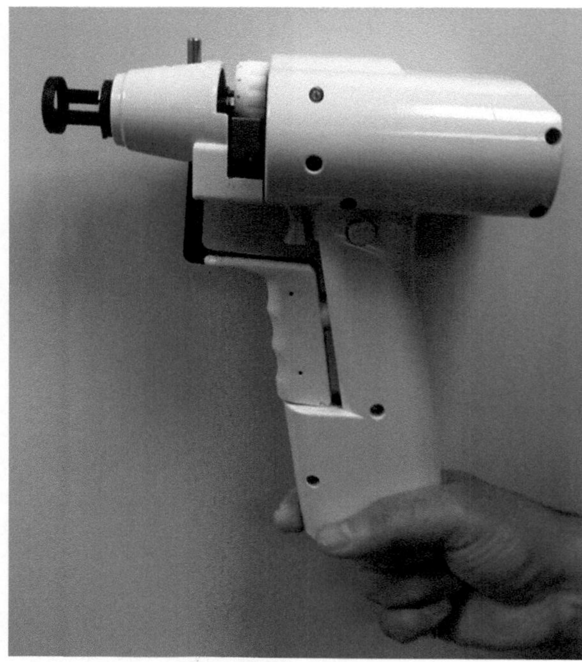

Fig. 1 Helios gene gun with a cartridge holder and a nosepiece in place. The diffuser is placed in the back of nosepiece (Image provided by Dr. Suneet Mehrotra)

nosepiece inserted into the Millicell insert (touching the filter but not the slices). Discharge once into each insert.

3. Check slices on an inverted microscope. Gold particles can be seen mainly at edges of the slices.
4. Immediately return the slices to the incubator. Slices can be imaged after 48–72 h of expression (Fig. 2).

3.6 Shooting of Fixed Sections with Dye-Coated Bullets

1. Prepare the gene gun by attaching the nosepiece with the diffuser screen in place. Insert cartridge holder and lock in place. Attach the gene gun to the helium gas and adjust the pressure to 150 psi (*see* **Note 7**). Discharge the gun first with no bullets to make sure the pressure is correct. Insert bullets into the cartridge holder.
2. Transfer sections into a petri dish and remove buffer until sections are almost dry. Arrange the sections so that they occupy an area smaller than the diameter of the gene-gun nosepiece. Hold the gun upright with the nosepiece touching the petri dish and discharge once (*see* **Notes 8** and **9**).
3. Check sections on an inverted microscope to ensure particles are seen on and around sections. If possible, sections should be viewed under a fluorescent dissecting microscope to assess the efficiency of delivery of the dye-coated particles (*see* **Notes 10–12**).

Fig. 2 A Purkinje neuron in a cerebellar organotypic slice from a postnatal day 19 mouse, 48 h after biolistic transfection with particles coated with GFP. Bar: 40 μm

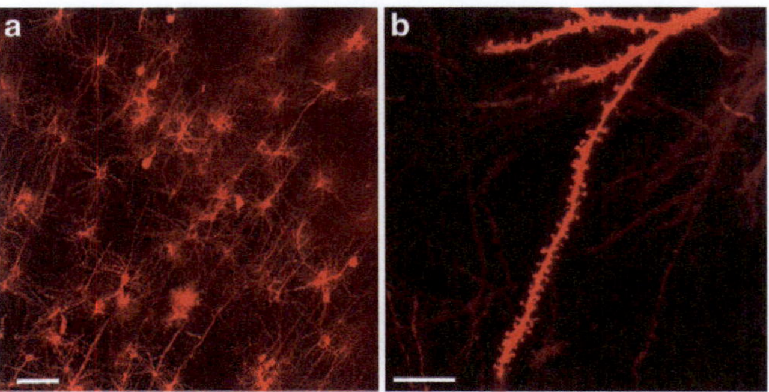

Fig. 3 (**a**) Fixed hippocampal sections diolistically labeled with particles coated with the dye DiI. Multiple cells and dendritic segments can be observed. Bar = 75 μm. (**b**) Dendritic spines can be seen at higher magnification. Bar = 10 μm

4. Rinse sections two times with buffer and make sure the sections are submerged. Incubate in buffer at 4 °C for at least 2–3 h before mounting.
5. Sections should be imaged within a few days of labeling (Fig. 3) to prevent excessive diffusion of the dye.

4 Notes

1. The size of the gold particles used for DNA transfection should be optimized for the specific preparation used and the cells targeted.
2. Spermidine should be stored at −20 °C and kept at 4 °C after thawing. Use within 3 weeks of thawing.
3. A fresh bottle of ethanol should be used for particle coating. Between washes, cover the cap with parafilm to ensure dryness of the ethanol.
4. The use of PVP is optional, and its effect on improving the coating should be determined by the user.
5. High-quality DNA needs to be prepared with a midi or maxi kit to ensure efficient transfection.
6. DNA concentration must be appropriate for efficient transfection and must be optimized for the specific system used. When more than one plasmid is used, these can be coprecipitated onto the gold particles.
7. The pressure used for shooting should be optimized for different thickness slices from different brain regions.
8. Different nosepieces and diffuser should be dedicated to particles coated with different plasmids and dyes. This will prevent inadvertent transfection or labeling with the wrong constructs and dyes.
9. The use of ear protection during gene-gun discharge is advisable.
10. If clumping of particles is observed in the diolistic approach, it is recommended to attach a filter membrane with 3 μm pore size (at the same place where the diffuser screen attaches to the nosepiece).
11. The transfection rate of neurons and glia in organotypic slices is moderate and variable depending on the preparation. The number of transfected cells tends to be higher with slices that are kept in culture for more than 1–2 days.
12. With diolistics in fixed sections, the same sections can be shot several times if the number of coated particles delivered is not sufficient. Although labeling of entire cells with the full arborization is possible, many cells are only partially labeled as some neurites are transected during sectioning.

References

1. Lordkipanidze T, Dunaevsky A (2005) Purkinje cell dendrites grow in alignment with Bergmann glia. Glia 51:229–234
2. Callaway EM, Borrell V (2011) Developmental sculpting of dendritic morphology of layer 4 neurons in visual cortex: influence of retinal input. J Neurosci 31:7456–7470
3. Dunaevsky A et al (1999) Developmental regulation of spine motility in the mammalian central nervous system. Proc Natl Acad Sci USA 96:13438–13443
4. Klein TM et al (1987) High-velocity microprojectiles for delivering nucleic acids into living cells. Nature 327:70–73
5. Lo DC, McAllister AK, Katz LC (1994) Neuronal transfection in brain slices using particle-mediated gene transfer. Neuron 13: 1263–1268
6. Luo L, Callaway EM, Svoboda K (2008) Genetic dissection of neural circuits. Neuron 57:634–660
7. Fu Y, Huang ZJ (2010) Differential dynamics and activity-dependent regulation of alpha- and beta-neurexins at developing GABAergic synapses. Proc Natl Acad Sci USA 107: 22699–22704
8. Gan WB et al (2000) Multicolor "DiOlistic" labeling of the nervous system using lipophilic dye combinations. Neuron 27:219–225
9. Benediktsson AM et al (2005) Ballistic labeling and dynamic imaging of astrocytes in organotypic hippocampal slice cultures. J Neurosci Methods 141:41–53
10. Harms KJ et al (2008) Transient spine expansion and learning-induced plasticity in layer 1 primary motor cortex. J Neurosci 28:5686–5690

Chapter 12

Lentiviral Vector Production, Titration, and Transduction of Primary Neurons

Baojin Ding and Daniel L. Kilpatrick

Abstract

Lentiviral vectors have become very useful tools for transgene delivery. Based on their ability to transduce both dividing and nondividing cells and to produce long-term transgene expression, lentiviruses have found numerous applications in the biomedical sciences, including developmental neuroscience. This protocol describes how to prepare lentiviral vectors by calcium phosphate transfection and to concentrate viral particles by ultracentrifugation. Functional vector titers can then be determined by methods such as fluorescence-activated cell sorting or immunostaining. Effective titers in the range of 10^8–10^9 infectious units/ml can be routinely obtained using these protocols. Finally, we describe the infection of primary neuronal cultures with lentiviral vectors resulting in 85–90 % cell transduction using appropriate multiplicities of infection.

Key words Lentivirus, Transfection, Titration, Immunofluorescence, Infection, Primary neuron cultures

1 Introduction

A key feature of lentiviruses as expression vectors is their capacity to transduce both proliferating and nonproliferating cells [1–3]. The development of replication-defective, self-inactivating lentiviruses has provided a safe and effective means for delivering transgenes both in vivo and in culture. These viruses lack dispensable genes from the HIV-1 genome and separate *cis*-acting sequences from *trans*-acting factors required for viral particle production, infection, and integration [4–6]. Second-generation vectors developed for producing self-inactivating lentivirus consist of three recombinant plasmids carrying sequences for the relevant transgene to be expressed, for viral packaging, and for envelope proteins. A third-generation packaging system provides maximal biosafety but requires the transfection of four separate plasmids [7, 8].

In this protocol, we describe the use of the second-generation packaging system together with calcium phosphate transfection

Renping Zhou and Lin Mei (eds.), *Neural Development: Methods and Protocols*, Methods in Molecular Biology, vol. 1018, DOI 10.1007/978-1-62703-444-9_12, © Springer Science+Business Media, LLC 2013

of HEK293T cells. Viral particles accumulate in the culture medium, and high-titer viral preparations are generated using ultracentrifugation. While several methods have been described for viral concentration, including filtration [9, 10], we have found that ultracentrifugation yields high concentrations and good retention of biological activity. Lentivirus amounts can be determined by nonfunctional or functional methods [11]. Examples of the former include detection of expressed virus proteins (such as p24) or their transcripts using ELISA or real-time PCR, respectively [12, 13]. Generally, these nonfunctional titration methods overestimate the functional vector titer since inactive viral particles contribute to their signal [14, 15]. Functional titration assays are typically based on vector-encoded reporter gene expression. For instance, vectors expressing green fluorescent protein (GFP) can be titrated by fluorescence-activated cell sorting (FACS) analysis [16]. For vectors that do not contain a reporter gene, the titers can be determined based on the integration of proviral or transgene copies into host cell genomic DNA using real-time PCR [17, 18]. In this protocol, we describe functional titration using immunofluorescence.

Lentiviral transgenes are extremely useful for exploring the function and development of primary neurons [19–22]. They can be used to express proteins in such cultures for multiple purposes, including to monitor changes in subcellular localization; to alter protein function using wild-type, constitutively active or dominant inhibitory isoforms and/or shRNAs; and also to express and study the regulation of gene promoters within a chromatin context as transgenes integrated into cellular DNA [16, 19, 20, 23]. Here, we outline the transduction of neuronal cultures using primary cerebellar granule neurons (CGNs) as an example [16, 23].

2 Materials

2.1 Lentivirus Preparation Reagents and Solutions

1. Plasmid preparation: A lentiviral vector containing the transgene of interest, a packaging vector such as psPAX2 (Addgene, 12260), and an envelope protein vector such as pCMV-VSVG (Addgene, 8454) (*see* **Note 1**).
2. Phosphate-buffered saline (PBS Buffer): 137 mM NaCl, 2.7 mM KCl, 8 mM Na_2HPO_4, 1.46 mM KH_2PO_4. Dissolve 8 g of NaCl, 0.2 g of KCl, 1.4 g of Na_2HPO_4, and 0.2 g of KH_2PO_4 into 800 ml distilled H_2O. Adjust pH to 7.4 and add distilled H_2O to 1 L. Sterilize by autoclaving.
3. Dulbecco's Modified Eagle's Medium (DMEM): Dissolve 13.4 g of DMEM powder (GIBCO® #12100-046) into 950 ml of ultrapure H_2O. Add 3.7 g of sodium bicarbonate powder (Mallinckrodt Chemicals, #7412-12). Adjust to pH 7.1–7.2 by adding 1N NaOH or 1N HCl with stirring. Bring final volume

to 1 L with water. Filter through a 0.2 μm filter and keep at 4 °C (*see* **Note 2**).

4. Complete DMEM: Just before use, supplement DMEM with 10 % fetal bovine serum (FBS, heat inactivated, GIBCO® #10082-147) and 1×Pen/Strep (Invitrogen, #15140).
5. 0.5 % trypsin-EDTA (GIBCO® #15400): Make tenfold dilution with PBS just before use.
6. HEK293T cells (ATCC, #CRL-11268) (*see* **Note 3**).
7. 100 mm cell culture dishes (Corning #430167).
8. ProFection Mammalian Transfection System-Calcium Phosphate (Promega #290710).
9. 0.45 μm syringe filter (Millipore, #SLHV033RS).
10. Beckman UltraClear ultracentrifuge tubes (25×89 mm, #344058).
11. 37 °C incubator with 5 % CO_2 humidified atmosphere.
12. A biosafety hood (BSL-2 compliant).

2.2 Lentivirus Titration and Transduction Components

1. Poly-D-lysine: Dissolve 250 mg of poly-D-lysine powder (Sigma, #P6407) into 50 ml of water to prepare 5 mg/ml stock. Store in aliquots at −20 °C for longer storage. Make 50-fold dilution of stock solution with water to make a 100 μg/ml working solution on the day of use (*see* **Note 4**).
2. Laminin (100× stock solution, Invitrogen, #23017-015). Make 100-fold dilution in PBS before use.
3. 24-well plate, 6-well plate, or 8-well chamber slides (BD Biosciences).
4. Hemocytometer (Hausser Scientific).
5. Neurobasal (NB) medium (GIBCO®, #21103).
6. Complete NB medium: NB medium supplemented with 1× B27 (50×, GIBCO® #17504-044), 1×L-glutamine (100×, GIBCO® #25030), 0.45 % D-glucose, 1×Pen/Strep (100×, Invitrogen, #15140). Filter, sterilize, and store at 4 °C.
7. Leica DM IRE2 fluorescence and phase contrast microscope or its equivalent.
8. Nylon mesh (74 μm; Small Parts, #CMN-0074-D).
9. 4 % paraformaldehyde: For 10 ml solution, add 6.6 ml of distilled water to a small beaker and warm with stirring. Add 0.4 g of paraformaldehyde (Sigma, #P6148) and 50 μl of 2N NaOH to the warming water while stirring. Once most of the paraformaldehyde powder is dissolved, remove solution from heat and cool to room temperature. Filter to remove undissolved material and keep in dark at 4 °C (*see* **Note 5**).

10. Normal goat serum (NGS) (Invitrogen, #PCN5000). Make 1 % and 5% dilutions with PBS.
11. Triton X-100 (Sigma, #9002-93-1). Make 1 % with water before use.
12. Bisbenzimide (H33258) (Sigma, #B2883). Prepare a 1 mg/ml stock solution in water and store in the dark at −20 °C. Dilute 1,000-fold with water just before use.
13. ProLong Antifade kit (Invitrogen, #P7481).

3 Methods

There are important biosafety issues that must be considered when working with self-inactivating lentiviral vectors. NIH Biosafety Level 2 criteria should be followed throughout, including (1) wear lab coat and gloves at all times, (2) avoid the creation of aerosols, (3) work with lentiviruses in a Class II laminar flow hood, (4) routinely decontaminate surfaces and immediately clean up and decontaminate any spills of virus material, and (5) decontaminate and dispose of all cultures, solutions, and other regulated wastes using approved procedures. For updated information on suitable laboratory biosafety practices, see the NIH website at http://oba.od.nih.gov/rdna_rac/rac_guidance_lentivirus.html. It is also important to check with the health and safety guidelines at your institution regarding the use of lentiviruses.

3.1 Lentivirus Preparation

1. ~24 h prior to transfection, plate $5–8 \times 10^6$ HEK293T cells (optimum cell number based on cell growth rate (*see* **Note 6**)) into each of four 100 mm dishes in 10 ml of fresh complete DMEM.
2. Approximately 3 h prior to transfection, replace medium with a fresh 10 ml of complete DMEM. At the time of transfection, the HEK293T cells should be ~80 % confluent and evenly distributed (Fig. 1) (*see* **Note 7**).
3. *$CaPO_4$ precipitation* (all volumes or amounts are for four 100 mm plates): In a 15 ml tube, prepare the transfection cocktail by adding the lentiviral transfer vector plasmid, psPAX2 packaging construct, and pCMV-VSVG viral envelope expression vector in a molar ratio of 1:1:1 with a total of 120 μg DNA. Combine with 248 μl of 2 M $CaCl_2$ (provided in kit) and add nuclease-free water to a final volume of 2 ml. While continuously vortexing the solution, dropwise add 2 ml of 2× HBS (pH 7.05) (*see* **Note 8**).
4. Leave the transfection cocktail in hood for ~5 min to permit precipitate formation. Then slowly add this solution to the plated cells dropwise, gently swirling the plate to ensure uniform dispersal. Return the cells to the incubator.

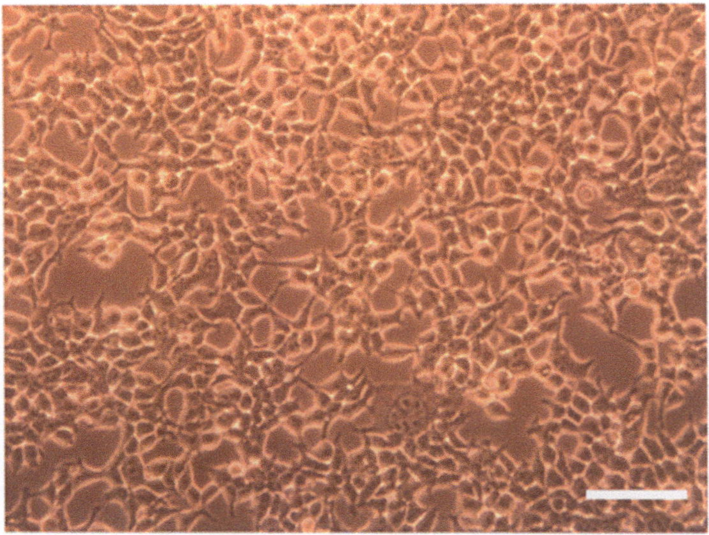

Fig. 1 Photomicrograph of HEK293T cells just prior to transfection. Bar = 100 μm

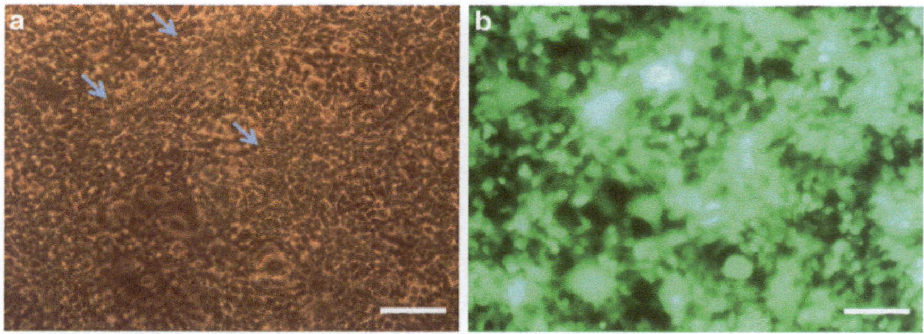

Fig. 2 HEK293T cells transfected with GFP-expressing lentivirus. (**a**) Bright-field image showing approximately confluent K293T cells at the time of virus harvest. Transfected cells tend to fuse into multinucleated cells (*arrow head*). (**b**) Fluorescence image of the same field as in (**a**). GFP(+) cells are typically >90 %. Fused HEK293T cells tend to give stronger signal. Bar = 100 μm

5. The next morning (~16 h after adding transfection cocktail), remove the medium and gently replace with 10 ml of fresh complete DMEM (*see* **Note 7**).
6. Harvest the virus-containing medium 48 h after medium replacement and store at 4 °C. Determine the transfection efficiency under a fluorescent microscope if appropriate, e.g., for viruses expressing GFP (Fig. 2) (*see* **Note 9**).
7. After the first collection, add 10 ml of fresh DMEM to each plate and collect the supernatant after an additional 24 h incubation.
8. Pool the two supernatants and centrifuge the virus-containing medium at 500 × *g* for 10 min at 4 °C. Filter medium through a 0.45 μm filter to remove debris.

9. Sterilize two or more 40 ml Beckman UltraClear tubes (25 × 89 mm) and adaptors by spraying with 70 % ethanol and allow them to dry in the biosafety hood (*see* **Note 10**).
10. In the hood, transfer the filtered lentivirus-containing medium to the ultracentrifuge tubes, completely filling them. Add additional DMEM as necessary to top off and balance them.
11. Centrifuge at 25,000 rpm in a Beckman SW 28 rotor for 100 min at 4 °C.
12. Carefully decant the supernatant and invert the tubes on a paper towel for 2 min to drain off the remaining liquid. A clear yellowish pellet should be visible (*see* **Note 11**).
13. Carefully suspend the virus pellet with complete NB medium (~100 μl per 10 cm plate) by first swirling medium over the pellet followed by gentle pipetting (*see* **Note 12**). Disperse into 20 μl aliquots in sterile 0.5 ml tubes and quick-freeze in a dry ice bath. Store at −80 °C.

3.2 Lentivirus Titration

For viruses expressing GFP (or any other fluorescent product):

1. Plate 7–10 × 10^4 HEK293T cells/well in a 24-well plate in 500 μl of complete DMEM as follows: one well to count cells prior to transduction; one well as a non-transduced control; six wells for serial dilutions of each virus (*see* **Note 13**). Incubate overnight in a 37 °C incubator with 5 % CO_2.
2. Cells should be ~60 % confluent at the time of transduction. Determine the preinfection cell number/well by harvesting cells in the counting well by trypsinization. Remove medium from the well and add 0.5 ml of 0.05 % trypsin in PBS. Incubate for 3 min in a 37 °C incubator and then neutralize the trypsin by adding 0.5 ml of complete DMEM. Transfer cells to a 1.5 ml conical tube and wash cells by centrifugation at 1,000 × *g* for 4 min, resuspension in PBS, and re-centrifugation. Suspend cells in 300 μl PBS and count using a hemocytometer.
3. Make tenfold serial dilutions of virus (10^{-2} to 10^{-7}) with complete DMEM as follows (*see* **Note 14**):

   5 μl stock virus + 495 μl media = 10^{-2}
   50 μl of 10^{-2} virus + 450 μl media = 10^{-3}
   50 μl of 10^{-3} virus + 450 μl media = 10^{-4}
   50 μl of 10^{-4} virus + 450 μl media = 10^{-5}
   50 μl of 10^{-5} virus + 450 μl media = 10^{-6}
   50 μl of 10^{-6} virus + 450 μl media = 10^{-7}
4. Remove all the medium from each well and add 500 μl of viral dilutions or medium alone for the non-transduced control. Place cells in the 37 °C incubator overnight.
5. The next morning, remove virus-containing medium and replace with 500 μl of complete DMEM. Incubate cells at 37 °C for an additional 48 h.

6. Visualize GFP expression for multiple fields under a fluorescence microscope and estimate infection based on % GFP(+) cells. For a good viral preparation, GFP expression should range from 100 to 0 % for successive viral dilutions and control wells.
7. Harvest cells from wells showing ~5–20 % GFP(+) for subsequent titering. Trypsinize, wash with PBS, and suspend cells in 300 μl of PBS (*see* **step 2** above).
8. Filter cells through ~80 μm nylon mesh (optional) and analyze cells by flow cytometry (*see* **Note 15**).
9. Calculate lentivirus titers using values from samples within a linear range of GFP expression based on flow cytometry results. Use the following formula:

$$T = (P \times N) / (D \times V)$$

T= titer (IFU/ml), P= % GFP positive cells (e.g., 0.2 for 2 % GFP (+) cells), N= number of cells determined at the time of viral transduction, D= dilution factor ($10^{-3} = 0.001$), and V= volume of viral inoculum (0.5 ml).

Titering by Immunofluorescence

In cases where an expressed protein is detectable by immunostaining, set up the titration as above using a chamber slide. Use 4×10^4 HEK293T cells/chamber and culture cells as before.

10. At the end of the experiment, fix the cells in each chamber with 200 μl of 4 % paraformaldehyde (*see* **Note 5**) by incubation at room temperature for 30 min.
11. Remove the paraformaldehyde, add 300 μl PBS per well, and incubate at room temperature for 5 min. Repeat PBS wash twice more (*see* **Note 16**).
12. Permeabilize cells with 200 μl of 1 % Triton X solution/well and incubate at room temperature for 10 min.
13. Remove liquid and wash wells with PBS three times (*see* **step 11** above).
14. Add 5 % NGS blocking solution and incubate at room temperature for 30 min.
15. Remove blocking solution and add an appropriate dilution of primary antibody in 1 % NGS (*see* **Note 17**). Incubate at room temperature for 1 h or at 4 °C overnight.
16. Remove primary antibody and wash wells with PBS four times at room temperature for 10 min each.
17. Add the appropriate fluorescently labeled secondary antibody in 1 % NGS. Incubate at room temperature for 30 min in the dark.
18. Remove the secondary antibody and wash with PBS three times at room temperature for 10 min in the dark.

19. Add nuclei stain such as bisbenzimide (1 μg/ml) in water. Incubate in the dark at room temperature for 1–2 min.
20. Wash twice with water, then air dry for 30 min in the dark.
21. In the meantime, prepare ProLong Antifade solution according to kit instructions.
22. Add a drop of ProLong solution to each chamber slide well, avoiding bubbles. Affix a cover slip.
23. Dry the slide in the dark for 30 min or overnight.
24. Capture multiple fields using a fluorescence microscope to determine the percentage of infected cells and calculate lentivirus titers as in step 9, Subheading 3.2 (*see* **Note 18**).

3.3 Transduction of CGNs

1. Plates to be used for CGN transduction should be treated with poly-D-lysine and laminin (*see* **Note 4**). Refer to Chapter 5 (Selvakumar and Kilpatrick).
2. Prepare *Percoll gradient-purified* CGNs for virus transduction as outlined in Chapter 5 (*see* **Note 19**).
3. Calculate how many lentivirus infectious units are needed for the experiment based on the CGN number/well, the number of wells, and the titer of the virus preparation being used. If near-quantitative cell transduction (85–90 %) is desired, we typically use MOI = 3 for virus titered by immunofluorescence or MOI = 5 for virus titered by flow cytometry (*see* **Note 20**).
4. Thaw virus vial and thoroughly mix the viral stock with culture medium by gently pipetting several times (*see* **Note 21**).
5. For cultures infected on the day of cell preparation (0 days in vitro (0 DIV)), diluted virus and cells in culture medium are first combined and then plated together (*see* **Notes 22** and **23**).
6. For infection of 1-DIV cultures, remove 50 % of culture medium from wells or plates and replace this with an equivalent volume of virus-containing medium at the appropriate MOI.
7. After 16–24 h, remove all virus-containing medium and replace with an equal volume of fresh complete NB medium. Continue culturing as outlined in Chapter 5 as dictated by the specific needs of the experiment.
8. Transduction efficiency of CGNs can be determined in chamber slides by direct fluorescence microscopy if the viruses express a fluorescent protein. Alternatively, degree of infection can be determined by immunofluorescence of the expressed protein as outlined above (Fig. 3).
9. Lentiviral transduction also can be performed using other neuronal cultures, including cortical neurons [16] and hippocampal neurons (data not shown) (*see* **Note 24**).

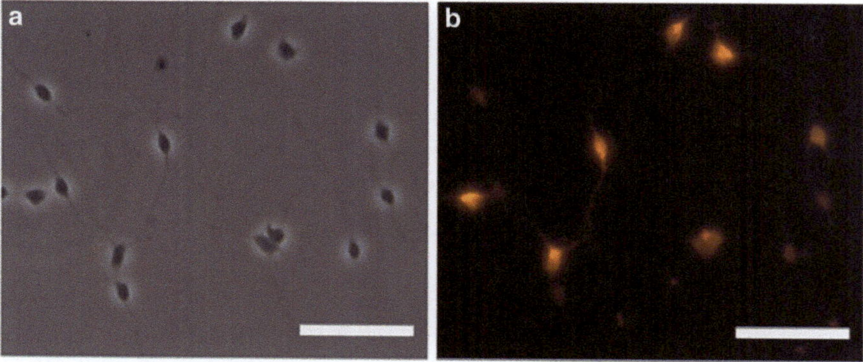

Fig. 3 Photomicrographs of lentivirus-transduced CGNs. (**a**) Bright-field image after transduction (MOI = 3) with virus expressing HA-tagged Nuclear Factor One dominant repressor. (**b**) Fluorescence image of the same field as in (**a**). CGNs were stained with anti-HA antibody (Rabbit mAB, Cell Signaling, #3724) and GtxRb IgG-Cy3 secondary antibody (Millipore, #AP132C). Bar = 50 μm

4 Notes

1. Purify plasmids using endotoxin-free Qiagen columns and suspend the precipitated DNA in TE buffer (pH 8.0) containing 0.2 mM EDTA to minimize endonuclease nicking that can reduce plasmid supercoiling (see below). Plasmid quality should be verified by optical reading (A260/A280 >1.8) and agarose gel electrophoresis to confirm mainly supercoiled DNA, which is optimal for cell transfection. Lentiviral transgene backbones containing central polypurine tract (cPPT) and woodchuck hepatitis virus posttranscriptional regulatory element (WPRE) sequences are recommended since these sequences can improve functional viral titers and enhance expression of transgenes [24, 25].
2. DMEM can be stored at 4 °C for 3–4 weeks.
3. HEK293T cells are maintained in complete DMEM. Depending on their rate of growth, cells are split at a ratio of 1:3 or 1:4 for passaging. It is best that cells be passaged at least three times after thawing frozen stocks before using for transfections. Some researchers recommend not using cells passaged more than 20 times.
4. Poly-D-lysine cannot be sterilized by filtration due to its high molecular weight. After poly-D-lysine treatment, we recommend washing the wells three times with 1×Pen/Strep in water to sterilize wells before plating cells.
5. 4 % paraformaldehyde can be stored at 4 °C in the dark for up to 1 week.
6. Be sure HEK293T cells are dispersed as single cells following trypsinization using gentle pipetting. Dilute cells in medium

and mix well prior to plating to ensure their even distribution on the plate surface. Transfection of four 100 mm plates typically generates lentivirus preparations in the range of $0.5–2 \times 10^8$ total infectious units. A larger number of plates can be used to generate more virus or to increase viral titers (e.g., if the lentiviral backbone being used tends to give lower viral yields).

7. Very carefully replace the medium to avoid disturbing attached cells.
8. It is important to *slowly* add 2×HBS buffer dropwise while vortexing to form a fine DNA-Ca_2PO_4 precipitate, which is critical for good transfection efficiency (*see* ref. 26). For multiple plates, preparing the DNA/Ca_2PO_4 precipitate as a single mixture is recommended. The larger volume facilitates the formation of a more homogenous precipitate, resulting in higher transfection efficiencies.
9. The transfection efficiency should be >90 % for a robust viral preparation.
10. To save time, sterilize the ultracentrifuge tubes and adaptors before virus collection. Keep sterile in the biosafety hood.
11. The pellet size is not necessarily well correlated with the virus yield. The pellet can also be difficult to detect due to its transparency.
12. Pipette gently and avoid bubbles to avoid virus inactivation. A separate small aliquot of a few microliters also can be prepared and frozen for titering purposes.
13. NIH3T3 cells also can be used for titration. Use duplicates or triplicates as preferred.
14. Depending on the virus concentration, extend the titration range from 10^{-2} to 10^{-12}, as required.
15. No fixation is required for GFP-expressing cells, and in fact fixation will reduce the fluorescent signal. For optimal sensitivity, we recommend analyzing the samples on the day of harvesting to minimize loss of signal with time.
16. The fixed cells can be stored at 4 °C with PBS for up to a week prior to staining.
17. If background signal is high for the no-virus control well, repeat the assay using antibody diluted with 2 % NGS.
18. Titers determined by immunofluorescence tend to be higher than those done by cytometry. Thus, when comparing results for multiple viruses in a single experiment, it is important to use the same titering method for the different virus preparations.
19. For optimal transduction efficiency, gradient-purified CGNs should be used.
20. To obtain a desired infection efficiency, the optimal MOI may vary depending on the quality of the virus particles, the

targeted cell type, and the titration method employed (e.g., viruses titered by flow cytometry of fluorescently labeled cells tend to require higher MOIs than viral preparations titered by immunostaining to obtain the same transduction efficiency). We recommend preliminary assays using different MOI (2–6) to transduce CGNs in chamber slides to determine an optimal MOI for >80 % infection using immunostaining or GFP fluorescence. Obviously, lower MOIs are suitable if a reduced % infection is desired. Also note that lentiviral infection can produce cytotoxicity at elevated MOIs, so it is important to find a balance between sufficient transduction (e.g., >80 %) and cell viability. Reduced viability is evident from increased numbers of floating cells or shrunken attached cell bodies. This can be further confirmed using appropriate cell viability (e.g., live/dead) assays.

21. Virus should be thawed only once since freeze-thaw cycles decrease virus titers. *See* **Note 12** regarding setting aside a virus aliquot specifically for titering purposes.

22. If there is a need to further enhance viral infection (e.g., for a low-titer virus), culture medium can be reduced by 50 % prior to adding virus to increase the effective virus concentration twofold.

23. Two alternative protocols for 0-DIV transduction are as follows: First, plate the CGNs in the absence of virus and then add virus 2–4 h later, as outlined for 1-DIV infections. In some instances, this can be advantageous by reducing cytotoxic effects of lentiviral infection (e.g., at higher MOIs) or of expressed proteins (e.g., those that interfere with cell cycle progression or exit). Second, under circumstances where early transgene expression is desired (e.g., to study cell proliferation or early differentiation events), CGN progenitors can be infected while they are maintained in the proliferative state to allow transgene expression prior to onset of growth arrest and differentiation. In this protocol, cell preparations are plated together with lentivirus and 2 μg/ml of the mitogen sonic hedgehog (Shh) (*see* Chapter 5 by Selvakumar and Kilpatrick). The medium is then replaced the following day to remove the virus using fresh complete NB medium either with or without sonic hedgehog (Shh), depending on the needs of the experiment. Replacement with medium without Shh induces onset of CGN differentiation. We typically culture with Shh for 1–2 days [23], depending on the length of time desired for prior transgene expression. Note that onset of lentiviral transgene expression is rapid, becoming substantial within 6 h following virus addition [16].

24. Lentiviral transduction of cortical and hippocampal neurons is typically done using poly-D-lysine-coated plates and does not require laminin substrate to promote neuronal survival of transduced cells, in our experience.

References

1. Lewis P, Hensel M, Emerman M (1992) Human immunodeficiency virus infection of cells arrested in the cell cycle. EMBO J 11(8): 3053–3058
2. Weinberg JB, Matthews TJ, Cullen BR, Malim MH (1991) Productive human immunodeficiency virus type 1 (HIV-1) infection of nonproliferating human monocytes. J Exp Med 174(6):1477–1482
3. Matrai J, Chuah MK, VandenDriessche T (2010) Recent advances in lentiviral vector development and applications. Mol Ther 18(3):477–490
4. Delenda C (2004) Lentiviral vectors: optimization of packaging, transduction and gene expression. J Gene Med 6(Suppl 1): S125–S138
5. Zaiss AK, Son S, Chang LJ (2002) RNA 3′ readthrough of oncoretrovirus and lentivirus: implications for vector safety and efficacy. J Virol 76(14):7209–7219
6. Zufferey R, Dull T, Mandel RJ, Bukovsky A, Quiroz D, Naldini L, Trono D (1998) Self-inactivating lentivirus vector for safe and efficient in vivo gene delivery. J Virol 72(12):9873–9880
7. Frecha C, Szecsi J, Cosset FL, Verhoeyen E (2008) Strategies for targeting lentiviral vectors. Curr Gene Ther 8(6):449–460
8. Tiscornia G, Singer O, Verma IM (2006) Production and purification of lentiviral vectors. Nat Protoc 1(1):241–245
9. Zimmermann K, Scheibe O, Kocourek A, Muelich J, Jurkiewicz E, Pfeifer A (2011) Highly efficient concentration of lenti- and retroviral vector preparations by membrane adsorbers and ultrafiltration. BMC Biotechnol 11(1):55
10. Ichim CV, Wells RA (2011) Generation of high-titer viral preparations by concentration using successive rounds of ultracentrifugation. J Transl Med 9:137
11. Geraerts M, Willems S, Baekelandt V, Debyser Z, Gijsbers R (2006) Comparison of lentiviral vector titration methods. BMC Biotechnol 6:34
12. Logan AC, Nightingale SJ, Haas DL, Cho GJ, Pepper KA, Kohn DB (2004) Factors influencing the titer and infectivity of lentiviral vectors. Hum Gene Ther 15(10):976–988
13. Scherr M, Battmer K, Blomer U, Ganser A, Grez M (2001) Quantitative determination of lentiviral vector particle numbers by real-time PCR. Biotechniques 31(3):520, 522, 524, passim
14. Ricks DM, Kutner R, Zhang XY, Welsh DA, Reiser J (2008) Optimized lentiviral transduction of mouse bone marrow-derived mesenchymal stem cells. Stem Cells Dev 17(3):441–450
15. Radcliffe PA, Sion CJ, Wilkes FJ, Custard EJ, Beard GL, Kingsman SM, Mitrophanous KA (2008) Analysis of factor VIII mediated suppression of lentiviral vector titres. Gene Ther 15(4):289–297
16. Wang W, Qu Q, Smith FI, Kilpatrick DL (2005) Self-inactivating lentiviruses: versatile vectors for quantitative transduction of cerebellar granule neurons and their progenitors. J Neurosci Methods 149(2):144–153
17. Lizee G, Aerts JL, Gonzales MI, Chinnasamy N, Morgan RA, Topalian SL (2003) Real-time quantitative reverse transcriptase-polymerase chain reaction as a method for determining lentiviral vector titers and measuring transgene expression. Hum Gene Ther 14(6):497–507
18. Kutner RH, Zhang XY, Reiser J (2009) Production, concentration and titration of pseudotyped HIV-1-based lentiviral vectors. Nat Protoc 4(4):495–505
19. Wang W, Stock RE, Gronostajski RM, Wong YW, Schachner M, Kilpatrick DL (2004) A role for nuclear factor I in the intrinsic control of cerebellar granule neuron gene expression. J Biol Chem 279(51):53491–53497
20. Wang W, Mullikin-Kilpatrick D, Crandall JE, Gronostajski RM, Litwack ED, Kilpatrick DL (2007) Nuclear factor I coordinates multiple phases of cerebellar granule cell development via regulation of cell adhesion molecules. J Neurosci 27(23):6115–6127
21. Kumar P, Woon-Khiong C (2011) Optimization of lentiviral vectors generation for biomedical and clinical research purposes: contemporary trends in technology development and applications. Curr Gene Ther 11(2):144–153
22. Osten P, Dittgen T, Licznerski P (2006) Lentivirus-based genetic manipulations in neurons in vivo. In: Kittler JT, Moss SJ, editors. The Dynamic Synapse: Molecular Methods in Ionotropic Receptor Biology. Boca Raton (FL): CRC Press; 2006. Chapter 13.
23. Wang W, Shin Y, Shi M, Kilpatrick DL (2011) Temporal control of a dendritogenesis-linked gene via REST-dependent regulation of nuclear factor I occupancy. Mol Biol Cell 22(6): 868–879
24. Zufferey R, Donello JE, Trono D, Hope TJ (1999) Woodchuck hepatitis virus posttranscriptional regulatory element enhances expression

of transgenes delivered by retroviral vectors. J Virol 73(4):2886–2892

25. VandenDriessche T, Thorrez L, Naldini L, Follenzi A, Moons L, Berneman Z, Collen D, Chuah MK (2002) Lentiviral vectors containing the human immunodeficiency virus type-1 central polypurine tract can efficiently transduce nondividing hepatocytes and antigen-presenting cells in vivo. Blood 100(3):813–822
26. Jordan M, Wurm F (2004) Transfection of adherent and suspended cells by calcium phosphate. Methods 33(2):136–143

Chapter 13

In Ovo Electroporation in Embryonic Chick Spinal Cords

Hui Wang and Michael P. Matise

Abstract

The developing spinal cord is a well-established model system widely used to study the signaling pathways and genetic programs that control neuronal/glial differentiation and neural circuit assembly. This is largely due to the relatively simple organization (compared to other CNS regions) and experimental accessibility of the neural tube, particularly in the chick embryo. In vivo transfection of cells within the developing chick neural tube using in ovo electroporation has emerged as a rapid and powerful experimental technique in that (1) transfected factors can be functionally tested in a spatially and temporally controlled manner and (2) the chick embryo provides a physiologically relevant in vivo environment to conduct biochemical studies such as dual-channel luciferase assay, co-immunoprecipitation (co-IP), and Chromatin Immunoprecipitation (ChIP). In this chapter, we will take an in-depth look at the in ovo electroporation system in embryonic chicken spinal cord. In the following chapter, we will continue by examining the use of in ovo electroporation in the dual-channel luciferase assay as an example of its biochemical application.

Key words In ovo electroporation, Hamburger and Hamilton (H&H) stage, Microcapillary injection, Vitelline membrane

1 Introduction

1.1 The Principle of Electroporation

Electroporation is among the most popular and efficient method to physically transfer foreign nucleic acid molecules into cells [1]. Under normal physiological conditions, cells do not internalize exogenous DNA/RNA molecules since the hydrophobic cell membrane provides a protective barrier that prevents passive diffusion of highly charged molecules into the cytoplasm. However, the application of a controlled low-current electric pulse can transiently increase membrane permeability by generating multiple conductive nanometer pores that contain a hydrophilic interface [2]. This physically induced membrane phenomenon permits charged molecules such as DNA or RNA to enter into the cell. As a result, highly negatively charged foreign nucleic acid molecules can be internalized through these conductive pores, which can then go on to be transcribed by the endogenous transcriptional machinery inside the cell if they contain the appropriate sequences that bind such factors.

Renping Zhou and Lin Mei (eds.), *Neural Development: Methods and Protocols*, Methods in Molecular Biology, vol. 1018, DOI 10.1007/978-1-62703-444-9_13, © Springer Science+Business Media, LLC 2013

The duration, timing, and voltage/current of each electrical pulse must be optimized in order to ensure that the majority of the cells induced to open membrane nanopores will recover once the electrical pulse is withdrawn rather than undergoing cell death.

1.2 In Ovo Chick Neural Tube Electroporation

Chicken embryos have been widely used as an in vivo model system for embryological studies for more than a century due to their ready availability and individual accessibility in ovo. The stages of chick embryonic development were catalogued in detail by Viktor Hamburger and Howard L. Hamilton in 1951. They divided chick embryonic development chronologically into 46 distinct stages [3]. These have become well known as Hamburger and Hamilton (H&H) stages and are based on significant developmental events such as primitive streak formation, neurulation, somitogenesis, as well as the development of external structures like limb buds and visceral arches. Because of its comprehensiveness, accuracy, and reliable representation of chick embryonic development, H&H staging has become the foremost method for determining and tracking the progress of embryonic development in chickens.

In ovo electroporation is a recent technique developed by Dr. Okumura and Dr. Nakamura [4, 5]. Subsequently, it has become widely utilized for functional gene overexpression and knockdown studies [6]. Prior studies have shown that in ovo electroporation provides the best transfection efficiency with chick embryos at a relative good survival rate compared to several other (non-virus-based) transfection methods, including lipofection (LP) and microparticle bombardment (MPB). The basic procedure of in ovo electroporation involves (1) injection of foreign gene/s into the patent lumen of the central canal of the developing chick neural tube and (2) application of an electric current across the mediolateral plane of the embryonic neural tube using laterally positioned microelectrodes. During current application, the net negative charge of DNA/RNA expression plasmids, which are loaded into the central canal, will migrate towards the positively charged cathode. The applied current is delivered in a series of controlled pulses that will simultaneously open nanopores in neural tube cells lining the central canal (dividing ventricular zone neural/glial precursors), allowing DNA to enter into neural progenitor cells on the cathode-side neural tube. A significant advantage of this procedure is that the contralateral, untransfected side of the neural tube positioned near the anode electrode does not take up injected DNA/RNA and thus serves as an internal control (*see* **Note 1**). Whether the foreign genes will be expressed transiently or stably in the descendants of transfected cells is dependent on the specific design and composition of the expression vectors. Conventional vertebrate plasmid vectors can be utilized for short-term gene expression (*see* Subheading 2.2.1), while provirus plasmid vectors and retrotransposons work well for stable gene expression by integrating the foreign genes into the host genome [7].

2 Materials

2.1 Animals

Specific pathogen-free (spf) fertilized White Leghorn chicken eggs (Aichi Line) are available from suppliers such as Charles River (MA, USA).

2.2 Plasmids

2.2.1 Gain of Function

Several conventional plasmid expression vectors are known to function well for transient foreign gene overexpression in the developing chick neural tube, such as pMiwSV (which contains a chicken beta-actin promoter and Rous Sarcoma Virus (RSA) enhancer) [8], pRc/CMV (which contains a cytomegalovirus (CMV) enhancer, Invitrogen, CA, USA), and pCAGGS (which has a combined CMV enhancer and chicken beta-actin (CAG) promoter; *see* **Note 2**) [9]. We routinely utilize the pCIG vector, a derivative of pCAGGS engineered to contain a nuclear-localized GFP driven by an internal ribosome entry site (IRES) sequence (constructed by A.P. McMahon lab) for transient gene overexpression [10]. GFP expression can be visualized within 3 h post-transfection (hpt), allowing for the rapid determination of transfection efficiency with single-cell resolution (*see* **Note 3**). With pCIG, robust foreign gene expression can be driven from 12 to 48 hpt.

2.2.2 Knockdown

There are several applicable methods to diminish specific endogenous target gene expression using in ovo electroporation. Firstly, short hairpin cDNA (containing an ~20 bp oligomer of target cDNA sequence) cloned downstream of a U6 promoter or H1 promoter contained shRNA expression vector, like pSilencer 1.0-U6 siRNA expression vector (Ambion, TX, USA), will be transcribed into small interfering RNA to block target gene expression [11]. Synthesized siRNA oligos (~20 bp oligomers of target mRNA sequence) also can be transfected directly into chick neural tube to transiently inhibit target gene translation (good for 24–48 hpt).

2.3 Reagents

Leibovitz's L-15, distilled deionized water (ddH_2O), plasmid elution buffer (10 mM Tris–Cl, pH 8.5), 4 % trypan blue solution, Dulbecco's Phosphate-Buffered Saline 1× (DPBS).

2.4 Other Materials

Borosilicate glass capillaries (WPI, 1.2 mm OD, 0.68 mm ID, 4 in. length), BD Vacutainer Blood Collection Set (Model 367251), cellophane tape, microscissors, micro-forceps, transfer pipettes, petri dishes, a small brush, 15 mL eppendorf tubes, 1.5 mL eppendorf tubes, 5 mL syringe, a large gauge needle, an appropriate platform for holding the egg during electroporation (e.g., we use ~2 cm high wax/paraffin mold of half egg), and pushpins.

3 Methods

3.1 Incubation of Fertile Eggs

1. Adjust forced-air incubator thermostat to 38 °C (100 °F) and >60 % humidity (*see* **Note 4**).
2. Incubate eggs horizontally (*see* **Note 5**). Make sure that the eggs are situated stably as they must remain horizontal.
3. Incubate eggs for an appropriate period of time to achieve the correct H&H stage (*see* **Note 6**). In this example, we will presume an incubation time of 53 h. At this time, over 70 % of incubated embryos will reach H&H stages 14–15 which is the ideal time window for the injection of plasmids into the developing spinal cord (*see* **Note 7**).

3.2 Preparation for Electroporation

1. About 30 min prior to the procedure, aliquot several milliliters of Leibovitz's L-15 media in 15 mL tubes and set them in a 37 °C water bath. Warmed L-15 will be used later during electroporation.
2. Prepare an appropriate amount of concentrated Maxiprep plasmids and dilute to the desired concentration with ddH_2O or plasmid elution buffer. Mix the diluted plasmid solution with 4 % trypan blue solution at a 7:1 to 10:1 ratio (*see* **Note 8**).
3. Fix a pair of electrodes (1 mm length, L-shape bent, BTX Model 516 Genetrode Kit, Harvard Apparatus, Massachusetts, USA) on a Genepaddle Holder (Harvard Apparatus, Massachusetts, USA) and adjust them to be parallel and spaced apart by ~4 mm (*see* **Note 9**). Connect lead wires from Genepaddle Holder to the electroporator power supply (BTX T820, San Diego, CA, USA). Use a small brush to clean electrodes with DPBS in a petri dish between each electroporation.
4. Generate glass capillary micropipette on vertical pipette puller (NARISHIGE, PC-10). Due to clogging of microcapillary tips, we generally avoid using pulled tips for more than four electroporations.
5. Remove the butterfly needle at one end of a BD Vacutainer Blood Collection tube and insert the glass capillary into the open end.
6. Remove the warmed Leibovitz's L-15 from the 37 °C water bath and place near the dissecting microscope for use during electroporation.

3.3 In Ovo Electroporation

1. At 53 h of incubation, remove the eggs from the incubator and wash the shell with 70 % ethanol. Place the egg against an Illuminator or fiber-optic light source to identify the position of the embryonic disc (candling) and circle with a pencil.
2. Use a pushpin to punch a small pole at the narrow end of each egg and carefully withdraw 4–5 mL albumen with a 5 mL

syringe and large gauge needle (*see* **Note 10**). Tape up the pinhole using cellophane tape.

3. Apply tape to the marked area overlying the embryonic disc. Using sharp scissors, open a small window (~2 cm diameter) in the shell overlying the embryonic disc, being careful to confine your cut to the taped portion so the eggshell does not shatter.
4. Take up a few microliters of plasmid solution with a glass microcapillary pipette and inject into the lumen of the spinal cord central canal (*see* **Note 11**). The trypan blue should fill up the space to indicate a good injection. This step is controlled by mouth pipetting.
5. Using a plastic transfer pipette, place a few drops of Leibovitz's L-15 on the top of embryonic disc (*see* **Note 12**). Gently place a pair of electrodes against the vitelline membrane about ~4 mm apart surrounding the neural tube, taking care to avoid damaging the paired vitelline vessels and any other extraembryonic tissues/membranes.
6. Deliver 5 square pulses (25 V, 50 ms each) using electroporator (*see* **Note 13**).
7. Place four drops of Leibovitz's L-15 on the top of chicken embryo (*see* **Note 12**) and remove electrodes from egg, taking care to avoid damaging membranes (*see* **Note 14**).
8. Seal the opened shell with cellophane tape. Record the time on the eggshell with a pencil and return egg to the incubator.

3.4 Collecting Embryos

1. Take the embryos out of incubator at desired H&H stage and remove tape. Excise the embryo from the yolk by cutting out a circle of vitelline membrane surrounding the chicken embryo with a microscissors.
2. Carefully transfer embryo into prechilled DPBS in a petri dish using curved forceps or a blunt transfer pipette cut to enlarge the tip diameter.
3. Using sharp forceps and microdissection scissors, dissect away the layers of membrane surrounding the embryo under a surgical microscope.

4 Notes

1. For certain applications, we also perform two-sided electroporations, i.e., after transfecting one side of the neural tube, we change the direction of the current (by switching the leads on the electroporator) and transfect the contralateral side. Based on our experience, the transfection efficiency for the primary transfected side is diminished by 10–20 % compared to the single-side electroporation, but the overall transfection efficiency for the entire spinal cord is increased as much as 50–70 %.

2. "Ubiquitous" expression vectors, like CAG, CMV, and PGK promoters, have tissue-specific and temporally restricted characteristics in the living chicken embryos. For example, the expression efficiency of CAG promoter is very good only in early embryonic stage of chick embryos (from E1.5 to E4) [12].
3. IRES-dependent GFP expression is much higher in an empty pCIG vector compared with pCIG vectors carrying a transgene.
4. More details for storing and incubating eggs refer to "Incubating and Hatching Eggs" (Texas A&M University) or "Care and Incubation of Hatching Eggs" (Mississippi State University).
5. Eggs will not develop properly if they are incubated vertically.
6. Several factors could influence the development rate of chicken embryos, such as the breed line, the temperature and humidity of incubation, and environmental weather and temperature (which can affect the eggs during shipping to the lab). For example, we always use White Leghorn chickens (Aichi Line) and the same incubator thermostat set to 38 °C (100 °F) all year long. During the warm seasons after 53 h incubation, approximately 70 % of incubated embryos will reach H&H stages 14–15; however, during the winter, it will take approximately half an hour more. Therefore, to create a standardized system, it is important to normalize the chicken embryos based on their morphology rather than chronological age.
7. During H&H stage 10 (ten somites, 29–33 h), three primary brain vesicles are well formed/closed, and it is considered as the earliest time to inject foreign DNA/RNA to examine their functions in anterior neural tube. During this stage, we injected plasmid from either the fourth ventricle or the telencephalon towards mesencephalon. Notably, since the anterior neuropore is closed at H&H stage 12, if we inject plasmid from the fourth ventricle towards mesencephalon, we make a small hole at the anterior tip of telencephalon to relieve the internal pressure as the plasmid solution is being injected. H&H stage 12 is considered the optimal stage to do anterior neural tube in ovo electroporation, since after that the head begins to turn (by the end of H&H stage 13), making the neural tube less accessible to injection. The direction of DNA movement is directly affected by the position of electrodes. DNA therefore will be transfected into ventral or dorsal side of brain vesicles but not left or right side, as electroporating DNA at H&H stage 13. H&H stages 13–16 is the best time window to conduct spinal cord transfections, since after H&H stage 16 the neural tube is thicker and completely closed, amnion covers most of the embryo, and the extraembryonic vitelline and yolk circulation becomes gradually more extensive, making secondary damage

to the fetal circulation during application of the electrical pulse, and embryo lethality, more likely.

8. Trypan blue is a vital stain that is not taken up by living cells. Here it is used to visualize the injection of DNA into the central canal.
9. The specific gap between electrodes is dependent on the size of embryo, with the goal being to avoid direct contact with the neural tube or body wall and vitelline vessels extending laterally into the yolk.
10. If doing late-stage in ovo electroporation after H&H stage 20, it is recommended to remove albumen at E2.0 to provide distance between the embryos and the shell and avoid damaging them when opening the egg.
11. It is not very difficult to localize the lumen/central canal of spinal cord at H&H stages 13–14 which is proximately at the midline between the left and right blood vessels.
12. Applying Leibovitz's L-15 medium prior to electroporation is to avoid embryo and membrane drying out and to increase conductivity during electroporation. Furthermore, L-15 medium is designed for supporting cell growth in environments without CO_2 equilibration. Based on our experience, using L-15 after electroporation can increase the survival rate of chick embryos.
13. Based on prior studies and our experience, 5 square pulses with 25 V 50 ms give the best transfection efficiency and good survival rate (more than 80 %) [5].
14. **Step 5** needs to be finished as soon as possible to avoid the diffusion of injected DNA.

References

1. Weaver JC (1995) Electroporation theory. Concepts and mechanisms. Methods Mol Biol 55:3–28, Epub 1995/01/01
2. Melikov KC, Frolov VA, Shcherbakov A, Samsonov AV, Chizmadzhev YA, Chernomordik LV (2001) Voltage-induced nonconductive pre-pores and metastable single pores in unmodified planar lipid bilayer. Biophys J 80(4):1829–1836, Epub 2001/03/22
3. Hamburger V, Hamilton HL (1992) A series of normal stages in the development of the chick embryo. 1951. Dev Dyn 195(4):231–272, Epub 1992/12/01
4. Funahashi J, Okafuji T, Ohuchi H, Noji S, Tanaka H, Nakamura H (1999) Role of Pax-5 in the regulation of a mid-hindbrain organizer's activity. Dev Growth Differ 41(1):59–72, Epub 1999/08/13
5. Muramatsu T, Mizutani Y, Ohmori Y, Okumura J (1997) Comparison of three nonviral transfection methods for foreign gene expression in early chicken embryos in ovo. Biochem Biophys Res Commun 230(2):376–380, Epub 1997/01/13
6. Wang H, Lei Q, Oosterveen T, Ericson J, Matise MP (2011) Tcf/Lef repressors differentially regulate Shh-Gli target gene activation thresholds to generate progenitor patterning in the developing CNS. Development 138(17):3711–3721, Epub 2011/07/22
7. Takeuchi JK, Koshiba-Takeuchi K, Matsumoto K, Vogel-Hopker A, Naitoh-Matsuo M, Ogura K

et al (1999) Tbx5 and Tbx4 genes determine the wing/leg identity of limb buds. Nature 398(6730):810–814, Epub 1999/05/11

8. Wakamatsu Y, Watanabe Y, Nakamura H, Kondoh H (1997) Regulation of the neural crest cell fate by N-myc: promotion of ventral migration and neuronal differentiation. Development 124(10):1953–1962, Epub 1997/05/01
9. Niwa H, Yamamura K, Miyazaki J (1991) Efficient selection for high-expression transfectants with a novel eukaryotic vector. Gene 108(2):193–199, Epub 1991/12/15
10. Megason SG, McMahon AP (2002) A mitogen gradient of dorsal midline Wnts organizes growth in the CNS. Development 129(9):2087–2098, Epub 2002/04/18
11. Das RM, Van Hateren NJ, Howell GR, Farrell ER, Bangs FK, Porteous VC et al (2006) A robust system for RNA interference in the chicken using a modified microRNA operon. Dev Biol 294(2):554–563, Epub 2006/04/01
12. Tanabe K, Takahashi Y, Sato Y, Kawakami K, Takeichi M, Nakagawa S (2006) Cadherin is required for dendritic morphogenesis and synaptic terminal organization of retinal horizontal cells. Development 133(20):4085–4096, Epub 2006/09/22

Chapter 14

Gene Transfer in Developing Chick Embryos: In Ovo Electroporation

Emma K. Farley

Abstract

In ovo electroporation is a popular technique to study gene function during development. This technique enables precise temporal and spatial genetic manipulation with the added advantages of being quick and inexpensive. In this chapter the transient transfection of a construct into the neural tube of a chicken embryo via in ovo electroporation is described. Modifications of this basic technique and methods to analyze the resulting electroporated embryos such as qPCR and microarray are also discussed.

Key words In ovo electroporation, Development, Gene expression, Chicken embryos, Vertebrate development, Neural tube

1 Introduction

During development temporally and spatially dynamic gene expression patterns instruct cells of an embryo to all the different cell fates that make up the adult organism. Therefore, determining the function of genes involved in embryonic development requires manipulation of gene expression in a spatially and temporally controlled manner. In ovo electroporation is a widely used technique to carry out this manipulation. Its advantages lie in the highly precise localization and timing of expression construct insertion, along with the speed and economy of the technique [1].

In ovo electroporation can be employed in both gain and loss of function studies, as well as reporter studies, to analyze gene function and regulation during development. This method has been applied to all areas of developmental biology including neurogenesis and neural crest differentiation [2, 3], axon outgrowth and guidance [4], axonal patterning in the developing limb [5], somitogenesis [6], skeletal muscle development [7], and eye development [8]. In ovo electroporation has been used to study development in mouse embryos [9], other vertebrates,

Renping Zhou and Lin Mei (eds.), *Neural Development: Methods and Protocols*, Methods in Molecular Biology, vol. 1018, DOI 10.1007/978-1-62703-444-9_14, © Springer Science+Business Media, LLC 2013

including *Danio rerio* [10] and *Xenopus laevis* [10, 11], and invertebrates such as Ascidiacea [12]. However, it has been most widely applied to avian embryos because the avian embryo develops in ovo and has a planar topology. These characteristics greatly facilitate injection of the DNA construct, placement of electrodes, and incubation of the electroporated embryo. Indeed in ovo electroporation is particularly suited to study neural development as the neural tube provides a body cavity which acts as a vestibule into which DNA can be injected.

In ovo electroporation involves microinjecting a gene expression construct typically into a natural body cavity such as the neural tube. Electrodes are placed flanking the site of injection and an electric current is applied across the embryo in the form of a rapid series of square wave pulses. This electric field transiently disrupts the stability of the plasma membrane, creating pores in the cell membrane. The negatively charged DNA constructs migrate towards the positive electrode and enter cells in their path via these pores. The tissue adjacent to the negatively charged electrode remains untransfected, providing an internal control. This method enables defined tissues to be targeted by location of DNA injection and positioning of electrodes, at precise times of development. (Fig. 1 outlines the procedure and shows examples of electroporated embryos.)

In ovo electroporation is most commonly applied to Hamburger–Hamilton (HH) stage (st) 10–20 embryos. Embryos older than HH st 20 have more compact tissues and increased tissue layers, making microinjection more difficult. For older embryos ex ovo explant electroporation is an alternative method [13].

Typically the construct used for in ovo electroporation contains a gene of choice or dsRNA along with green fluorescent protein (GFP), or another marker, in order to identify the cells which have taken up the construct. A popular construct to obtain transient transfection is pCAβ-IRES-GFP containing a β-actin promoter, cytomegalovirus (CMV) enhancer, a polylinker for inserting your desired gene followed by an internal ribosomal entry site (IRES) and GFP [14]. Expression of the translated product can be detected 2.5 h after electroporation and peaks around 20–24 h [15]. Expression of transient construct can be maintained for 3–11 days, but weakens as cells divide [1, 16].

Constitutive expression can be obtained by integrating plasmids into the genome using methods such as transposon-mediated gene transfer [17]. Using constructs with inducible promoters for example the tetracycline on tetracycline off system enables further control over the timing of exogenous DNA expression [18]. It is also possible to use cell-type-specific enhancers.

Conversely loss of function studies can be carried out using RNA interference [19], or dominant negative constructs [20]. Constructs containing the gene of interest linked to either a repressor or an activator enable investigation of the transcriptional activity of genes in vivo [21]. As well investigating gene function in

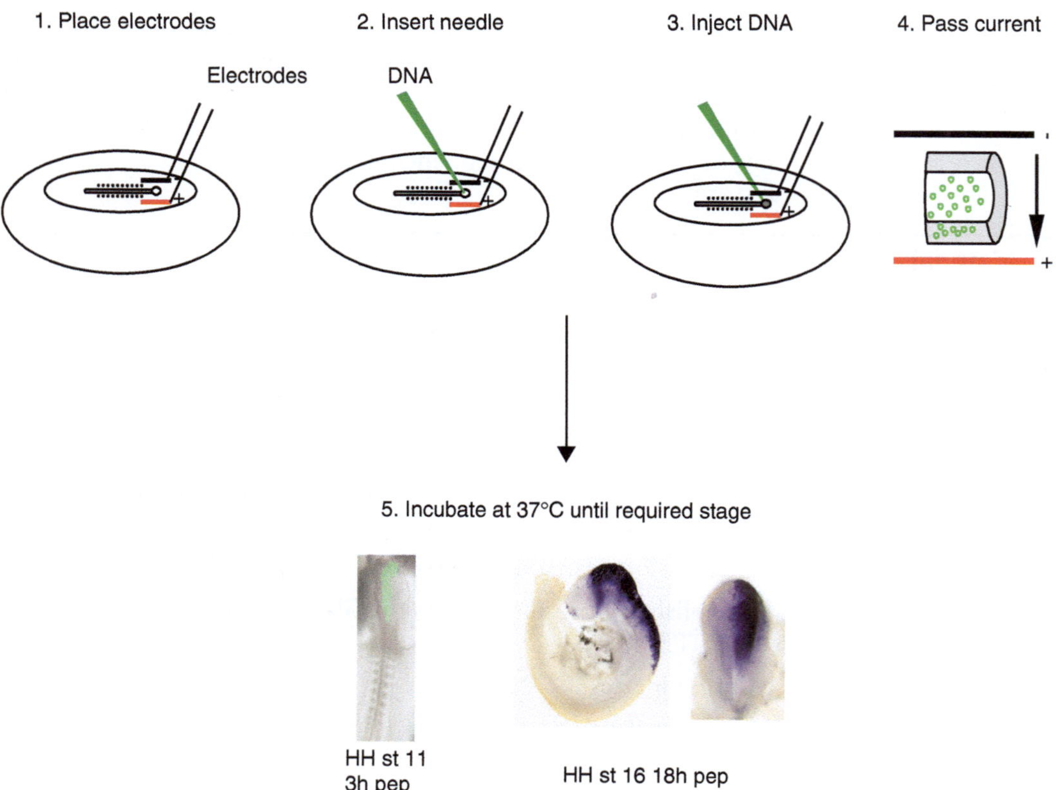

Fig. 1 Schematic of electroporation technique. Fertilized egg incubated until the required developmental stage, e.g., HH st 10. (*1*) Egg is windowed to visualize embryo and electrodes are placed across the region one wishes to electroporate. (*2*) Needle containing DNA is inserted into the neural tube. (*3*) DNA is injected into this region. (*4*) A series of square wave electrical pulses (5× 12 V of 50 ms duration at 100 ms intervals) are applied across the electrodes such that the DNA is taken up by cells adjacent to the positive electrode. (*5*) Eggs are sealed and incubated until required developmental stage. For example until HH st 11, 3 h post electroporation (pep), electroporation is shown by expression of GFP. Another example HH st 16, 18 h pep, embryo electroporated with *pcaβ-Dmrt5-IRES-GFP* and then expression visualized by in situ for exogenous *Dmrt5* (using mouse *Dmrt5* probe). In both examples the over-expression is only on one side of the embryo

ovo electroporation can be used to study the activity of enhancers during development in vivo using reporter constructs [22].

Traditionally, the downstream effects of in ovo electroporation of an investigator's gene of interest have been analyzed using in situ hybridization and antibody staining of a few select gene products. This approach has the major disadvantage of surveying only a limited number of genes for qualitative transcriptional changes. The sequencing of the chicken genome and the availability of a highly representative chicken genome microarray (Affymetrix GeneChip Chicken Genome Array) now enables the coupling of in ovo electroporation to genome-wide analysis. This strategy provides an opportunity to use in ovo electroporation to investigate the role of genes during the embryonic development in a fast, economical, and comprehensive manner. A good description of this procedure has been described by Chambers and Lumsden [23].

This protocol explains in ovo electroporation of pCAβ-IRES-GFP into the neural tube of HH st 10 embryos. Once electroporated embryos can be collected and the effects of this genetic manipulation studied by in situ, antibody staining, qPCR, microarray, or next generation sequencing.

2 Materials

2.1 Electroporation Equipment

Electrodes—Sonidel Ltd has a variety of electrodes for different requirements; we use CUY610P1.5-1.

ECM830 Electro-S Square Porator BTX Inc. 45-0052.

Picospritzer III INTRACEL 051-0500-900 0-100 psi.

Needle puller—we use model P-30/P from INTRACEL.

Tungsten probes 0.25 mm diameter 50 mm long. We use World Precision Instruments 501317.

Fire polished Borosilicate glass for Glass needles B100-75-10 from INTRACEL.

Mouth pipettes.

18 G microlance.

30 G microlance.

Cello tape (Niceday).

2.2 Reagents

2 U/ml dispase dissolved in Tyrode's.

Absolutely RNA Miniprep Kit, Stratagene.

GeneChip Chicken Genome Array, Affymetrix.

Fast Green.

Tyrode's: 137 mM NaCl, 2.7 mM KCl, 1 mM $MgCl_2$, 1.8 mM $CaCl_2$, 0.2 mM Na_2HPO_4, 12 mM $NaHCO_3$, 5.5 mM D-glucose.

India Ink.

Polyvinyl alcohol (PVA).

3 Methods

Carry out all procedures at room temperature unless otherwise stated. Eggs can be ordered on a weekly basis and kept at 14 °C (*see* **Note 1**). This protocol outlines the basic method of in ovo electroporation; modifications of this technique are discussed in **Note 2**. There are several critical steps to ensure effective electroporation and viable embryos. High efficiency electroporation needs a clean preparation of the construct and adjustment of the electroporation parameters for your chosen tissue type and development stage. Sterile working conditions are required to reduce death due to bacterial infection (*see* **Note 3**).

3.1 Preparation of Eggs

1. Fertilized eggs are placed at 37 °C in a humidified incubator (*see* **Note 4**) until the required developmental stage has been reached; we electroporated embryos at Hamburger–Hamilton (HH) [24] stage 10, which require 36 h incubation. Eggs should be incubated on their sides. The top of the egg should be marked with a pencil and this is where the embryo lies (*see* **Note 5**).
2. Cello tape is placed over this pencil mark in the location where a window will be cut and a small piece of tape is also put at the rounded pole of the egg (*see* **Note 6**).
3. A hole is made in this pole of the egg which contains the air sack (this is the more rounded end) using clean scissors. 2 ml of albumin is removed from this hole using an 18 G attached to a syringe (*see* **Note 7**).
4. Once albumin is removed seal this hole with a small piece of Cello tape (*see* **Note 8**).
5. A window is cut in the top of the egg using curved scissors and being careful not to disrupt the embryo by keeping scissors as parallel to the shell as possible. This window needs to be large enough to visualize the embryo and place the electrodes and needle.
6. To visualize the embryo 10 % India ink/Tyrode's solution is injected under the embryo using a 30 G needle attached to a 1 ml syringe (*see* **Note 9**).

3.2 Electroporation of Embryo

1. Put a few drops of Tyrode's solution on the embryo (*see* **Note 10**).
2. Place electrodes across the region you wish to electroporate and push down so that there is tension across the embryo and vitelline membrane (*see* **Note 11**).
3. Fill a freshly pulled needle with DNA construct/Fast Green mix using a mouth pipette (*see* **Note 12**). The Fast Green is added in order to visualize the DNA filling the region of interest. 0.5 μl of 2 % Fast Green can be added to 5 μl volume of DNA solution. The concentration of your plasmid varies depending on electroporation site, embryo age, and experimental aims. We typically use 4 μl of 5 μg/μl DNA and 0.5 μl Fast Green. 1 μl of 10 % PVA in Tyrode's can be added if ventral electroporation is required (*see* **Note 13**).
4. Attach the needle to the Picospritzer and a micromanipulator. Insert the needle into the region you wish to electroporate, e.g., neural tube (*see* **Note 14**).
5. Using the Picospritzer inject the DNA mix into the region until dye fills the entire space. If you want to get ventral expression wait for a minute for the injected DNA to settle before **step 6**.

6. Use a BTX electro square Porator machine or similar to produce five pulses of 50 ms duration at 100 ms intervals (*see* **Note 15**).
7. Tyrode's solution is placed on embryo to cool the embryo; the needle and electrodes should be carefully removed from the embryo.
8. Replace the egg shell and cover with Cello tape to seal the egg (*see* **Note 16**). Place the egg back at 37 °C in the humidified incubator and leave until developmental stage required is reached.
9. The electrodes should be cleaned between electroporations (*see* **Note 17**).

3.3 Collection of Material for Microarray/qPCR

1. Once the required developmental stage has been reached remove embryos from the egg by cutting around the membrane on which the embryo is sitting. The embryo can be picked up using tweezers and placed in Tyrode's solution.
2. Membranes are removed in this dish and embryos placed in another dish containing clean Tyrode's.
3. It is important to check that you have good electroporation efficiency by looking at GFP expression; collect only embryos that have a high GFP for qPCR or microarray. I typically use only tissue with 70 % of cells electroporated or greater for qPCR/microarray experiments.
4. The region for collection is dissected, unwanted mesodermal tissue may be removed by addition of 2 U/ml dispase in Tyrode's; this causes mesodermal tissue to detach from neural ectoderm. The two can then be separated by manipulation with tungsten needles (*see* **Note 18**).
5. The electroporated region and contralateral control regions can then be dissected using tungsten needles.
6. Place the isolated tissue in 100 μl lysis buffer provided in the Absolutely RNA Miniprep Kit (Stratagene). In our experiments we collect tissue from six stage matched embryos in 100 μl lysis buffer.
7. This is vortexed for 1 min and stored at −80 °C material can be stored here for several months prior to further processing.
8. For purification of such small samples of RNA we find Stratagene Absolutely RNA micro Kit to be very good. Once isolated the RNA can be used to make cDNA for qPCR analysis or can be used for microarray analysis—a good review on this can be found in Chambers and Lumsden [23].

3.4 Collection of Embryos for In Situ or Antibody Staining

1. Embryos are removed from their membranes and placed in clean Tyrode's. A small hole is made in the top of the midbrain and other body cavities of interest to ensure penetration of PFA and proper fixation. Place embryos in 4 % PFA for 1 h on

a roller at room temperature (*see* **Note 19**). Embryos can then be kept in PBS at 4 °C and the protocols for antibody or in situ can be carried out.

4 Notes

1. Eggs should be stored at 14 °C for at most 1 week. Eggs stored for longer are less viable. Transportation of eggs should be gentle. Shaking can cause deformation and less viable embryos.
2. This protocol could be further refined by using fluorescence-activated cell sorting to ensure that pure populations of cells containing only the electroporated construct are obtained for analysis by microarray, qPCR, or next-generation sequencing. Other modifications to this basic technique include use of beads soaked in the DNA construct, which are micro-surgically implanted into the embryo instead of injection of DNA. This allows for focal electroporation and electroporation of different constructs in close proximity, thus increasing possible precision and complexity of in ovo electroporation studies [25].
3. Some people use Penicillin–Streptomycin to reduce the likelihood of bacterial infection; we find that this is not required if clean practice is undertaken. Some labs wipe the eggs with ethanol; however, as the egg shell is porous, wiping with products such as ethanol is not advisable. We again find that if clean practices are maintained including the placement of Cello tape on the egg to prevent pieces of shell falling into the egg then cleaning the outside of the egg is not required.
4. Incubation time and developmental progress of the embryo are dependent on the temperature and humidity; any incubator set to 37 °C can be used as long as humidity of at least 45 % and good air circulation can be achieved. To reach 45 % humidity it is usually sufficient to place a tray of distilled water at the bottom of the incubator.
5. During incubation eggs should be kept on their sides such that the embryo will be correctly positioned for electroporation. It is useful to mark the top of the egg with a pencil so that you know where to cut a window in the shell; the embryo sitting on the top of the yolk lies directly below this point. For best development we suggest taking each egg out one at a time; time spent outside the 37 °C incubator at room temperature slows down the development of the embryos. For working with the egg it is best to make an egg holder—using 3 cm plate and some paraffin wax, melt the wax pour into the dish and then when it is almost cooled press an egg sideway into the paraffin to create a dimple for the eggs to sit in.

6. We have found certain types of Cello tape toxic to the development of chicken embryos we use Niceday; however, we suggest that you test several brands to ensure that the tape is nontoxic.
7. The syringe should be angled down to avoid damaging the yolk. It is important not to damage the yolk or embryo as this affects survival.
8. Failure to seal this hole can lead to the egg drying out and bacterial infection.
9. We have found certain inks toxic; we suggest testing different types of India ink from local suppliers—we buy ours from a local market. When injecting the ink under the embryo put the needle into yolk and then lift needle upwards so that the ink is located directly under embryo; this improves visualization of the embryo and reduces the amount of ink required.
10. Addition of Tyrode's lowers the electric resistance, provides a uniform solution through which the current can pass, and prevents overheating of the embryo.
11. Creating tension across the membrane aids insertion of the glass needle. This method prevents the need to break the membranes which can increase risk of bacterial contamination and damage to embryo. Some people remove or create a hole in the vitelline membrane with forceps. We have always found this unnecessary; however, those working with older embryos or starting out may find it easier to break vitelline membrane for electroporation.
12. This is done by sucking the DNA from the tube up the pulled needle. If you do not wish to use a mouth pipette, an alternative, although slower, method is to pipette a drop of the solution at the thick end of the needle and have the needles facing tip down; capillary action will cause the DNA to move into the tip of the needle.
13. The efficiency of electroporation greatly depends on the quality of DNA preparation; we use QIAGEN maxiprep kits but do not use fast filter at end; rather, precipitate DNA using traditional methods. We also advise against freeze–thaw of aliquots. We aliquot 5 μl of highly concentrated 5–10 μg DNA construct and keep them at −20 °C using one per experiment and discarding. Transfection efficiency also depends on time point of injection, concentration and electroporation settings. Addition of PVA helps in making the solution heavier and thus is good when ventral electroporation is needed.
14. During electroporation contact between electrodes and major blood vessels, as well as with embryos, should be avoided to prevent severe damage and death. It is also advised to keep electrodes away from the heart.

15. You will see bubbles if the electroporation worked correctly. Any other electroporator that generates square wave pulses can be used. Different voltages may be required depending on the stage of the embryo and tissue type you wish to electroporate. For older embryos use electrodes which are not fixed so that you can adjust the distance between them; higher voltages will also be required. Electroporation settings should be chosen according to embryonic stages, distance between electrodes, and tissue being electroporated.
16. Ensure that the window is completely sealed. Dehydration of the embryo after electroporation leads to loss of embryos.
17. Cleaning the electrodes between electroporations is important to obtain efficient electroporation as remaining proteins on electrodes interfere with efficient electroporation. This can be done by placing some Tyrode's in a small dish and placing the electrodes there and pulsing a current; a pipette can then be used to squirt liquid at the electrodes to remove the plaque on the electrodes. It is important not to damage the electrodes by wiping them or scratching them with tweezers.
18. It is important to work out how long it takes for the dispase to break down the connections between mesenchyme and epithelium.
19. Do not freeze thaw PFA; thaw freshly for each fixation

References

1. Itasaki N, Bel-Vialar S, Krumlauf R (1999) 'Shocking' developments in chick embryology: electroporation and in ovo gene expression. Nat Cell Biol 1(8):E203–E207
2. Sandberg M, Kallstrom M, Muhr J (2005) Sox21 promotes the progression of vertebrate neurogenesis. Nat Neurosci 8(8):995–1001
3. Lee S et al (2009) Retinoid signaling and neurogenin2 function are coupled for the specification of spinal motor neurons through a chromatin modifier CBP. Neuron 62(5): 641–654
4. Islam SM et al (2009) Draxin, a repulsive guidance protein for spinal cord and forebrain commissures. Science 323(5912):388–393
5. Luria V et al (2008) Specification of motor axon trajectory by ephrin-B:EphB signaling: symmetrical control of axonal patterning in the developing limb. Neuron 60(6):1039–1053
6. Watanabe T et al (2009) EphrinB2 coordinates the formation of a morphological boundary and cell epithelialization during somite segmentation. Proc Natl Acad Sci USA 106(18): 7467–7472
7. Gros J, Serralbo O, Marcelle C (2009) WNT11 acts as a directional cue to organize the elongation of early muscle fibres. Nature 457(7229):589–593
8. Skowronska-Krawczyk D et al (2009) Conserved regulatory sequences in Atoh7 mediate non-conserved regulatory responses in retina ontogenesis. Development 136(22):3767–3777
9. Garcia-Moreno F et al (2010) A neuronal migratory pathway crossing from diencephalon to telencephalon populates amygdala nuclei. Nat Neurosci 13(6):680–689
10. Hendricks M, Jesuthasan S (2007) Electroporation-based methods for in vivo, whole mount and primary culture analysis of zebrafish brain development. Neural Dev 2:6
11. Wizenmann A et al (2009) Extracellular Engrailed participates in the topographic guidance of retinal axons in vivo. Neuron 64(3): 355–366
12. Irvine SQ et al (2008) Cis-regulatory organization of the Pax6 gene in the ascidian Ciona intestinalis. Dev Biol 317(2):649–659
13. Pu HF, Young AP (1990) Glucocorticoid-inducible expression of a glutamine synthetase-CAT-encoding fusion plasmid after transfection of intact chicken retinal explant cultures. Gene 89(2):259–263

14. Yaneza M et al (2002) No evidence for ventrally migrating neural tube cells from the mid- and hindbrain. Dev Dyn 223(1):163–167
15. Nakamura H, Watanabe Y, Funahashi J (2000) Misexpression of genes in brain vesicles by in ovo electroporation. Dev Growth Differ 42(3): 199–201
16. Luo J, Redies C (2004) Overexpression of genes in Purkinje cells in the embryonic chicken cerebellum by in vivo electroporation. J Neurosci Methods 139(2):241–245
17. Sato Y et al (2007) Stable integration and conditional expression of electroporated transgenes in chicken embryos. Dev Biol 305(2): 616–624
18. Watanabe T et al (2007) Tet-on inducible system combined with in ovo electroporation dissects multiple roles of genes in somitogenesis of chicken embryos. Dev Biol 305(2): 625–636
19. Pekarik V et al (2003) Screening for gene function in chicken embryo using RNAi and electroporation. Nat Biotechnol 21(1):93–96
20. Lee SK, Pfaff SL (2003) Synchronization of neurogenesis and motor neuron specification by direct coupling of bHLH and homeodomain transcription factors. Neuron 38(5): 731–745
21. Araki I, Nakamura H (1999) Engrailed defines the position of dorsal di-mesencephalic boundary by repressing diencephalic fate. Development 126(22):5127–5135
22. Uchikawa M et al (2003) Functional analysis of chicken Sox2 enhancers highlights an array of diverse regulatory elements that are conserved in mammals. Dev Cell 4(4):509–519
23. Chambers D, Lumsden A (2008) Profiling gene transcription in the developing embryo: microarray analysis on gene chips. Methods Mol Biol 461:631–655
24. Hamburger V, Hamilton HL (1992) A series of normal stages in the development of the chick embryo. 1951. Dev Dyn 195(4):231–272
25. Simkin JE, McKeown SJ, Newgreen DF (2009) Focal electroporation in ovo. Dev Dyn 238(12):3152–3155

Chapter 15

In Utero Electroporation in Mice

Chunlei Wang and Lin Mei

Abstract

In utero electroporation has been extensively used to study a variety of developmental questions in the developing brain. This protocol aims to provide the basic knowledge for a beginner to get familiar with the technique. Basically, by electroporating a DNA construct into a subpopulation of progenitor cells in the ventricular zone of embryonic brain, the progenitor cells carrying the DNA will undergo neurogenesis, migration, and final differentiation to become mature neurons positioned in distinct cortical layers according to their birth date. In addition, by controlling the direction of electroporation, a specific cortical area can be targeted. Thus, in utero electroporation allows gene modification in a specific cortical layer in a specific cortical area.

Key words In utero electroporation, Plasmid DNA, Mouse embryos, Gene modification, Brain development

1 Introduction

It has been more than a decade since in utero electroporation was first developed to introduce DNA into mouse embryos [1, 2]. Before that, electroporation had been successfully used to deliver DNA into cultured cells, chick embryos, and cultured mouse embryos [3–5]. The application of electroporation in mouse embryos in utero opened a new window to look at neural development in mammals. Not only it is capable of labeling the developing cells by expressing a fluorescent protein (e.g., EGFP), but it can also over-express or mis-express a gene of interest into the cells to achieve the gain-of-function of the gene. In addition to that, it can also be used to study the loss-of-function of a gene by expressing the shRNAs or microRNAs which target the gene or introducing a cre expression to conditionally knock out the gene previously floxed genetically [6]. The convenience of this technology led to a large body of studies on the gene functions and networks regarding early neural developmental processes such as neurogenesis and neuronal migration [7, 8]. More recently, it has also been used to address late-phase developmental questions such as dendritic

Renping Zhou and Lin Mei (eds.), *Neural Development: Methods and Protocols*, Methods in Molecular Biology, vol. 1018,
DOI 10.1007/978-1-62703-444-9_15, © Springer Science+Business Media, LLC 2013

patterning and axonal path-finding when neurons have already migrated to their final positions [9, 10]. Due to its ability to temporally and spatially target a specific population of neurons in the brain (e.g., layer 2/3 pyramidal neurons in the somatosensory cortex), it has also caught the attention of physiologists and behaviorists [11, 12]. The long-term expression of a protein such as ChR2 channels by in utero electroporation enabled the analysis of the network connectivity at behavior level in adult animals.

To understand how in utero electroporation targets neurons in the brain, it is important to know how electroporation works to allow DNA entering the neurons. First, transient electropores are formed in the cell membrane immediately after electrical pulses, and then the negatively charged DNA outside of the cell is moved in through electropores by electroporation field [13, 14]. We use somatosensory cortex as an example to see how DNA (e.g., GFP expressing construct) is introduced into layer 2/3 neurons in the region (Fig. 1). When DNA is injected into the lateral ventricle (LV) of the mouse embryo (at embryonic day (E) 15.5), DNA particles diffuse to ventricular zone (VZ) surface of cortical plate in the somatosensory cortical region where the endfeet of radial glia cells (shown in grey in Fig. 1c), the progenitor cells of cortical neurons, will uptake DNA upon electrical pulses were applied. After the uptake of DNA, the radial glial cells will undergo cell mitotic changes to become neurons which continue carrying the DNA and migrate to layer 2/3 in the somatosensory cortex.

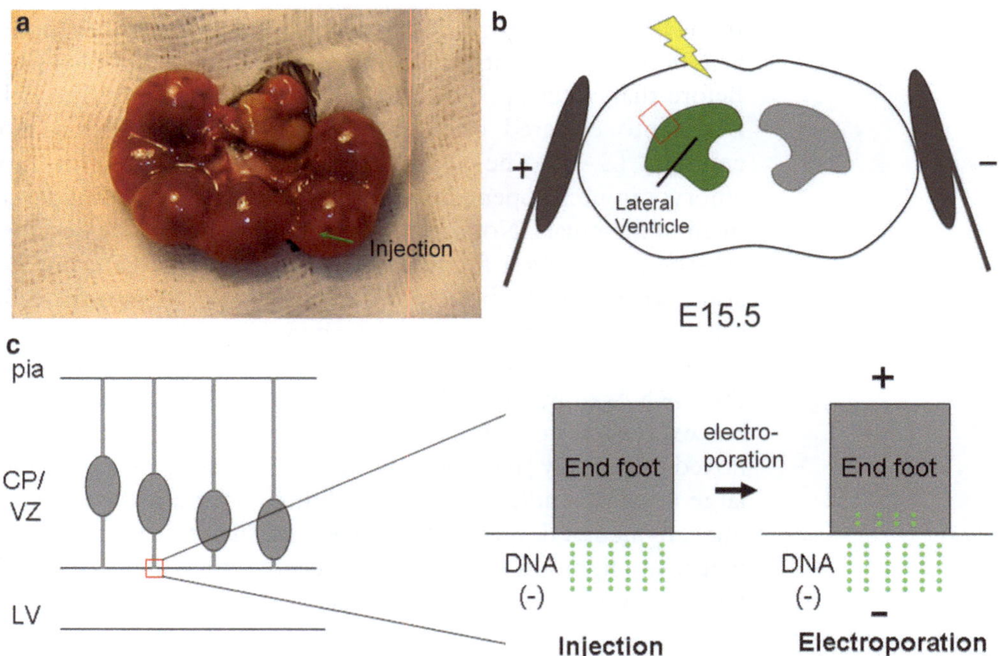

Fig. 1 The principle of in utero electroporation (**a**) Embryos at embryonic day (E) 15.5. Arrow points to the ventricle of the brain. (**b** and **c**) Diagrams illustrating the working principle of in utero electroporation

In this protocol, we describe the details of the in utero electroporation method in targeting layer 2/3 neurons in the somatosensory cortex. Briefly, after anesthesia with isoflurane/O_2 inhalation, pregnant mice at E15.5 (or E14.5) are subjected to abdominal incision to expose the uterus. Through the uterine wall, embryos are visualized and plasmids DNA are injected into the lateral ventricle through a glass micropipette. Electric pulses are then delivered into embryos by gently clasping their heads with forceps-shaped electrodes (positive side on the injected brain side). Five 33–36 V pulses of 50 ms are applied at 1 s intervals. Uterine horns are repositioned into the abdominal cavity before the abdominal wall and the skin are sutured.

2 Materials

2.1 Pregnant Mice

CD1 mice (Charles River, USA) are used in most cases, because they usually bear more than ten pups and feed well after pups are born. Other mouse strains, such as C57BL/6, can also be used. The noon of a day when a vaginal plug is found is designated as embryonic day (E) 0.5. The day of birth is designated postnatal day (P) 0.

All experiments should be performed in accordance with the protocols approved by the institutional animal care and use committee.

2.2 Plasmid DNA Preparation

Plasmids are purified using the Qiagen EndoFree Maxiprep Kit according to the manufacturer's protocol. Use PBS or purified water to collect DNA and measure the concentration with NanoDrop. Before in utero electroporation, dilute the DNA solution with H_2O or PBS to a final concentration of 1–1.5 μg/μl (pCAG-IRES-EGFP is used). Add fast-green (final concentration at 0.1 %) into the DNA solution. pCAG-IRES-EGFP carries the EGFP (enhanced green fluorescent protein) gene downstream of the CAG promoter with an IRES sequence in between. Purified plasmids should be handled on a clean bench to avoid pathogenic contamination.

2.3 Micropipettes for DNA Injection

Borosilicate Glass tubes with filament are used in this protocol (Fig. 2a, b). Pull glass capillary tubes using the micropipette puller, P-87 (Sutter Instrument Co.), under the following conditions: 001 heat, 885; pull, 45; velocity, 28; time, 200; 002 heat, 750; pull, 50; velocity, 50; time, 200 (Fig. 2c, d). Note that the pulling is two steps instead of one step in order to make the tip not too long. Usually the tip length is 0.8–1 cm (Fig. 2e). Sterilize the pipettes under UV light on a clean bench for about15 min.

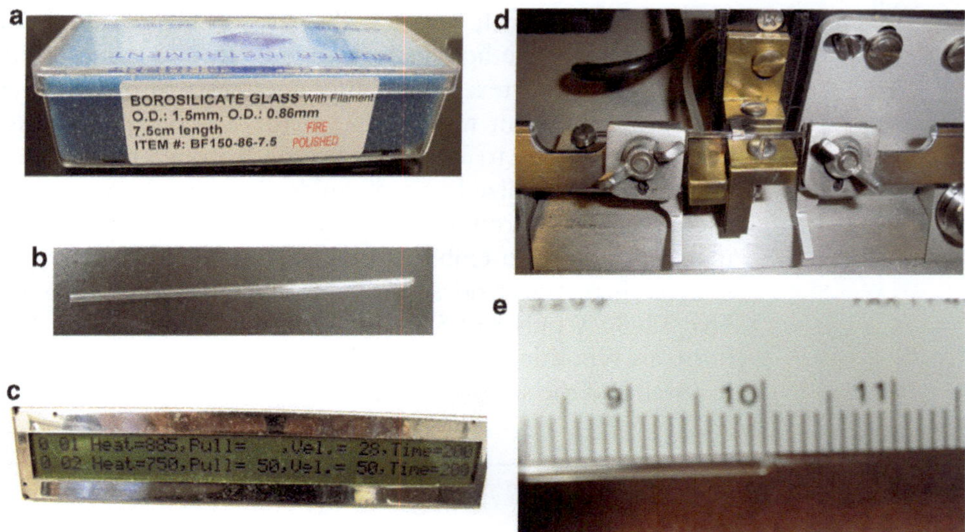

Fig. 2 The making of micropipette (**a** and **b**) Borosilicate glass tubes used in the protocol. (**c** and **d**) The program used for making glass capillary tubes. (**e**) An example of the glass tube with the appropriate tip size

2.4 Mouth Pipette System Sigma-Aldrich, A5177-5EA (Sigma)

2.5 Electroporator (Fig. 3a)

ECM 830 (BTX) connected to a 7 mm electrode and a foot-controlled pedal.

2.6 Vaporizer for Isoflurane Anesthesia (Fig. 3b)

Isoflurane is used for anesthetizing the animal. Set oxygen flow at 1 l/min and isoflurane at 3–5 % for induction and at 1.5–2 % for maintenance.

2.7 Hot Bead Sterilizers (Fig. 3c)

When temperature reaches 250 °C, instruments require approximately 20 s of contact with the beads for complete sterilization.

2.8 Surgical Tools (Fig. 3d)

Different tools including straight and curved forceps with blunt ends, two scissors with different sizes, suturing needle holder, and wound clip applier for 7 or 9 mm clips.

2.9 Skin Cleaner (Fig. 3e)

PDI Povidone-Iodine Swabsticks and Alcohol Swabsticks.

2.10 Sterilized Gauze

size 4″ × 4″.

2.11 Sterilized PBS

500 ml pre-warmed.

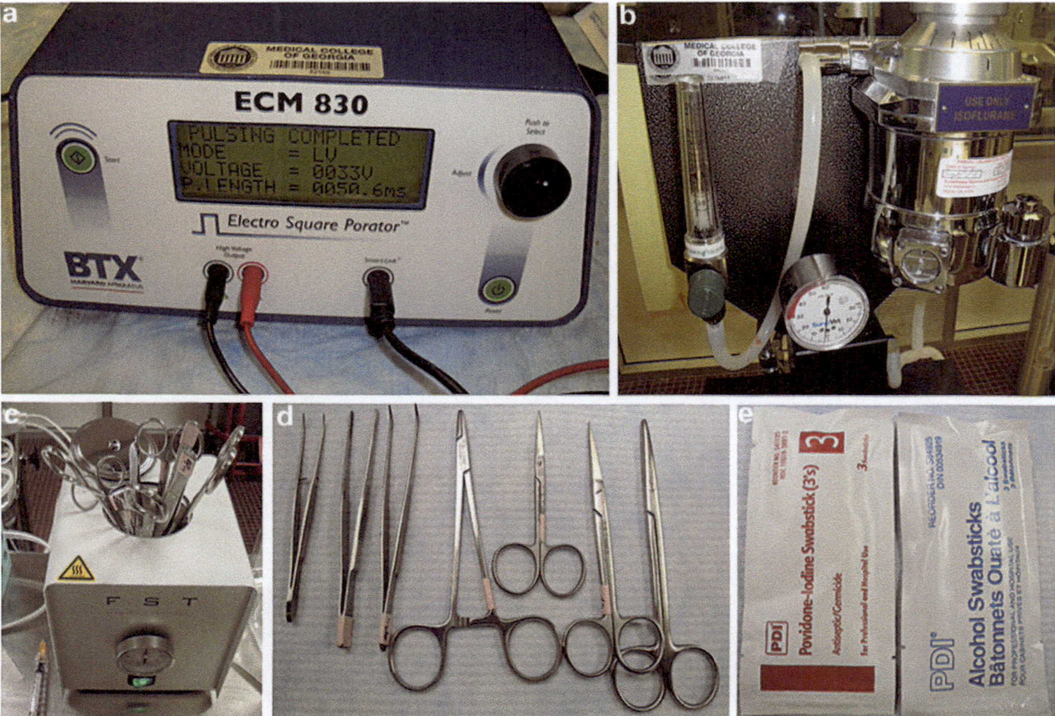

Fig. 3 Materials and tools for surgery (**a**) Electroporator. (**b**) Isofurane machine. (**c**) Sterilizer. (**d**) Surgical tools. (**e**) Skinner cleaner

3 Surgical Procedures

First, set up the surgical area as shown in Fig. 4.

Next, follow Subheadings 3.1–3.11 as below.

3.1 Anesthesia of the Animal (Gas Anesthesia)

1. Induce anesthesia of the mouse in a chamber with high flow of isoflurane (3–5 %) mixed with O_2 (1 l/min).
2. Transfer the mouse to surgical area with continuous flow of isoflurane (1.5–2 %) mixed with O_2 (1 l/min) (Fig. 5a).
3. Inject carprofen (5 mg/kg) subcutaneously for pain relief.
4. Make sure that the animal is fully anesthetized by observing the disappearance of toe pinch reflex.

3.2 Preparation of the Skin

1. Shave the abdominal skin to remove the fur (Fig. 5b).
2. Disinfect the abdominal skin with povidone-iodine and alcohol.

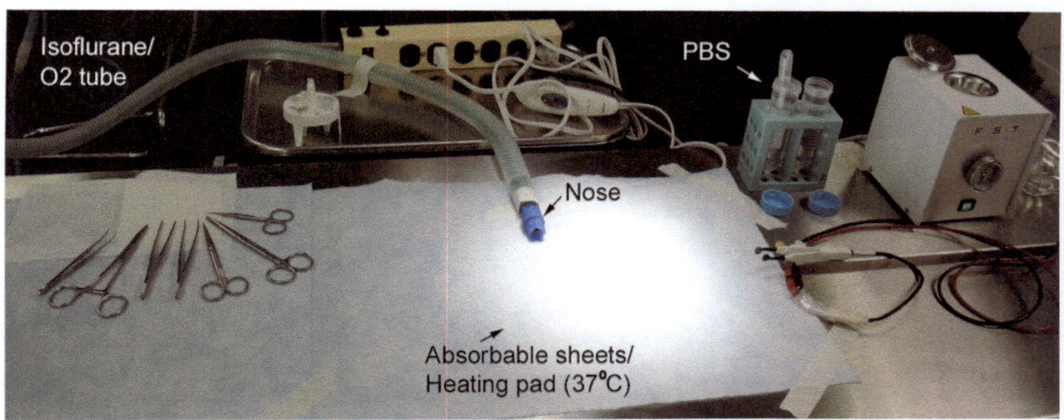

Fig. 4 Surgical setup

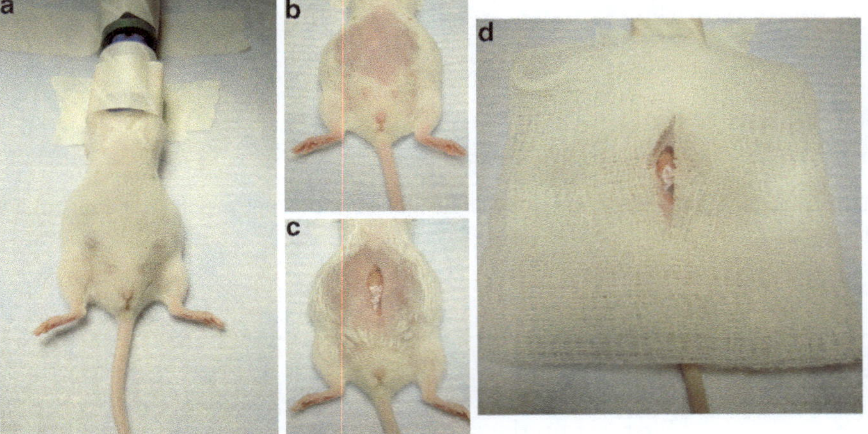

Fig. 5 Preparation of the skin (**a–d**) Four continuous steps (described in the text) to prepare the skin for further embryo manipulation

3.3 Exposure of the Embryos

1. Make an incision along the midline of the skin for about 2–2.5 cm in length (Fig. 5c).
2. Cover the skin with a sterilized gauze (4″×4″) with an open window (4–5 cm) in the middle (Fig. 5d). Drop some PBS in the cut area.
3. Cut the muscle wall underneath the skin along the midline for about 2–2.5 cm in length (Fig. 6a–c). Drop PBS to wet the gauze near the cut area.
4. Gently pull the embryos out on the gauze (Fig. 6d). Drop some PBS to moisturize the embryos.

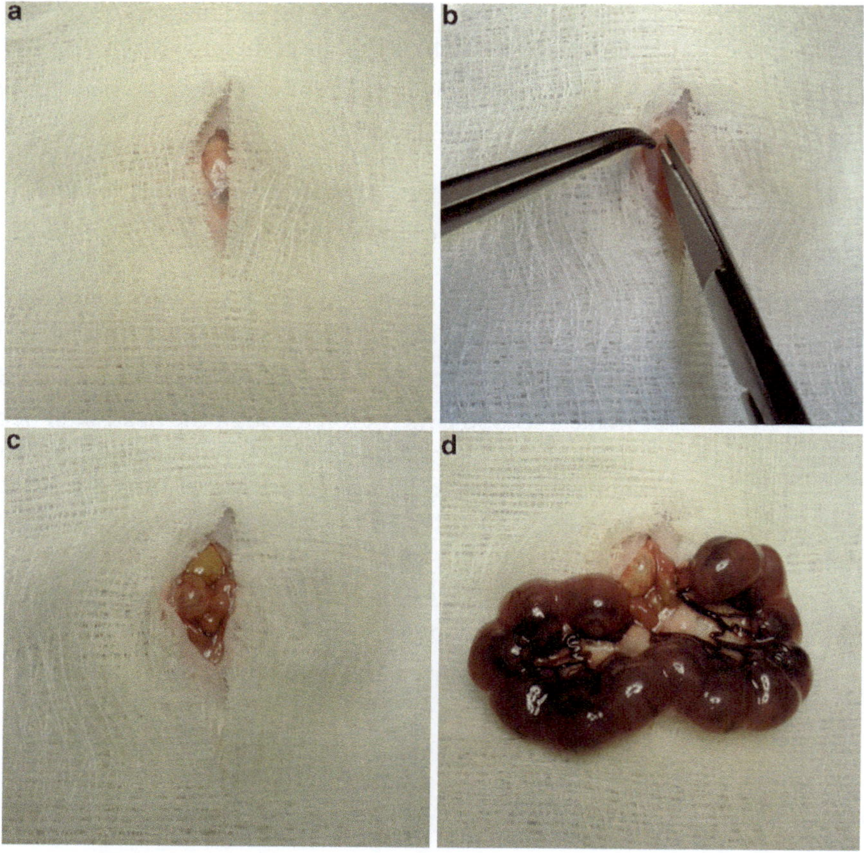

Fig. 6 Exposure of the embryos (**a–d**) Four continuous steps (described in the text) to expose the embryos

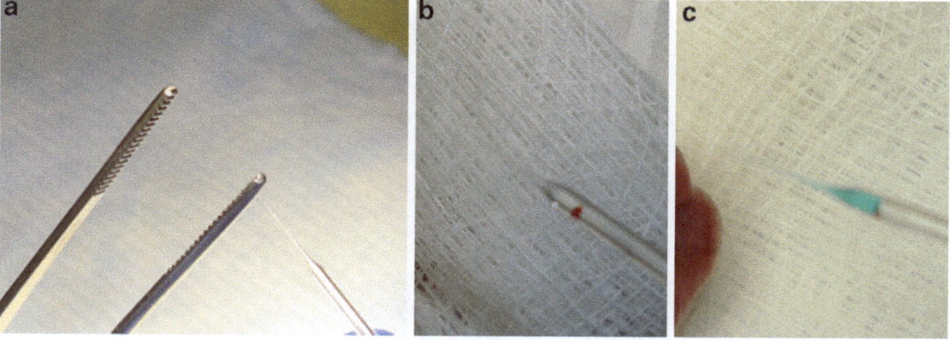

Fig. 7 Preparation of micropipette tip (**a–d**) Three continuous steps (described in the text) to prepare the micropipette tip for injection

3.4 Preparation of Glass Pipette Tip

1. Carefully pinch the glass pipette tip against the forceps (Fig. 7a).
2. Fill 1–2 μl PBS into the pipette tip and inject out with mouth pipette to test whether the tip size is suitable for injection.
3. Repeat steps (1) and (2) until comfortable to inject the PBS out of the pipette tip.

4. Fill 2 μl PBS into the pipette tip and mark the position (Fig. 7b), and then inject the PBS out completely.
5. Put the mouth pipette system with the glass pipette somewhere clean for next step.

3.5 Filling of DNA Solution

1. Find an embryo whose head is facing outside or visible.
2. Fill the DNA solution to the pre-marked position (Fig. 7c).

3.6 Microinjection of DNA into Brain Ventricles

1. Fix the position of the embryo head with a curved blunt forceps by pushing against the uterus on the head side (Fig. 8a).
2. Make a first penetration through the uterus wall (Fig. 8a).
3. Make a second penetration through the head skull between the eye and the lambdoid suture (in the middle of the head), directly into the lateral ventricle on one side of the brain (Fig. 8a). This step will need practice to get the sense of penetration into the ventricle.
4. Inject some DNA first to test whether the pipette tip is in the ventricle (Fig. 8b).
5. Continue to inject the remaining DNA into the ventricle (Fig. 8c).

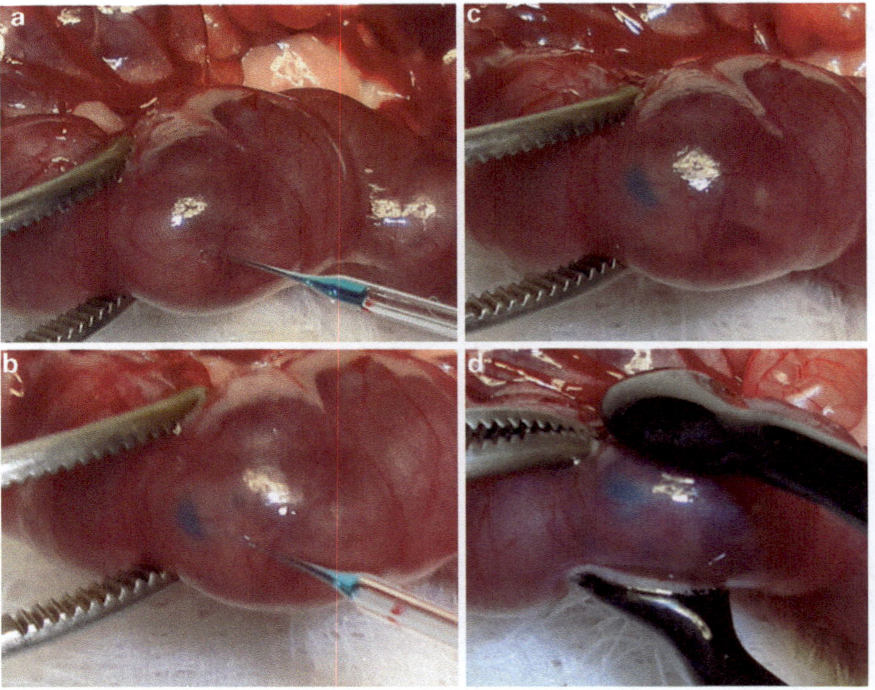

Fig. 8 Injection of DNA and electroporation (**a–c**) Three continuous steps (described in the text) to inject DNA to embryonic brain ventricles followed by electroporation (**d**)

3.7 Clasping of the Embryonic Head with Electrodes

1. Drop some PBS on electrode surfaces and the uterus wall containing the embryo.
2. Clasp the uterus with electrodes (Fig. 8d). The positive electrode should be placed on the injected side. Usually use the curved blunt forceps to push the uterus wall so that the embryonic head can be positioned in between the two electrodes.
3. Make sure the electrode surfaces and uterus wall are contacted.

3.8 Delivery of Electric Pulses

1. Make sure the contact surface of the electrodes and uterus wall is filled with PBS.
2. Deliver the electric pulses with a foot-controlled pedal (Fig. 8d). The bubble will be generated on the negative electrode with each pulse. The embryo will have some stretching response during electroporation. Sometimes the head may escape from the electrodes, so it is important to use a forceps to stabilize the embryo by pushing on the uterus wall during electroporation.

3.9 Return of Embryos Back into the Abdominus

1. Cut the upper part of the gauze to make the window bigger in order to relieve the restrain.
2. Drop PBS on all embryos and the abdominal cavity. This is to make all surfaces as smooth as possible.
3. Lift the embryos on one body side with one blunt forceps and lift the abdominal skin with another blunt forceps (Fig. 9a).
4. Gently insert the embryos back to the abdominal cavity (Fig. 9b).
5. Repeat steps (3) and (4) for the embryos on the other body side (Fig. 9c).
6. Drop PBS into the abdominal cavity to allow embryos positioned more naturally (Fig. 9d).

3.10 Suturing of the Abdominal Wall

1. Suture the muscle wall first. Use absorbable suture (Fig. 10a).
2. Suture the Skin with wound clips (9 or 7 mm) (Fig. 10b). Tighten the wound clips with needle holder.
3. Dry the skin with tissue paper.

3.11 Recovery from Anesthesia

1. Turn off the isoflurane machine.
2. Release the mouse from the isoflurane tube nose.
3. Recover the mouse for 5 min at 37 °C until it turns over.
4. Put the mouse back to the cage and to the animal room.

3.12 Post-surgery

1. It is recommended to watch closely on the mouse for the first 2 days. Analgesic drugs are provided during the first 3 days

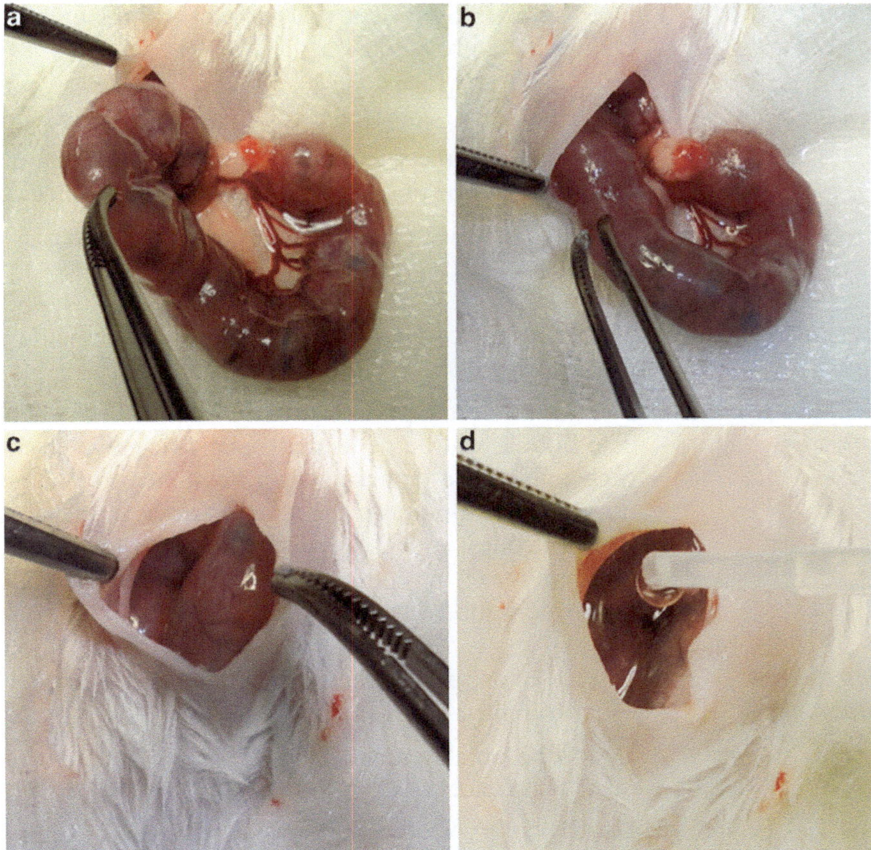

Fig. 9 Sending back of embryos (**a–d**) Four continuous steps (described in the text) to send embryos back

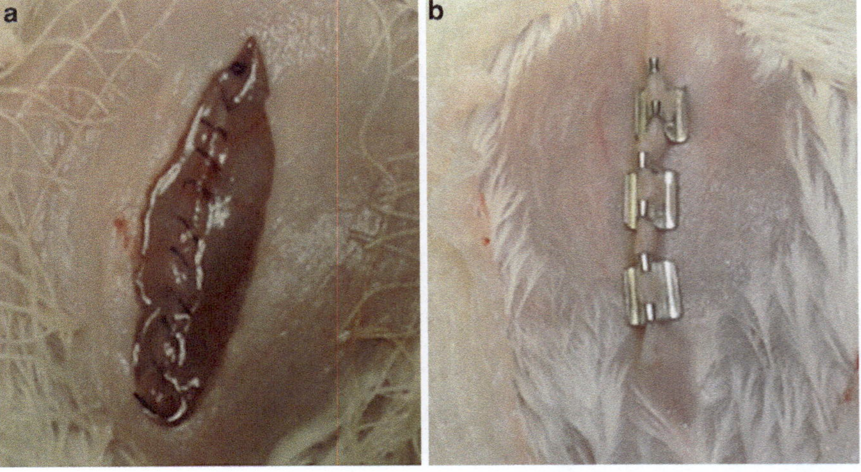

Fig. 10 Suturing of muscle and skin (**a,b**) Two steps (described in the text) to close the abdominal wall

after surgery to relieve the pain from the surgery. Abortion of the embryos or death of the mother mouse usually happens during the first two post-surgery days. If the mouse is in stress, euthanize it. If the mouse survives without abortion during the first 2 days, it usually will give birth as normal.

2. The fluorescence can be detected in live pups during the first 3 days after they are born, under stereo-microscope or by using a portable fluorescent detector.

4 Notes

1. Isoflurane is recommended as anesthetics, but intraperitoneal injection of the cocktail of ketamine (80–100 mg/kg) and xylazine (5–10 mg/kg) can also be used. The use of isoflurane inhalation is safer than the cocktail injection in terms of accidental death of the animals during anesthesia. In addition, the isoflurane anesthesia makes the mouse recover quickly (usually it turns over within 5 min).
2. Make sure that the toe pinch reflex completely disappears before making an abdominal incision. If the mouse shows some response during skin or muscle cutting, wait for a few minutes after the response disappears.
3. The abdominal incision should be made as small as possible. Usually the size of the incision is between 2 and 2.5 cm in length. A smaller incision less than 2 cm may squeeze the embryos and may be harmful for their survival.
4. During the pulling of the embryos out onto gauze with blunt-ended forceps, be gentle and do not use extra strength to pull embryos. Usually pull embryos of one body side to the same side of gauze as shown in Fig. 6d.
5. Pinch the glass pipette tip to an appropriate size (8–10 μm of the inner diameter). Smaller than 8 μm may cause difficulty to inject DNA out and bigger than 10 μm is not good for penetration through the uterus wall and the embryo head. This step can be done immediately after pipette pulling or during surgery.
6. The heads of embryos are not always visible. If embryo heads are hiding inside, I usually rotate the embryos with the blunt forceps until I can see them.
7. To penetrate the micropipette into the brain ventricle, make a first penetration through the uterus wall and the second penetration through the brain head. Do not do single penetration directly to the brain ventricle otherwise the pipette tip may most likely reach other brain regions. Use the blunt forceps to push on the uterus wall on the brain side so the embryo can be fixed, otherwise the embryo would escape during penetration of the brain head.

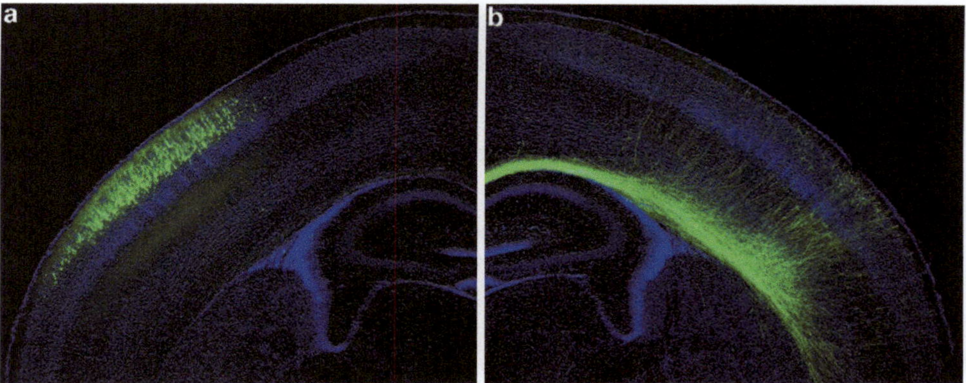

Fig. 11 Electroporation of neurons in somatosensory cortex. This image shows cortical neurons at postnatal day (P) 9 electroporated with EFPG construct at E15.5. Neurons are distributed in layers 2/3 in the somatosensory cortex (**a**) and their callosal axons project to the contralateral cortex (**b**)

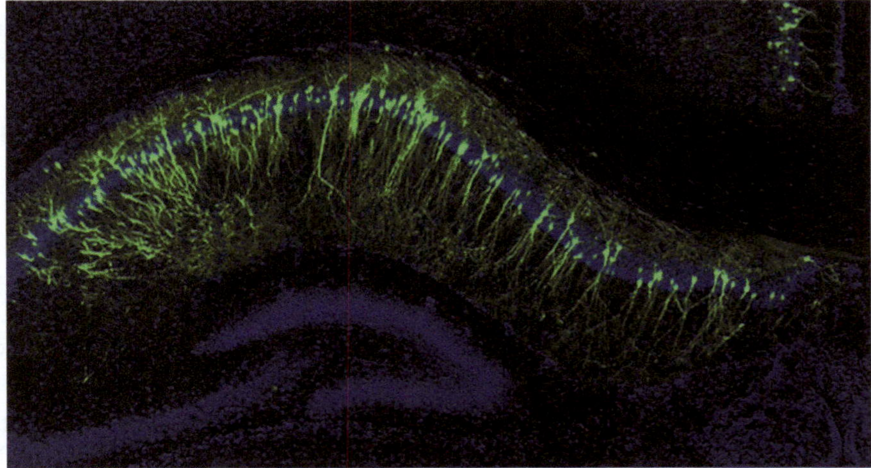

Fig. 12 Electroporation of neurons in hippocampus. This image shows hippocampal CA1 neurons at P10 electroporated with EGFP construct at E15.5. Note that the negative electrode should be put on the injected side of the embryo during in utero electroporation [15]

8. During DNA injection, if the DNA (colored with fast green) is seen in other regions, adjust the position of the pipette tip by pulling back or penetrating a little further until DNA can be injected in the ventricle. A correct injection will show a clearly defined triangle-shaped ventricle (Fig. 8).

9. To target cortical neurons in the somatosensory cortex (Fig. 11), the positive electrode should be put on the injected side of DNA. To target cortical neurons in the hippocampal neurons (Fig. 12), negative electrode should be put on the injected side of DNA. Ideally, the DNA can be delivered to the progenitor cells in any area of ventricular surface by setting the direction of electrodes.

10. The voltage for E15.5 embryos is 33–36 V and a higher voltage is not necessary. For younger embryos, use lower voltages. Usually drop 2 V when the embryo is 1 day younger.
11. After in utero electroporation, gently send the embryos back to the abdominal cavity. Drop sufficient PBS on embryos so that they can be smoothly sent back. It is a big challenge for a big litter (>12 embryos), but always be patient and be gentle for this step. Do not use too much strength.
12. Always suture the muscle and skin separately. After suturing the skin with wound clips, always tightened clips. This is important because the mouse tends to bite the clips after recovery. If the clips are not tightened, they could be gone the wound is not healed.
13. The total surgery time is best kept within 1 h but one and a half hours may also work. The shorter the time, the better the recovery.

References

1. Saito T, Nakatsuji N (2001) Efficient gene transfer into the embryonic mouse brain using in vivo electroporation. Dev Biol 240: 237–246
2. Tabata H, Nakajima K (2001) Efficient in utero gene transfer system to the developing mouse brain using electroporation: visualization of neuronal migration in the developing cortex. Neuroscience 103:865–872
3. Muramatsu T, Mizutani Y, Ohmori Y, Okumura J (1997) Comparison of three nonviral transfection methods for foreign gene expression in early chicken embryos in ovo. Biochem Biophys Res Commun 230:376–380
4. Itasaki N, Bel-Vialar S, Krumlauf R (1999) 'Shocking' developments in chick embryology: electroporation and in ovo gene expression. Nat Cell Biol 1:E203–E207
5. Swartz M, Eberhart J, Mastick GS, Krull CE (2001) Sparking new frontiers: using in vivo electroporation for genetic manipulations. Dev Biol 233:13–21
6. Matsuda T, Cepko CL (2007) Controlled expression of transgenes introduced by in vivo electroporation. Proc Natl Acad Sci USA 104:1027–1032
7. Chen G, Sima J, Jin M, Wang KY, Xue XJ, Zheng W, Ding YQ, Yuan XB (2008) Semaphorin-3A guides radial migration of cortical neurons during development. Nat Neurosci 11:36–44
8. Sanada K, Tsai LH (2005) G protein betagamma subunits and AGS3 control spindle orientation and asymmetric cell fate of cerebral cortical progenitors. Cell 122:119–131
9. Cancedda L, Fiumelli H, Chen K, Poo MM (2007) Excitatory GABA action is essential for morphological maturation of cortical neurons in vivo. J Neurosci 27:5224–5235
10. Wang CL, Zhang L, Zhou Y, Zhou J, Yang XJ, Duan SM, Xiong ZQ, Ding YQ (2007) Activity-dependent development of callosal projections in the somatosensory cortex. J Neurosci 27:11334–11342
11. Huber D, Petreanu L, Ghitani N, Ranade S, Hromadka T, Mainen Z, Svoboda K (2008) Sparse optical microstimulation in barrel cortex drives learned behaviour in freely moving mice. Nature 451:61–64
12. Petreanu L, Huber D, Sobczyk A, Svoboda K (2007) Channelrhodopsin-2-assisted circuit mapping of long-range callosal projections. Nat Neurosci 10:663–668
13. Neumann E, Schaefer-Ridder M, Wang Y, Hofschneider PH (1982) Gene transfer into mouse lyoma cells by electroporation in high electric fields. EMBO J 1:841–845
14. Shimogori T, Ogawa M (2008) Gene application with in utero electroporation in mouse embryonic brain. Dev Growth Differ 50:499–506
15. Navarro-Quiroga I, Chittajallu R, Gallo V, Haydar TF (2007) Long-term, selective gene expression in developing and adult hippocampal pyramidal neurons using focal in utero electroporation. J Neurosci 27: 5007–5011

Chapter 16

General Introduction to In Situ Hybridization Protocol Using Nonradioactively Labeled Probes to Detect mRNAs on Tissue Sections

Daehoon Lee, Shan Xiong, and Wen-Cheng Xiong

Abstract

In situ hybridization (ISH) is a type of hybridization that uses a labeled complementary DNA or RNA strand (i.e., probe) to localize a specific DNA or RNA sequence in a portion or section of tissue (In Situ) or in the entire tissue (whole mount ISH). Localization of endogenous transcripts is a desirable approach for confirming expression patterns. This is distinct from immunohistochemistry, which usually localizes proteins in tissue sections. DNA ISH can be used to determine the structure of chromosomes. However, RNA ISH (hybridization histochemistry) is used to measure and localize mRNAs and other transcripts within tissue sections or whole mounts. RNA–RNA hybrids approach may offer increased sensitivity, which is more stable than that of DNA–RNA hybrids. Here we describe the efficient ISH protocol for nonradioactive (i.e., in direct methods using digoxigenin (DIG) system) RNA probes, and it can be performed in less than 3 days.

Key words In situ hybridization, Tissue section, Nonradioactive probes, Digoxigenin (DIG) labeling

1 Introduction

In situ hybridization (ISH) techniques allow specific nucleic acid sequences to be detected in morphologically preserved chromosomes, cells, or tissue sections. In combination with immunocytochemistry, ISH can relate microscopic topological information to gene activity at the DNA, mRNA, and protein level. The technique was originally developed at 1969 [1, 2]. At first, the radioactive nucleotides were mainly used in the ISH experiment. The advantage of radiolabeled probes is their ability to detect very low levels of transcripts. However, they also have some major limitations including a very long exposure time and a relatively poor resolution, depending on the radioisotope used. Moreover, the photographic emulsion revealing the hybridization signals is not at

Renping Zhou and Lin Mei (eds.), *Neural Development: Methods and Protocols*, Methods in Molecular Biology, vol. 1018, DOI 10.1007/978-1-62703-444-9_16, © Springer Science+Business Media, LLC 2013

the same focus as the tissue section, which hampers their solution and microscopic observation [3]. The most recent use of nonradioactively labeled nucleotides (digoxigenin (DIG) or fluorescently labeled probes) has considerably improved ISH with both a shortening of the development time and an excellent histological resolution [4–6]. Of all the methods developed, the DIG based detection has been proved to be the most appropriate one for infrequent transcripts [7]. Furthermore, it opens new opportunities for combining different labels in one experiment. In addition, there are many sensitive antibody detection systems available for such probes that further enhance the flexibility of this method [8]. In this chapter, therefore, we describe nonradioactive alternatives for ISH.

2 Materials

Lab area for RNA extraction should be RNase free and preferably a separate area. All the solutions have to be prepared using 0.1 % diethylpyrocarbonate (DEPC) water [9]. To prepare 0.1 % DEPC water: add 1 mL DEPC to 1 L water, stir overnight using a stir bar, and autoclave it the next day. There should be no smell of DEPC in the treated water after autoclaving. The water and other materials should be stored in an RNase-free place or drawer.

2.1 *In Situ* Hybridization Reaction Components

1. 50× Denhardt's Reagent: 1 % (w/v) Ficoll, 1 % (w/v) Polyvinylpyrrolidone (PVP), 10 mg/mL RNase-free bovine serum albumin. Adjust the volume to 50 mL with DEPC-treated water then 1 mL aliquots and store at −20 °C until use.
2. Hybridization Buffer (without salmon sperm DNA): 50 % Formamide, deionized, 0.3 M NaCl, 10 mM Tris–HCl (pH 8.0), 1 mM 0.5 M EDTA (pH 8.0), 0.5 mg/mL Yeast tRNA, 10 % Dextran sulfate in formamide, 10 mM DTT, 1× Denhardt's reagent. Adjust the volume to 20 mL with DEPC-treated water then 1 mL aliquots and store at −20 °C until use.
3. Triethanolamine-HAc: dissolve 0.1 M Triethanolamine in 200 mL DEPC-treated water. Immediately add 500 μL (0.25 %) acetic anhydride then mix well.
4. Prehybridization Buffer: 50 % Formamide deionized in 4× saline–sodium citrate (SSC) Buffer.
5. Buffer 1: 100 mM Tris–HCl (pH 7.5), 150 mM NaCl. Add autoclaved, certainly used deionized water to final volume of 500 mL.
6. Buffer 1 T: 0.1 % Triton X-100 in Buffer 1.
7. Blocking solution: 1 % blocking reagent for nucleic acid hybridization and detection in Buffer 1 T.

8. Buffer 2: 100 mM Tris–HCl (pH 9.5), 150 mM NaCl, and 50 mM $MgCl_2$. Add autoclaved, certainly used deionized water to final volume of 200 mL. Store at room temperature in a bottle wrapped with aluminum foil.
9. TE Buffer: 100 mM Tris–HCl, 50 mM EDTA (pH 8.0). Make up to 500 mL with DEPC-treated water.
10. RNase Buffer: 500 mM NaCl, 10 mM Tris–HCl (pH 8.0), and 1 mM EDTA (pH 8.0). Make up to 500 mL with DEPC-treated water.
11. Nitro Blue Tetrazolium (NBT): NBT is resuspended to 100 mg/mL in 70 % dimethylformamide (DMF) and stored in aliquots at −20 °C.
12. 5-Bromo-4-Chloro-3-Indolyl Phosphate (BCIP): BCIP is resuspended to 50 mg/mL in anhydrous DMF and stored in aliquots at 20 °C.
13. Stop Reaction Buffer: 10 mM NaCl (pH 8), 1 mM EDTA (pH 8). Make up to 500 mL with DEPC-treated water and filter through a 0.45 μm Corning filter.
14. 20× saline–sodium citrate (SSC): Weigh 175.3 g of NaCl and 88.9 g sodium citrate. Make up to 1 L with distilled DEPC-treated water.
15. 10× Phosphate buffered saline (PBS): A 1 L stock of 10× PBS can be prepared by dissolving 80 g NaCl, 2 g KCl, 14.4 g $Na_2HPO_4 \cdot 2H_2O$, and 2.4 g KH_2PO_4 in 800 mL of distilled water, and topping up to 1 L. The pH is ~6.8, but when diluted to 1× PBS it should change to 7.4.
16. 10× PBST: 0.1 % Triton X-100 in 10× PBS.

2.2 Probe Design and Synthesis

Antisense RNA probes are created from in vitro transcription of short (150–400 bp) PCR products carrying the T7 or SP6 RNA polymerase recognition sequence at one end. Probes longer than 500 bases may not penetrate tissue. If probes are longer than 500 bases, shorten them by alkaline hydrolysis accordingly. The most convenient template is cloned cDNA, using primers that will amplify as a high proportion of exon-containing sequence as possible. Alternatively, a template genomic DNA can be used. Some may choose to also synthesize the complementary (sense) strand as a control probe, or use one of the suggested probes for specificity (Fig. 1).

2.2.1 PCR to Generate Probe Synthesis Template

1. Assemble the following in a PCR tube: Forward primer (25 pmol/μL) 1 μL, reverse primer (25 pmol/μL) 1 μL, dNTPs (10 mM) 0.5 μL, and PCR buffer (10× stocks) 2.5 μL, cloned cDNA (200 ng/μL) 1 μL, Taq polymerase 0.5 μL, and add ddH_2O to final volume of 20 μL. Perform a standard PCR reaction, for example 95 °C for 5 min with repetition of (95 °C

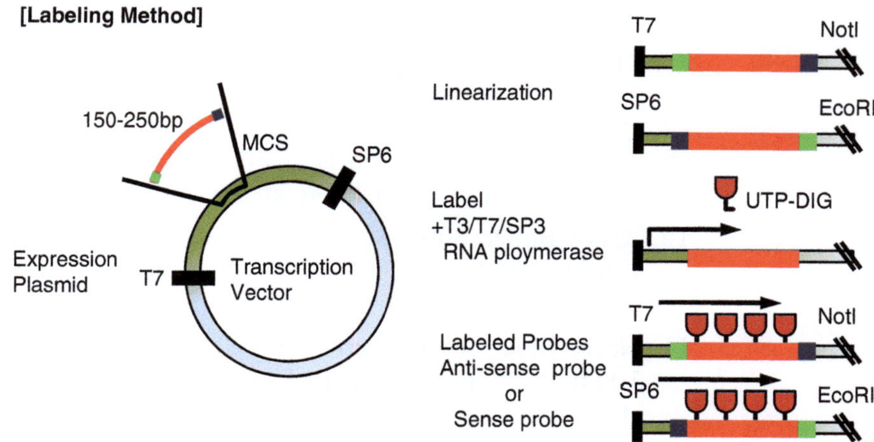

Fig. 1 Schematic drawing of a principle for In vitro transcription. The plasmid containing a portion of probe template is linearized by restriction enzyme in two separate restriction digest reactions. The template for probe synthesis is shown as a *red line* with and indicating the direction of transcription. In vitro transcription is used polymerase (T7 or SP6) and the PCR product as a template. Both linearized DNA is used as the template to generate DIG labeled probes, respectively, sense and antisense probe. In this example, the vector is first cleaved with NotI and then in vitro transcribed with T7 RNA polymerase to generate the antisense probe used for ISH. Opposite, EcoRI and SP6 RNA polymerase are used to synthesize a sense probe, which can be used as a negative control

for 30 s, 56 °C for 30 s, 72 °C for 30 s) for 35 cycles, then 72 °C for 10 min. Check on an 0.1 % agarose gel to make sure that the PCR product is of the expected size.

2. The RNA to be transcribed should be cloned into the multiple cloning site of a transcription vector which contains a promoter for SP6, T7, or T3 RNA Polymerase.
3. Linearization of the cDNA-containing RNA expression vector [10] (Generate blunt ends or 5′ overhang): 1 μg Plasma DNA, 1 μL restriction enzyme, 2 μL 10× restriction buffer, 2 μL BSA (if required), add RNase-free ddH_2O to final volume of 20 μL. Put in an appropriate water bath (usually 37 °C) for 2–5 h.
4. Clean up linear DNA: PCR product cleanup by Kit. Follow manufacturer's instructions.

2.2.2 DIG-Labeled Probe Synthesis (In Vitro Transcription)

1. In Vitro Transcription Reaction: 1 μg DNA, 2 μL 10× Transcription Buffer, 2 μ DIG labeled NTP mix, 1 μL RNase inhibitor (20 units), 2 μL RNA polymerase (20 U/μL, T7 or Sp6), and RNase-free ddH_2O to final volume of 20 μL. Put in an appropriate PCR equipment for 2 h at 37 °C.
2. Add 50 μL ddH_2O and mix well.

3. Run 5 μL on a mini-gel (0.8 % gel, 60 V, 15 min), place 25 ng template DNA in neighboring lane. RNA band should be ten times stronger than template DNA.
4. Stop the reaction by adding. Add 1 μL tRNA (20 μg/μL), 25 μL 3.2 M LiCl (Lithium Chloride), mix well, and add 400 μL EtOH.
5. Precipitate overnight at −20 °C. Centrifuge 20 min at 15,000 rpm at 15,000 × *g*; pellet should be well visible.
6. Wash pellet with 500 μL 70 % EtOH. Dry in SpeedVac for 5 min.
7. Dissolve in 100 μL H_2O (DEPC-treated) or not dissolve in EtOH. Aliquot labeled probes and store them at −80 °C for at least 1 year.

2.3 Preparation of Tissue Sections

Keep the time between perfusion or killing of animal and fixation as short as possible, for best preservation of morphology. Place tissue in freshly prepared RNase-free 4 % para-formaldehyde (PFA). Fix the tissue for at least 6 h overnight (but no more than 24 h) at 4 °C. After completed decalcification, leave tissue in 15 % sucrose for 2–3 h after removing in 30 % sucrose overnight. Freeze the tissue in liquid nitrogen or hexane on dry ice. Store in a −80 °C freezer [11].

1. For the processing of tissue, clean all knives and other materials with RNase remover (i.e., RNase ZAP; AMBION) and work as aseptically as possible.
2. Remove the tissue from the animal, immediately snap-freeze the tissue, and store it in liquid nitrogen. Work quickly to avoid degradation of RNA.
3. Cut 10–30 μm thick frozen sections. To get a higher signal, cut sections thicker than 10 μm.
4. Mount tissue directly on slides to prevent detachment of the section (*see* **Note 1**).
5. Dry slides in an oven at 40 °C overnight.
6. Circle the sections with a PAP pen (silicone pen) to prevent smudging of the substrate.
7. If the target tissue has lipid vesicles, that will interfere with the RISH (Reflectance ISH) detection. To minimize nonspecific background caused by lipid vesicles, do the following (all at room temperature). Delipidize the sections by extracting them for 5 min in chloroform. Dry the section to evaporate the chloroform.
8. If the target tissue does not have lipid vesicles that interfere with the RISH detection.

3 Methods

Prepare all solutions for procedures below (probe labeling through posthybridization) with distilled, deionized water (ddH_2O) that has been treated with 0.1 % DEPC. To avoid RNase contamination, wear gloves and mask throughout the procedures and use different glassware for prehybridization and posthybridization steps. Perform all procedures below at room temperature unless a different temperature is stated.

3.1 Prehybridization

1. To rehydrate, incubate sections for 5 min with RNase-free 95 % EtOH then rinse with RNase-free 70 % EtOH.
2. Circle the sections one more time with a PAP pen (silicone pen) to prevent smudging of the substrate.
3. 2× with DEPC-treated PBS, 5 min each time. Immediately followed by 2× with DEPC-treated PBS containing 0.2 % glycine, 5 min each time.
4. Wash for 10 min with DEPC-treated PBS.
5. Permeabilize sections for 30 min at 37 °C with TE Buffer containing 1 μg/mL RNase-free Proteinase K. Permeabilization is the most critical step of the entire ISH procedure. Alternative permeabilization protocols for improved efficiency of digestion include incubation of slides for 20–30 min at 37 °C with 0.1 % pepsin in 0.2 M HCl (*see* **Note 2**).
6. Post-fix sections for 5 min at 4 °C with DEPC-treated PBS containing 4 % PFA.
7. 2× with DEPC-treated PBS, 5 min each time.
8. To acetylate sections, place slide containers on a rocking platform and incubate slides twice for 5 min with 0.1 M triethanolamine Buffer, containing 0.25 % (v/v) acetic anhydride. Acetic anhydride is highly unstable. Add acetic anhydride to each change of triethanolamine–acetic anhydride solution immediately before incubation (*see* **Note 3**).
9. 2× rinse with DEPC-treated PBS. Immediately followed twice by 2× saline–sodium citrate (SSC) Buffer, 5 min each time.
10. Incubate sections at 37 °C for at least for 30 min with prehybridization Buffer [9].

3.2 In Situ Hybridization

1. Denature the hybridization buffer by incubating it at 95 °C in a water bath for 10 min and then adding 1 mg/mL salmon sperm DNA.
2. Drain prehybridization Buffer from the slides and overlay each section with 100 μL of hybridization Buffer containing 10–50 ng of DIG-labeled RNA probe (*see* **Note 4** and Fig. 2a).

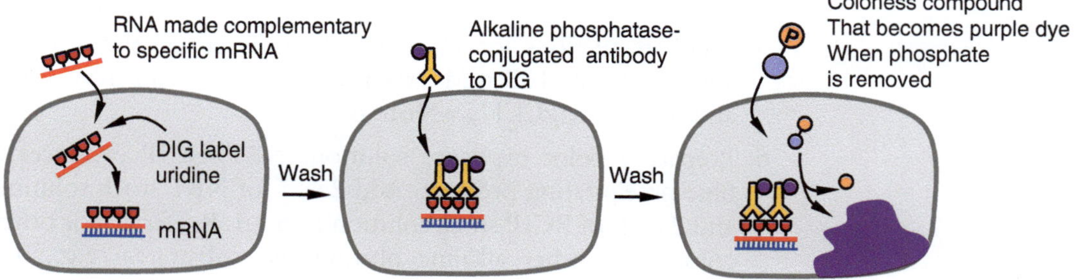

Fig. 2 Schematic of the procedure. A digoxigenin (DIG)-labeled antisense probe hybridizes to a specific mRNA. Alkaline phosphatase-conjugated antibodies to DIG recognize the DIG-labeled probe. The enzyme is able to convert a colorless compound into a dark purple precipitate

3. Cover samples with hydrophobic plastic cover slip (*see* **Note 5**).
4. Incubate sections at 42 °C overnight in a humid hybrid chamber.

3.3 Posthybridization

Be particularly careful with RNase. This enzyme is extremely stable and difficult to inactivate. Use a separate set of glassware for posthybridization and prehybridization procedures to avoid RNase contamination in prehybridization steps.

1. Remove coverslips from sections by immersing slides for 10 min in 4× SSC.
2. In a shaking water bath at 37 °C, wash sections for 2 × 15 min with 2× SSC.
3. In a shaking water bath at 37 °C, wash sections for 2 × 15 min with 1× SSC.
4. To digest any single-stranded (unbound) RNA probe, incubate sections for 30 min at 37 °C in RNase Buffer containing 20 μg/mL RNase A.
5. In a shaking water bath at 37 °C, wash sections for 2 × 30 min with 0.1 SSC. Or if sample has high background or nonspecific binding, try post-hybridization washings at 52 °C with 2× SSC containing 50 % formamide.

3.4 Immunological Detection

1. Using a shaking platform, wash sections for 2 × 10 min with Buffer 1.
2. Cover samples for 30 min with blocking solution (1 % blocking reagent for nucleic acid hybridization and detection in Buffer 1 T) in humid vertical slide mailers.
3. Decant blocking solution and incubate sections for overnight at 4 °C in a humid chamber with blocking solution, and a suitable dilution of alkaline phosphatase-conjugated anti-DIG antibody. For optimal detection, incubate several sections from the same sample with different dilutions of the antibody (Fig. 2b).

4. Using a shaker to wash sections for 2 × 10 min with Buffer 1.
5. Incubate sections for 10 min in humid vertical slide mailers with Buffer 2 [100 mM Tris–HCl (pH 9.5), 100 mM NaCl, and 50 mM $MgCl_2$] (*see* **Note 6**).
6. Prepare a color exposure solution. NBT/BCIP produces a blue precipitating product. Add 27 μL of NBT stock solution and 21 μL of BCIP stock solution to 6 mL Buffer 2. For other colors, use other alkaline phosphatase substrates (e.g., use INT/BCIP for brown precipitates).
7. Cover each sample slide with approximately 300 μL drops of NBT/BCIP color exposure solution for 30 min to 16 h (Fig. 2c) in a humid chamber (in the dark) until desired color has developed (depending on abundance of transcript).
8. When color development is optimal, stop the color reaction by incubating the slides in Stop reaction Buffer [10 mM Tris–HCl (pH 8.1), 1 mM EDTA].
9. Dip slides briefly in distilled water.
10. Counter-stain sections for 2–5 min with 0.1 % Nuclear Fast Red or 0.001 % Fast green FCF in ddH_2O.
11. Wash with ddH_2O for 10 min.
12. Mount sections using an aqueous mounting solution (Fig. 3). Do not use xylene-based mounting solutions. These lead to crystal formation of color precipitates.

3.5 Controls for In Situ Hybridizations

Adequate controls must always be included to ensure the particularity of detection signals. Controls must include negative and positive specimens as well as technical controls to detect positive and negative results [12].

3.5.1 Negative Controls

We suggest that ISH experiments should have at least two negative specimens to ensure negative results. (1) Hybridization with sense probe. (2) Digestion of mRNA with RNase prior to ISH. (3) Hybridization with irrelevant probe (e.g., probe for viral sequences). (4) Hybridization in the presence of excess unlabeled antisense probe.

3.5.2 Positive Controls

If using tissue or cell line known to contain mRNA of interest, those specimens are efficient for positive controls. Antisense RNA probes can be used to test quality of tissue mRNA for abundant "housekeeping genes."

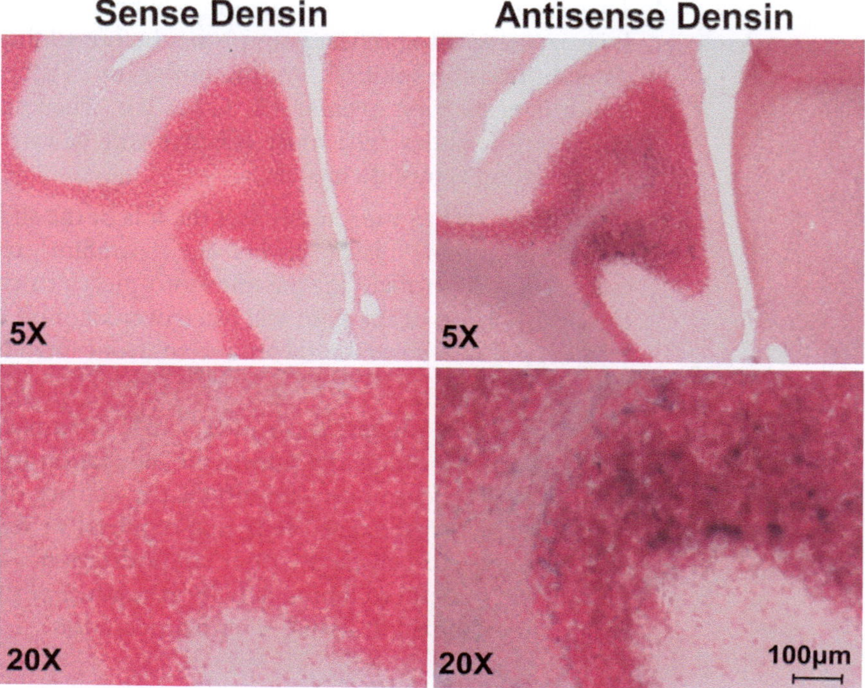

Fig. 3 Use of a DIG-labeled antisense RNA probe and NBT/BCIP detect formation of colored spots of Densin mRNA. In situ hybridization analysis of Densin expression in granular cell layer of wild-type cerebellum at 2 month. The experiments were repeated three times, and representative images are shown. Note that only antisense, but not sense, probes showed signals, demonstrating the specificity

4 Notes

1. Gelatin-coated slides are excellent for large tissue sections obtained from frozen specimens. Alternatively, silane-coated slides can be used, but are better suited for small tissue samples or cell preparations.
2. Keep your own RNase-free Proteinase K (PK) for in situ. Sometimes higher PK concentrations may be used. Deproteinization with PK makes the tissue more accessible to the probe. But 1 and 10 μg/mL are suggested for frozen sections in other literatures.
3. Acetylation of the positively charged amino groups in the proteins of the tissue decreases background binding of the negatively charged RNA probe. Some protocols suggest putting in the slides, then adding acetic anhydride, stirring until beading disappears, stopping stirring, and waiting for 10 min.

4. These steps should be performed in a hybridization chamber. It is a good idea to preheat your probes before adding them to the hybridization solution. Put the buffer into a tube and heat it in a hybridization oven at 42 °C for 1 h. Then add heated buffer to the slides. The most effective way to carry out the hybridization is in slide mailers.
5. It is a good idea to thoroughly seal the lids of the slide mailers with hydrophobic plastic covers (i.e., parafilm) to prevent evaporation of probe. Cut parafilm into pieces to the size of the clear portion of the slide and place over solution using forceps. Try to use coverslip technique to avoid introduction of bubbles.
6. Some literatures suggest to include 5 mM Levamisole in the Buffer 2 when washing and during the incubation with NBT/BCIP because Levamisole can inhibit the activity of endogenous alkaline phosphatase. It is not required in fetal tissues (rat, mouse, or chicken) after hybridization. However, Levamisole is useful to inhibit activity at residual of the endogenous enzyme in adult tissues after hybridization.

References

1. Mary LP, Gall JG (1969) Molecular hybridization of radioactive DNA to the DNA of cytological preparations. Proc Natl Acad Sci USA 64:600–604
2. John HA, Birnstiel ML et al (1969) RNA-DNA hybrids at the cytological level. Nature 223:582–587
3. Olivier B, Walter W (1998) A simplified in situ hybridization protocol using non-radioactively labeled probes to detect abundant and rare mRNAs on tissue sections. Biochemica 1:10–16
4. Singer RH, Ward DC (1982) Actin gene expression visualized in chicken muscle tissue culture by using in situ hybridization with a biotinated nucleotide analog. Proc Natl Acad Sci USA 79:7331–7335
5. Farquharson M, Harvie R, McNicol AM (1990) Detection of messenger RNA using a digoxigenin end-labeled oligonucleotide probe. J Clin Pathol 43:423–428
6. Morris RG, Arends MJ, Bishop PE et al (1990) Sensitivity of digoxigenin and biotin-labeled probes for detection of human papillomavirus by in situ hybridization. J Clin Pathol 43: 800–805
7. Science RA (2008) DIG application manual for nonradioactive in situ hybridization, 4th edn. Roche Diagnostics GmbH, Mannheim, Germany
8. Sugimoto N, Nakano S, Katoh M et al (1995) Thermodynamic parameters to predict stability of RNA/DNA hybrid duplexes. Biochemistry 34:11211–11216
9. Sambrook J, Fritsch E, Maniatis T (1989) Molecular cloning: a laboratory manual. Cold Spring Harbor Laboratory Press, Cold Spring Harbor, NY
10. Valentino KL, Eberwine JH, Barchas JD (1987) In situ hybridization: applications to neurobiology. Oxford University Press, Oxford
11. Naber S, Smith L, Wolfe HJ (1992) Role of the frozen tissue bank in molecular pathology. Diagn Mol Pathol 1:73–79
12. Herrington CS, McGee JO (1992) Principles and basic methodology of DNA/RNA detection by in situ hybridization. In: Herrington CS, McGee JO (eds) Diagnostic molecular pathology: a practical approach, vol 1. IRL, New York, pp 69–102

Chapter 17

Immunofluorescence Staining with Frozen Mouse or Chick Embryonic Tissue Sections

Hui Wang and Michael P. Matise

Abstract

Immunofluorescence (IF), a form of immunohistochemistry (IHC) with specific applications, is commonly used for both basic research and clinical studies, including diagnostics, and involves visualizing the cellular distribution of target molecules (e.g., proteins, DNA, and small molecules) using a microscope capable of exciting and detecting fluorochrome compounds that emit light at specific, largely nonoverlapping wavelengths. The procedure for carrying out IF varies according to the tissue type and methods for processing and preparing tissue (e.g., fixative used to preserve tissue morphology and antigenicity). The protocol presented here provides a general guideline for multichannel IF staining using frozen embryonic mouse or chicken tissue sectioned on a cryostat. In general, the procedure involves the following: (1) fixing freshly dissected tissues in a 4 % paraformaldehyde solution buffered in the physiological pH range, (2) cryopreservation of tissue in a 30 % sucrose solution, (3) embedding and sectioning tissue in Optimal Cutting Temperature (OCT) matrix compound, (4) direct or indirect detection of the target antigen/s using fluorochrome-conjugated antibodies.

Key words Immunofluorescence, Immunohistochemistry, Cryostat tissue section, Antigen, Antibody, Paraformaldehyde, OCT

1 Introduction

Immunofluorescence (IF) involves the detection of target antigens through binding of specific fluorochrome-conjugated primary or secondary antibodies (immunohistochemistry, or IHC) in situ, allowing visualization of the cellular localization with fluorescence microscopes equipped with epifluorescence. In many cases, direct labeling fluorescence assays (such as labeling DNA with DAPI, or detecting proteins containing fluorescent domains) can be combined with IF-based IHC. The basic principle behind successful IF is to optimize the specificity of binding of the antibody reagents to the target antigens with minimal background (i.e., nonspecific binding) signal. The primary antibody refers to the reagent that will bind directly to the target antigen of interest, which can then

Renping Zhou and Lin Mei (eds.), *Neural Development: Methods and Protocols*, Methods in Molecular Biology, vol. 1018,
DOI 10.1007/978-1-62703-444-9_17, © Springer Science+Business Media, LLC 2013

be visualized either directly or indirectly. The secondary antibody refers to the reagent whose antigen is the primary antibody, and is used for indirect IF and signal amplification.

1.1 Prepare Frozen/Cryostat Tissue Sections

Well-prepared tissues are a prerequisite for high-quality IF outcomes. For the preparation of frozen tissue sections, approximately five steps are necessary: tissue dissection, tissue fixation, antigen retrieval (optional), OCT embedding, and cryostat sectioning. Of these steps, tissue fixation is generally considered the most critical. For most histological and cytological studies, formaldehyde-based organic compound that cross-link proteins within the cell/tissue are used to preserve the overall tissue morphology through subsequent antibody processing stages as well as to immobilize the target antigen/s within the cell [1, 2]. However, such cross-linking fixatives also can produce changes that are detrimental to IHC in general, including limiting cell and nuclear membrane penetration that can negatively impact the trafficking of large molecules (like antibodies) during post-fixative processing, and the induction of conformational changes in the structure of proteins which might alter or mask certain epitopes (antigen determinants) on antigens [3]. Thus, although adequate fixation is required for good cytological preservation, overfixation can lead to "antigen masking" which can block or prevent antibodies from binding to the protein epitope/s to which they were generated. The procedure presented below is for 4 % w/v paraformaldehyde to preserve embryonic tissue.

Formaldehyde-based fixatives, including 4 % v/v formaldehyde/10 % v/v formalin and 3.7–4 % w/v paraformaldehyde (*see* **Note 1**), are widely used to stabilize histological/cytological morphology and proteins by cross-linking nearby basic amino acids (primarily the residues of lysine) with methylene ($-CH_2-$) bridges [4, 5]. Notably, freshly made 3.7–4 % w/v paraformaldehyde is superior to 4 % v/v formaldehyde/10 % v/v formalin, especially for immunostaining purposes, in that it contains pure formaldehyde without by-products and additives such as methanol and formic acid, both of which can influence antigen immunoreactivity and staining outcomes (*see* **Notes 1** and **2**). In general, for embryonic tissue fixation, dissected tissues should be immersed in the fixative solution immediately after collection, and the volume of fixative solution should be at least 20 times that of the tissue sample. To generate well-prepared tissue, several specific characters of paraformaldehyde-dependent fixation should be noted. First, the penetration of formaldehyde is relatively good (about 0.5 mm/h in tissue blocks; *see* **Note 3**) but fixed tissues present a low permeability environment. Certain detergents (like Triton X-100 or Tween-20) therefore are applied to facilitate antibody penetration. Second, it is necessary to minimize antigen masking for immunostaining purposes. Therefore, it is preferable to fix tissues at the minimum time necessary to prevent cytological autolysis and preserve the structural integrity.

To prevent uneven or over-fixation, tissues can be trimmed into small blocks in which the thickness in at least one dimension should be less than 4 mm.

Many primary antibodies, especially those raised against specific protein structural motifs (e.g., monoclonal antibodies) can be sensitive to formaldehyde fixation even with optimal procedures because the cross-linking induces protein conformational changes that can mask the target epitope/s. In such cases, antigen retrieval techniques might help to partially rescue the immunoreactivity. Heat-induced antigen retrieval (HIAR) is commonly used for aldehyde-fixed paraffin-embedded tissue sections [6, 7]. Although the mechanisms are not fully characterized, it has been demonstrated that heating can break methylene bridges to unmask epitopes [8, 9]. However, such procedures should be used with caution on fragile frozen tissue sections since they can negatively impact overall tissue morphology. Recently, several independent groups have attempted to address this limitation by developing SDS-dependent antigen retrieval methods for frozen tissue sections; these are not described in this chapter (*for details, see* [10]).

1.2 Immunostaining Frozen/Cyrostat Tissue Sections

The common reagents of all IHC procedures are antibodies, immunoglobulin (Ig) proteins that can bind to one unique (monoclonal antibodies) or distinct (polyclonal antibodies/antisera) epitopes on target antigens (*see* **Notes 4** and **5**). Antibody selection, preparation, and incubation are all critical for specific antigen detection, signal amplification, and background signal reduction. Briefly, the unique structural elements of the hypervariable (HV) regions, which are located in the N-terminal variable domains of the light and heavy chains (V_L and V_H) of antibodies, are essential to recognize specific antigens. Accordingly, antibodies with high specificity and affinity are preferred for various immuno-detection techniques. Generally, to achieve optimal immunostaining outcomes (high-quality specific staining with minimal background signal), lower titer is required for the antibody with higher affinity; with the same titer, and the antibody with higher affinity needs shorter incubation time to reach equilibrium.

Unlike enzyme-conjugated (e.g., peroxidase and phosphatase) antibodies utilized in conventional IHC, IF employs antibody molecules chemically linked to specific fluorochromes that can be detected with an epifluorescent microscope set up to excite/visualize a range of fluorochromes as needed. The basic principle behind epifluorescence is that different fluorochrome molecules absorb a specific range of wavelengths of photons (the excitation spectrum), followed by rapid emission of a specific range of wavelengths (emission/radiation spectrum). Thus, detection instruments should be equipped at minimum with both an epifluorescent illuminator with a set of narrow band-pass excitation filters and objective lenses with a similar matched set of emission filters.

One important application of the IF technique is for protein co-localization studies, since multiple target antigens can be labeled with different antibodies conjugated with fluorochromes that absorb/emit at distinct, nonoverlapping or minimally overlapping wavelengths (like Cy3 with ~570 nm excitation, Fluor 488 with ~499 nm excitation, and Fluor 350 with ~343 nm excitation; more information can be found at http://flowcyt.salk.edu/fluo.html). Dual- or multi-labeled tissue can be visualized using a properly equipped fluorescent microscope/digital imaging system capable of exciting/detecting the different channels and the raw image data is processed through commercially available software packages to determine co-localization.

One of the major problems with the IF technique is irreversible photobleaching, a dynamic process of photo-induced fluorophore destruction. Although the mechanism is still unclear, it is likely caused by photo-induced oxygen-dependent or -independent redox reaction between proximal excited triplet dyes (Dye-to-Dye mechanism) or between excited triplet dye and singlet oxygen molecules (Dye-to-Oxygen mechanism) [11]. Two methods commonly used to attenuate photobleaching are (1) reducing intensity and duration of light exposure, and (2) mounting tissues with anti-fading coverslipping medium containing singlet oxygen scavengers. Overall, for IF staining, an optimal fluorochrome dye should have at least three properties: (1) bright luminescence, (2) narrow excitation and emission spectrum, and (3) exceptional photostability.

IF staining methods can be divided into two types: direct detection and indirect detection. The former uses fluorochrome-conjugated antibodies to directly detect target antigens. This approach has the general advantage of being quick and having relatively low background signals due to nonspecific binding of the primary antibody to nontarget antigens that may be structurally similar. One of the major disadvantages of direct IF is low sensitivity, since the amount of primary bound in tissue sections is limited by the abundance of the antigen. Indirect IF methods employ an additional step using secondary antibodies raised against the specific Ig subclass of the primary antibody. In this approach, many fluorochrome-conjugated secondary Ab molecules can bind to the primary and thereby amplify the signal (*see* **Note 6**). One of the major benefits of using indirect IF is that it provides a wider choice of fluorochrome used with the secondary antibody (i.e., its emission spectrum), allowing greater flexibility in double- or multi-labeling experiments involving multiple distinct detection channels. However, to avoid cross-reactivity in multiple-labeling, it is necessary to select primary antibodies raised from distinct species. It should be noted that the use of an Ig-directed secondary antibody increases the likelihood of nonspecific binding to endogenous Ig-like antigens in situ, creating a higher background signal than direct methods. Furthermore, secondary antibodies cannot distinguish between endogenous Ig proteins and the primary antibody if the primary antibody is generated from a host animal of the same species as the tissue being stained. For example, anti-mouse secondary antibodies can detect both endogenous antibodies and the experimental primary antibody in mouse embryonic tissues older than E12.5, resulting in nonspecific staining.

Therefore, one of the important goals in indirect IF experiments is to reduce the background signal. Hydrophobic and ionic interactions are primary contributors not only for the specific antibody–antigen recognition but for background signals as well. Hydrophobicity is shared, more or less, by most proteins. Particularly, formaldehyde-based fixation augments the hydrophobicity of tissue proteins [12]. Thus, in addition to specific signal reduction, over fixation also causes the enhancement of background signals. To reduce the nonspecific signals caused by hydrophobic interactions, several methods are commonly used in immunostaining: (1) blocking proteins, like normal animal sera and BSA, are utilized prior to or during antibody incubation (*see* **Note 7**); (2) several nonionic detergents (e.g., Tween-20 and Triton X-100) are added to blocking solutions, antibody diluents, and even washing buffers (*see* **Note 8**). In aqueous media, hydrophobic interactions are generally stronger with higher ionic strength. However, high ionic strength reduces nonspecific signals due to ionic interactions. The dilemma is that conditions reducing one kind of nonspecific signals might enhance the other. In general, phosphate buffer (PB) in physiological pH range is widely used throughout the procedure (including fixative preparation, sucrose infiltration, washing buffer, and polyclonal antibody diluents; *see* **Note 9**) because this ionic buffer solution provides ideal aqueous media to reduce nonspecific background caused by hydrophobic and ionic interactions, and many other reasons (like autofluorescence *see* **Note 10**). But this solution is not recommended as a monoclonal antibody diluent (*see* **Notes 5, 11** and **12**). Moreover, all Ig proteins contain one C-terminal crystalline fragment (Fc) which can be recognized by distinct classes of Fc receptors (FcRs). To this end, it is theorized that background signals might be elicited from the nonspecific interactions between antibody molecules and endogenous FcRs as well [13, 14]. It is also thought that pre-incubation with blocking proteins (*see* **Note 7**) could reduce this nonspecific signal. However, this idea has been challenged by recent data suggesting that with fixation in formaldehyde, acetone, or alcohol, endogenous FcRs could no longer interact with the Fc fragment of antibodies [15].

In summary, there is no universal optimal protocol for tissue processing and immunostaining in that desirable conditions will vary depending on the specific tissues and/or antibodies being used. To maximize the signal-to-noise ratio, it is thus necessary to optimize staining conditions for each individual experiment through a series of pilot studies where some of the key variables can be tested.

2 Materials

2.1 Animals

Specific pathogen free (spf) fertilized White Leghorn chicken eggs (Aichi Line) and experimental mice are available from suppliers such as Charles River (MA, USA).

2.2 Reagents

Distilled deionized Water (ddH_2O; *see* **Note 13**) Commercial Cell Culture Grade Water is available from suppliers such as Cellgro (VA, USA).

Dulbecco's Phosphate-Buffered Saline 1× (DPBS; *see* **Note 9**; [16]). Commercial DPBS is available from suppliers such as Cellgro (VA, USA). Formulation for self-made (g/L): $CaCl_2$ (anhydrous) 0.1, KCl 0.2, KH_2PO_4 0.2, $MgCl_2 \cdot 6H_2O$ 0.1, NaCl 8, $Na_2HPO_4 \cdot 7H_2O$ 2.1716.

0.05 M Tris Buffer in physiological pH range (7.0–7.6), diluted from commercial concentrated Tris Buffer (available from suppliers such as Teknova, CA, USA) (*see* **Notes 5, 11** and **12**).

20 % w/v paraformaldehyde (pH 7.4) in ddH_2O: (1) Weigh 20 g paraformaldehyde powder and transfer to a Wheaton glass bottle with screw cap. (2) Add 80 mL ddH_2O and 75 μL 10 N NaOH. (3) Incubate within a 60 °C water bath, stirring occasionally until powder dissolves into solution. (4) Adjust final volume up to 100 mL and pH value to 7.4. (5) Filter through 0.22 μm filters and dispense into 10×10 mL aliquots with 50 mL Falcon tubes. (6) Store at −20 °C (can be kept for a couple of weeks without significant loss of potency).

4 % w/v paraformaldehyde in DPBS (buffered in the physiological pH range): (1) Thaw 20 % paraformaldehyde in 60 °C water bath stirring (occasionally by hand). (2) Add 40 mL DPBS into each tube (make up to 50 mL final volume). (3) Keep on ice up to 1 day.

30 % w/v sucrose in DPBS (buffered in the physiological pH range): freshly made and stored on ice.

Optimal Cutting Temperature compound (OCT) is available from suppliers such as SAKURA (OH, USA).

Others: unconjugated primary antibodies and fluorochrome-conjugated secondary antibodies (for important recommendations on antibody storage and preparation, *see* **Note 14**), Fluoro-Gel (Electron Microscopy Sciences, PA, USA; *see* **Note 15**), Hoescht/DAPI/TOPRO nucleic acid/DNA dyes (to visualize cell nuclei) (*see* **Note 16**), normal animal serum, and Triton X-100.

2.3 Others

Superfrost Plus and ColorFrost Plus Microscope Slides (Fisher Scientific, PA, USA), Microscope Cover Glass (Fisher Scientific, PA, USA), 0.22 μm vacuum filter system (Corning, NY, USA), a Wheaton glass bottle with screw cap, disposable embedding molds, Falcon tubes, 14 mL polypropylene round-bottom tubes, Coplin jars, slide boxes, slide holders, eppendorf tubes, transfer pipettes, petri dishes, and syringes.

3 Methods

3.1 Prepare Frozen/Cyrostat Tissue Sections

1. Prepare 4 % w/v paraformaldehyde and 30 % w/v sucrose with DPBS and store them on ice.
2. Collect whole embryos from the fertilized eggs or the uterus of a pregnant mouse, and wash them once with prechilled DPBS in a large petri dish (*see* **Note 17**).
3. Transfer whole embryos into a clean petri dish containing pre-chilled DPBS. Clean up the amnion and yolk sac, and dissect out the desired tissues under a stereomicroscope (*see* **Note 18**).
4. Immerse tissue blocks with prechilled 4 % paraformaldehyde/DPBS solution in 14 mL polypropylene round-bottom tubes on ice (*see* **Note 19**). Fixation time will vary depending on the age/size of the tissue blocks (*see* **Note 20**).
5. Transfer used paraformaldehyde into a hazardous waste disposal bottle (*see* **Note 21**). Wash tissue blocks quickly with one change of prechilled DPBS to stop fixation reaction, and then wash three times for 15 min each in prechilled DPBS on ice.
6. Equilibrate tissue blocks with prechilled 30 % w/v sucrose DPBS solution on ice either overnight or until tissue sinks to the bottom of tube (*see* **Note 22**).
7. Infiltrate tissue blocks with OCT in a small petri dish on ice about 1 h.
8. Fill the labeled embedding mold with OCT and transfer tissue blocks into it using curved forceps or plastic transfer pipettes cut to an appropriate size to avoid damaging tissue. Pipette air bubbles out carefully or allow to rest undisturbed until bubbles disappear. Gently and slowly adjust the position of each tissue block on the bottom of mold with a clean syringe.
9. Freeze embedding mold in an ice bucket filled with pulverized dry ice (*see* **Note 23**). Frozen tissue blocks can be wrapped with parafilm and keep in −20 or −80 °C for long-term storage.
10. Section the frozen OCT block using a cryotome cryostat (e.g., HM505E MICROM/Thermo Scientific, Walldorf, Germany), typically at 10–16 μm. Cold sections will readily stick onto the room temperature Superfrost Plus microscope slides.
11. Completely air dry sections in room temperature (for at least 30 min). Dried sections are ready for immunostaining or can be stored at −20 °C in a slide box sealed with parafilm.

3.2 Immunostaining of Frozen/Cyrostat Tissue Sections

1. If slides were stored at −20 °C, warm slides in room temperature for half an hour prior to proceeding.
2. Wash slides with DPBS three times for 10 min each in a Coplin jar (*see* **Note 24**).
3. Add 150–200 μL/slide of primary antibody/s suspended in antibody dilution buffer (generally, for polyclonal antibody using DPBST: DPBS with 4 % Triton X-100 and 1 % normal animal serum or BSA, and for monoclonal antibody using 0.05 M Tris buffer with 4 % Triton X-100 and 1 % normal animal serum or BSA) (*see* **Notes 11, 25** and **26**).
4. Cover each slide with a parafilm coverslip trimmed to fit over sections and incubate in a sealed humidified box at 4 °C for overnight (*see* **Notes 27** and **28**).
5. Take humid box back to room temperature. Float off parafilm with DPBS and place slides in a Coplin jar. Wash with DPBS three times for 10 min each. Protect slides from overhead/ambient light in the following steps.
6. Add 150–200 μL/slide of secondary antibody/s in antibody dilution buffer: DPBST (DPBS with 4 % Triton X-100 and 1 % normal animal serum or BSA). Cover each slide with a parafilm coverslip and incubate in a humid box at room temperature for 1 h. (To protect from light, cover the humid box with foil or place it in a drawer.)
7. Wash slides in a Coplin jar with DPBS three times for 10 min each. (Cover the Coplin jar with foil or place it in a drawer.)
8. (Optional step, only required to label cell nuclei): Dilute Hoescht dye at 1:10,000 with DPBS and incubate slides in a Coplin jar for 5–10 min.
9. Apply several drops of Fluoro-Gel on one side of slide. Carefully cover the slides with microscope cover glasses from the Fluoro-Gel applied side to the other (*see* **Note 29**).
10. Dry slides for 30 min in a drawer or a slide holder at room temperature. Slides are ready to be observed under epifluorescent or confocal microscope, or can be stored in a slide box at 4 °C for a couple of months.

4 Notes

1. Even through 10 % v/v formalin and 3.7–4 % w/v paraformaldehyde are both formaldehyde-containing solutions, they could be quite different in ingredient and usage. Formaldehyde (HCHO) is a gas that can be saturated in water at 37 % w/v or 40 % v/v. The saturated formaldehyde solution is also called 100 % formalin or strong formalin. Thus, 10 % formalin is actually 4 % v/v (3.7 % w/v) formaldehyde. Two chemical

reactions, namely the polymerization and Cannizzaro reaction of formaldehyde, slowly happen in formaldehyde/formalin solutions. Specifically, these reactions produce white polymerized precipitates, methanol, and formic acid that therefore coexist in aged formaldehyde/formalin solutions. Moreover, unless specific labeled, most manufactures add methanol deliberately to inhibit these reactions. In contrast, paraformaldehyde solutions, freshly made from polymerized formaldehyde white powder, contain pure formaldehyde. Thus, the freshly made paraformaldehyde solution is actually the "methanol-free" or "ultrapure" formaldehyde solution.

2. Instead of cross-linking, methanol will clump and precipitate proteins. Because of the inconvenience of preparing paraformaldehyde solutions, another option for short-time use is to purchase methanol-free formaldehyde.
3. Formaldehyde is small molecule; therefore its penetration is good, which however is not applicable for tissues covered with dense membrane capsules, like kidney.
4. Antigen–antibody interaction is similar to substrate–enzyme interactions that involve non-covalent binding.
5. *Polyclonal* antibodies are produced from the immune response elicited in antigen injected animals. Specifically, they are generated by diverse plasma cells and can target various epitopes of the given antigen, i.e., they are a group of antibodies with different specificities and affinities. Accordingly, polyclonal antibodies have a good tolerance for tissue fixation and other immunostaining conditions (like antibody titration/dilution, incubation time, temperature, and pH), while they have obvious disadvantages for immunohistochemical purpose including batch-to-batch variability and higher chance for immunochemical cross-reactivity (common or similar epitope/s shared by different antigens). Furthermore, polyclonal antibodies are produced in animals and each set of host animals can produce slightly different immune responses; thus, polyclonal antibodies are typically a more scarce resource. In contrast, *monoclonal* antibodies are secreted by cultured mouse (or rabbit) hybridoma cells that are derived from single cell clones. Thus, they are homogeneous, immunochemical identical, and can be produced as long as the parent cell line is maintained. To this end, it is imperative to screen high-quality plasmocyte clones that can give rise to monoclonal antibodies with high affinity for specific epitope recognition without cross-reactivity. There are two general limitations for using monoclonal antibodies: (1) when using formaldehyde-based fixatives, optimal fixation is mandatory to prevent alteration or masking of target epitope by cross-linking; and (2) they are generally more sensitive to specific immunostaining conditions due to their molecular

homogeneity, especially for pH value and the presence of certain ions, e.g., it has been systemically determined that the presence of Na^+ ions in antibody diluents affect the immunoreactivity of many, possibly most, monoclonal antibodies [17, 18].

6. Since the primary antibody can only detect unique (monoclonal antibodies) or a limited number of epitopes (polyclonal antibodies) on the target antigen, the direct method can result in low levels of specific signal. In contrast, a secondary antibody can recognize a larger number of epitopes (on the primary antibody) thus amplifying the specific signals dramatically. Sometimes two rounds of signal amplifications can be carried out, i.e., after incubating with fluorochrome-linked secondary antibody, another fluorochrome-linked antibody is used for further signal amplification by interacting against the secondary antibody. The three-step indirect method should only be employed in cases where amount of functional epitopes are very low since it can increase the level of background signal. The protocol presented here only provides the details of secondary indirect method.
7. Blocking proteins are commonly used to reduce background signals by competitively inhibiting nonspecific hydrophobic interactions between antibody and tissue proteins. In general, if using normal animal sera as blocking proteins, the animal species of serum should match that of secondary antibody, e.g., normal goat serum is preferred when the secondary antibody is generated in goat. However, most recent findings suggested that this routine protein blocking step is dispensable for immunostaining properly fixed cell and tissue samples [15].
8. Both Tween-20 and Triton X-100 work for IF staining with the very subtle difference that the aromatic ring of Triton has some absorption under UV light. Some protocols call for the use of Triton X-100 in blocking solutions and antibody diluents and Tween-20 only for washing buffers. In our experiences, these differences do not seem to have an impact on the outcome of IF experiments.
9. DPBS maintains cell tonicity by buffering reagents in the physiological pH range (from 7.0 to 7.6). Ionic strength provided by DPBS also helps to reduce background signals caused by nonspecific ionic interactions.
10. Some cellular organelles, such as mitochondria and lysosomes, could produce autofluorescence. This luminescence is aroused mostly from endogenous fluorochromes, such as flavin coenzymes (FAD and FMN) and reduced pyridine nucleotides (NADH and NADPH) [19]. If these interfere with the detection of fluorescent signals of interest (i.e., if the target signal is weak), autofluorescence should be minimized. For this, fixed

samples can be washed in a PBS solution containing 0.1 % sodium borohydride ($NaBH_4$).

11. DPBS is still widely used for tissue preparation and immunostaining. However, it has been determined that the presence of Na^+ ions in this solution can impair the immunoreactivity of many of monoclonal antibodies (*see* **Note 5**). For this reason, it is recommended to use 0.05 M Tris buffer as diluents of primary monoclonal antibodies.
12. Given that the isoelectric points (pI) of individual monoclonal antibodies are quite different from one another, there is no common pH value that is favorable for all monoclonal antibodies. In general, buffers near physiological pH value are commonly used as antibody diluents.
13. ddH_2O is used to prepare all solutions which are not purchased from commercial suppliers.
14. It is important to store and prepare antibodies appropriately. Upon receipt, keep the antibodies at manufactory recommended temperature immediately. Generally, highly concentrated antibodies, which need to be diluted at a ratio higher than 1:200, should be dispensed into small volume aliquots and store at −20 °C (or even lower) to avoid repetitive freezing and thawing. It is also recommended to make a prediluted aliquot for each highly concentrated antibody to a 1:100 ready-to-use ratio (e.g., for an antibody which is recommended to be used at 1:6,000 ratio, to make its prediluted aliquot is to dilute it at 1:60 with water, which is then ready to be used at 1:100), and store the ready-to-use aliquot at 4 °C. Next, keep all thawed antibodies on ice when in use. Notably, all fluorochrome-conjugated antibodies need to be protected from light by storing them in foil-covered tubes in opaque boxes. Moreover, antibodies that have been stored for over 1 year may contain insoluble aggregates and/or polymerized precipitates. These aggregates and polymers have diminished immunoreactivity but can dramatically increase background signals by nonspecific hydrophobic interacting with tissue proteins. These aggregates can be removed prior to staining by centrifugation at top speed for at least 10 min at 4 °C.
15. Fluoro-Gel is a water-based anti-fading mounting medium.
16. Hoechst (~480 nm emission) and TOPRO (~675 nm emission) dyes can be used to label nucleic acids/nuclei. They may be preferable to DAPI due to their less-toxic and better penetration characteristics.
17. Prechilling DPBS is required for good tissue preservation.
18. In this example, we will dissect the developing spinal cords of mouse or chick embryos. From stage E8.0 (E1.5 for chick) to

Table 1
Recommended fixation time for tissue blocks in different ages/sizes

Stage	Tissue processing	Recommended fixation time
E8.0–E9.5	Decapitation	No more than 40 min
E10.5	Decapitation and evisceration	40 min to 1 h
E11.5	Decapitation and evisceration	1–1.5 h
E12.5	Decapitation and evisceration	1.5–2 h
E13.5–E18.5	Dissecting spinal cord out	2 h

E10.0 (E2.0 for chick), we first collect the whole embryos and cut off the head at the cervical level rostral to the forelimb buds. From stage E10.5 (E2.5 for chick) to stage E12.5 (E3.5 for chick), embryos are also eviscerated. From stage E13.5 (E4.0 for chick) to E18.5 (E20.5 for chick), we will dissect the spinal cord out of the embryo for even fixation.

19. Compared to room temperature, the rates of penetration and cross-linking for formaldehyde are slower on ice. Thus, pre-chilled paraformaldehyde solution serves for better tissue preservation and optimal fixation.
20. The penetration of formaldehyde solution into tissue is approximately 0.5 mm/h. To calculate the time required for fixative to fully penetrate tissue, divide 0.5 mm/h by the thickness of tissue. For spinal cord tissues prior to E13.5, the lumen of the transected neural tube allows for the free flow of fixative. Therefore, immersion in fixative will allow for penetration of fixative into the tissue from both the lumen side and the outside of the spinal cord. In this case, take the calculated time for full penetration above and divide by two. The recommended fixation time for tissue blocks in different ages/sizes is listed in Table 1. We recommend changing to fresh paraformaldehyde solution after the first hour of fixation for longer times.
21. Formaldehyde solution is dangerous and highly toxic. Inappropriate manipulation can cause unexpected damage to both humans and the environment. Please follow the standard regulations of United States Occupational Safety and Health Administration (OSHA) stringently.
22. Sucrose solution functions as cryoprotectant that is used at 15–30 % for immunostaining purpose. Some protocols suggested using a two-step infiltration process, namely sink in 15 % sucrose solution first and then sink in 30 %, instead of the one-step 30 % infiltration for embryonic tissues. However, based on our experience, there is no obvious difference.

23. Slowly and carefully place the embedding mold into the crushed dry ice without disturbing the position/orientations of tissue blocks. Alternative methods for freezing tissue blocks include using liquid nitrogen, precooled isopentane with liquid nitrogen, or a freeze medium mixed with dry ice and methanol/ethanol. OCT tissue blocks should be wrapped in parafilm to avoid drying out in the freezer.
24. Pour DPBS out carefully along the side of Coplin jar to avoid damaging tissue sections. Once washed by DPBS, do not let the sections dry through the remainder of the procedure.
25. Triton X-100 functions by both increasing tissue permeability and reducing background signals (*see* **Note 8**).
26. As using normal animal serum as blocking proteins, the animal species should be compatible with that of the secondary antibody that will be used (*see* **Note 7**).
27. A humid environment is used to prevent evaporation of the antibody staining solution and drying out of tissue sections. To set up a humid box, place several pieces of damp paper towels (soaked in DPBS) on the bottom, and set a suitable holder (e.g., an inverted tube rack) on the wet paper towels and place slides on the top of the holder.
28. Primary antibody incubation could be keep at 4 °C up to 48 h. An alternative method for primary antibody incubation is to incubate slides in a humid box at room temperature for a couple of hours.
29. Repel DPBS away and avoid generating air bubbles.

References

1. Feldman MY (1973) Reactions of nucleic acids and nucleoproteins with formaldehyde. Prog Nucleic Acid Res Mol Biol 13:1–49
2. Fraenkel-Conrat H, Olcott HS (1948) J Am Chem Soc 70(8):2673–2684
3. Battifora H, Kopinski M (1986) The influence of protease digestion and duration of fixation on the immunostaining of keratins. A comparison of formalin and ethanol fixation. J Histochem Cytochem 34(8):1095–1100
4. Fox CH, Johnson FB, Whiting J, Roller PP (1985) Formaldehyde fixation. J Histochem Cytochem 33(8):845–853
5. Puchtler H, Meloan SN (1985) On the chemistry of formaldehyde fixation and its effects on immunohistochemical reactions. Histochemistry 82(3):201–204
6. Krenacs L, Krenacs T, Stelkovics E, Raffeld M (2010) Heat-induced antigen retrieval for immunohistochemical reactions in routinely processed paraffin sections. Methods Mol Biol 588:103–119
7. Shi SR, Key ME, Kalra KL (1991) Antigen retrieval in formalin-fixed, paraffin-embedded tissues: an enhancement method for immunohistochemical staining based on microwave oven heating of tissue sections. J Histochem Cytochem 39(6):741–748
8. Yamashita S (2007) Heat-induced antigen retrieval: mechanisms and application to histochemistry. Prog Histochem Cytochem 41(3):141–200
9. Yamashita S, Okada Y (2005) Mechanisms of heat-induced antigen retrieval: analyses in vitro employing SDS-PAGE and immunohistochemistry. J Histochem Cytochem 53(1):13–21
10. Brown D, Lydon J, McLaughlin M, Stuart-Tilley A, Tyszkowski R, Alper S (1996) Antigen retrieval in cryostat tissue sections and cultured cells by treatment with sodium dodecyl sulfate (SDS). Histochem Cell Biol 105(4):261–267

11. Song L, Hennink EJ, Young IT, Tanke HJ (1995) Photobleaching kinetics of fluorescein in quantitative fluorescence microscopy. Biophys J 68(6):2588–2600
12. Ramos-Vara JA (2005) Technical aspects of immunohistochemistry. Vet Pathol 42(4): 405–426
13. Bussolati G, Leonardo E (2008) Technical pitfalls potentially affecting diagnoses in immunohistochemistry. J Clin Pathol 61(11): 1184–1192
14. Daneshtalab N, Dore JJ, Smeda JS (2010) Troubleshooting tissue specificity and antibody selection: procedures in immunohistochemical studies. J Pharmacol Toxicol Methods 61(2): 127–135
15. Buchwalow I, Samoilova V, Boecker W, Tiemann M (2011) Non-specific binding of antibodies in immunohistochemistry: fallacies and facts. Sci Rep 1:28
16. Dulbecco R, Vogt M (1954) Plaque formation and isolation of pure lines with poliomyelitis viruses. J Exp Med 99(2):167–182
17. Boenisch T (1999) Diluent buffer ions and pH: their influence on the performance of monoclonal antibodies in immunohistochemistry. Appl Immunohistochem Mol Morphol 7(4):300–306
18. Boenisch T (2001) Formalin-fixed and heat-retrieved tissue antigens: a comparison of their immunoreactivity in experimental antibody diluents. Appl Immunohistochem Mol Morphol 9(2):176–179
19. Monici M (2005) Cell and tissue autofluorescence research and diagnostic applications. Biotechnol Annu Rev 11:227–256

Chapter 18

β-Galactosidase Staining of LacZ Fusion Proteins in Whole Tissue Preparations

Margaret A. Cooper and Renping Zhou

Abstract

The lacZ gene product, β-galactosidase, has classically been used as a reporter of gene expression. β-Galactosidase activity can be detected using a chromogenic substrate, X-gal, which leaves an intense blue precipitate when cleaved by the enzyme. Insertion of the lacZ coding DNA targeted into a specific gene creates a β-galactosidase-tagged fusion protein that is expressed under the endogenous promoter. Analysis of the hybrid protein takes advantage of the chromogenic detection system, as the distribution and relative abundance of the expressed protein can be efficiently visualized.

Key words β-Galactosidase, LacZ, X-gal, Embryonic development, Histochemistry, Cryosectioning, Recombinant protein, Protein localization

1 Introduction

The use of the bacterial lacZ gene, which encodes β-galactosidase, is a well-established method in molecular biology for reporting gene expression [1, 2]. By constructing a targeted insertion of the lacZ gene, in frame, at the genetic locus for a protein of interest [3–5], the intact promoter of the altered gene will drive the expression of the resulting hybrid protein containing β-galactosidase [3, 6–8]. The expression pattern and intensity of β-galactosidase can then be used as a readout for the transcription of the targeted gene.

The LacZ construct can be designed to specifically interrogate the regulation and function of the targeted protein. LacZ can be inserted to replace an entire protein coding region [9], to replace a single domain [4, 8], or to express alongside the wild type protein without disrupting function [10]. With the endogenous promoter intact, β-galactosidase staining reflects the normal level and pattern of transcription in a cell but can also report subcellular localization if the fusion protein retains its localization properties [3, 6, 7].

Renping Zhou and Lin Mei (eds.), *Neural Development: Methods and Protocols*, Methods in Molecular Biology, vol. 1018, DOI 10.1007/978-1-62703-444-9_18, © Springer Science+Business Media, LLC 2013

β-Galactosidase staining is particularly useful in embryos for viewing the tissue specificity and timing of protein translation of developmentally regulated genes. Embryos or small pieces of tissue expressing the hybrid protein are first lightly fixed and then permeabilized to allow a chromogenic substrate to penetrate the tissue. Staining with 5-bromo-4-chloro-indolyl-β-d-galactopyranoside (X-gal), a substrate analogue synthesized for use in 1964 [11], produces a colorless indoxyl monomer when cleaved by the β-galactosidase enzyme [12]. The ferric and ferrous ions in the development buffer act as electron acceptors to facilitate the dimerization and oxidation reactions to produce the stable, and water insoluble, blue precipitate [13, 14]. The stained tissue can be imaged whole with a dissecting scope or sectioned onto slides to reveal staining in cross section.

β-Galactosidase staining of whole tissue is a versatile method to observe the expression of a protein of interest [6, 15] (Fig. 1). In whole tissues, differential staining patterns and intensities can be conveniently observed across the tissue. Similarly, staining differences can also be compared across developmental stages to analyze temporal changes in protein expression. Unlike in situ hybridization studies, which detect mRNA expression, β-galactosidase staining detects the localization of the translated protein, which does not always coincide spatially or temporally with mRNA [16]. Protein localization can also be visualized by immunostaining; however, the specificity and availability of antibodies can limit its potential for use. β-Galactosidase staining is very sensitive and has an advantage for being directly visualized in the tissue and long-lasting.

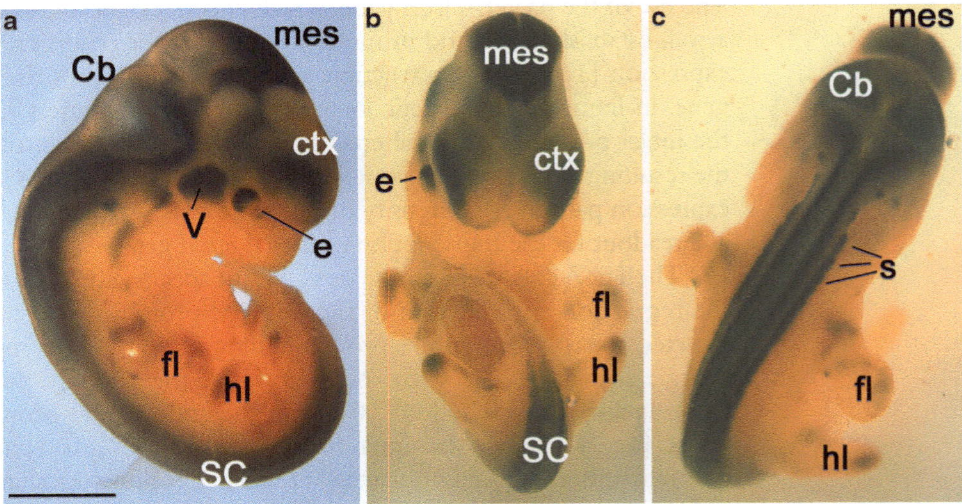

Fig. 1 Whole tissue staining of an E13 embryo. (**A**) Side view of the embryo expressing an EphA5 receptor:LacZ fusion protein. (**B**) Frontal view. (**C**) Dorsal view. Note that the β-galactosidase detection is highly discrete. Expression gradients are detectable along the spinal cord and in the eye. *Cb* cerebellum, *ctx* cortex, *e* eye, *fl* forelimb, *hl* hindlimb, *mes* mesencephalon, *s* somites, *SC* spinal cord, *V* fifth cranial nerve. Scale bar = 1 mm

2 Materials

Solutions should be prepared with deionized water and stored at room temperature unless otherwise specified. Refer to extra details provided in the notes when indicated.

2.1 β-Galactosidase Development Solutions

1. Phosphate buffered saline (PBS, 10×): Weigh 80 g of NaCl, 2.0 g of KCl, 14.4 g of Na_2HPO_4, and 2.4 g of KH_2PO_4. Place solutes in a 1 L bottle with a stir bar and add 800 mL of water. Adjust the pH to 7.4. Transfer to a graduated cylinder and adjust volume to 1 L with water. Transfer back to the bottle and sterilize by autoclaving.
2. Paraformaldehyde: 4 % solution in 1× PBS (*see* **Note 1**). Weigh 8 g of paraformaldehyde and transfer to a 250 mL bottle. Add one pellet of sodium hydroxide to bring up the pH and help dissolve the solute. Add 170 mL of water and 20 mL of 10× PBS. Place the bottle on a heating block with a stir bar and thermometer. Bring the temperature up to 60 °C. Adjust the pH of the fixative to approximately 7.4 with HCl using pH indicator strips or a pH meter. Transfer the solution to a graduated cylinder and bring the volume up to 200 mL with water. Place a fluted filter paper circle into a funnel and filter the paraformaldehyde back into the bottle. Store at 4 °C up to 1 week.
3. Glutaraldehyde: Commercially available as a 25 % solution (Sigma-Aldrich, MO). Store at 4 °C.
4. Fixation buffer: 2 % paraformaldehyde/0.5 % glutaraldehyde in 1× PBS. Store at 4 °C (*see* **Note 1**).
5. 1 M Magnesium chloride ($MgCl_2$) solution in water.
6. 1 % Sodium deoxycholate (Sigma-Aldrich, MO) solution in water.
7. 2 % Nonidet P-40 (NP-40) (Sigma-Aldrich, MO) solution in water (*see* **Note 2**).
8. Potassium *ferri*cyanide ($K_3Fe(CN)_6$, (J. T. Baker, NJ), 0.5 M): Weigh 8.23 g of solute (dark red crystals) and transfer to a 50 mL conical tube. Dissolve in water up to 50 mL. Wrap tube with foil and protect from the light.
9. Potassium *ferro*cyanide ($K_4Fe(CN)_6$, (J. T. Baker, NJ), 0.5 M): Weigh 9.21 g of solute and transfer to a 50 mL conical tube. Please note that its molecular weight is different from that of potassium ferricyanide, so read and label tubes carefully. Dissolve solute (light yellow crystals) in water up to 50 mL. Wrap tube with foil and protect from the light. The solution may be stored for up to 1 year.

10. X-gal stock: Dissolve 40 mg of X-gal in 1 mL of dimethylformamide (DMF) (*see* **Note 3**). The solution may be stored for up to 1 year.
11. Wash buffer: 0.02 % NP-40 in 1× PBS.
12. Equilibration buffer: 5 mM $K_3Fe(CN)_6$, 5 mM $K_4Fe(CN)_6$, 2 mM $MgCl_2$, 0.01 % sodium deoxycholate, 0.02 % NP-40, in 1× PBS (*see* **Note 4**). Pre-warm the solution in a 37 °C water bath.
13. Development buffer: Equilibration buffer with 1 mg/mL X-gal (*see* **Notes 4** and **5**). Pre-warm the solution in a 37 °C water bath.

2.2 Surgical and Staining Equipment

1. Surgical tools: one pair of large scissors and two forceps.
2. Dulbecco's Modified Eagle Medium (DMEM) (Invitrogen, CA): Store at 4 °C.
3. Sterile 100 mm Petri dishes (Corning, NY).
4. Tray with ice, big enough to hold two or three, 100 mm Petri dishes.
5. 50 mL polypropylene conical centrifuge tubes.
6. Water bath set to 37 °C.
7. Sodium azide (Sigma-Aldrich, MO).

2.3 Tissue Embedding and Sectioning Equipment

1. A graded series of ethanol dilutions in water: 50, 70, 80, 90, 95, 100 %.
2. Xylene (Reagent grade, J.T. Baker, NJ).
3. Paraffin (Tissue-Prep 2) or Paraplast (Fisher Scientific, NJ): Melted in a glass beaker covered with foil. Store in an oven heated to 58 °C.
4. Embedding molds (Sakura, Japan).
5. Plus-charged glass slides (VWR, PA).
6. Dry ice, powdered.
7. Tissue-Tek O.C.T. Freezing Compound (Sakura, Japan).

3 Methods

3.1 Whole Embryo or Tissue Staining

1. Dissect embryonic day (E)9 up to E15 mouse embryos or small pieces of tissue (*see* **Note 6**). Timed pregnant dams are terminally anesthetized before dissection when embryos are ready. A surgical incision is first made horizontally across the lower abdomen and a second incision is made vertically up the midline with scissors. Remove the uterus sacs with forceps and

place in a Petri dish on ice filled with cold 1× DMEM. At one pole of each embryo sac, use two sharp forceps to pull open the sac and remove the embryo into a clean Petri dish on ice filled with cold 1× DMEM until all the tissue is collected.

2. Transfer embryos to a 50 mL tube with 20 mL of cold 1× PBS and rinse to remove any remaining blood or media. Approximately five embryos or pieces of tissue can be processed together in one tube.
3. Change the solution to 20 mL of cold fixation buffer and lightly fix for 30 min (*see* **Note 7**).
4. Remove fixative and permeabilize the tissue with three washes in wash buffer, 30 min each. Keep at room temperature.
5. Remove wash buffer and incubate the tissue with two washes in equilibration buffer, 30 min each. The buffer should be prewarmed to 37 °C. Incubate tubes in a 37 °C water bath.
6. Remove equilibration buffer and add the development buffer containing 1 mg/mL X-gal. Allow tissue to develop for 18 h in development buffer at 37 °C (*see* **Notes 3** and **8**).
7. After the desired level of development has been achieved, rinse tissue three times for 30 min each with 1× PBS or until the solution no longer appears yellow. This step stops the reaction which can continue to develop within the tissue if not properly washed away.
8. Tissue can be imaged with a dissecting scope equipped with a camera right away or stored in 1× PBS at 4 °C for later use (*see* **Note 9**).

3.2 Sectioning Stained Embryos

1. To generate sections: (a) Dehydrate stained embryos in stepped concentrations of ethanol solution from 50 to 100 %. (b) Repeat the last wash in 100 % ethanol one more time to ensure that water is out of tissue. (c) Incubate in a 1:1 ethanol–xylene solution. (d) Follow with two washes in 100 % xylene two washes with xylene. Each incubation step is 30 min (*see* **Note 10**).
2. Warm the embryos briefly before transferring to melted paraffin to prevent hardening of the wax. Change the melted paraffin or Paraplast four times, every 30 min. Keep all materials in the 58 °C oven.
3. Allow the tissue to harden into paraffin embedding blocks and cut 5–10 μm sections onto glass "plus" charged slides with a microtome.
4. To coverslip: (a) place slides into an oven for 1 h to soften the paraffin, (b) soak in xylene twice for 15 min each, (c) mount coverslip with Permount. Sectioned tissues allow better resolution of the staining (Fig. 2).

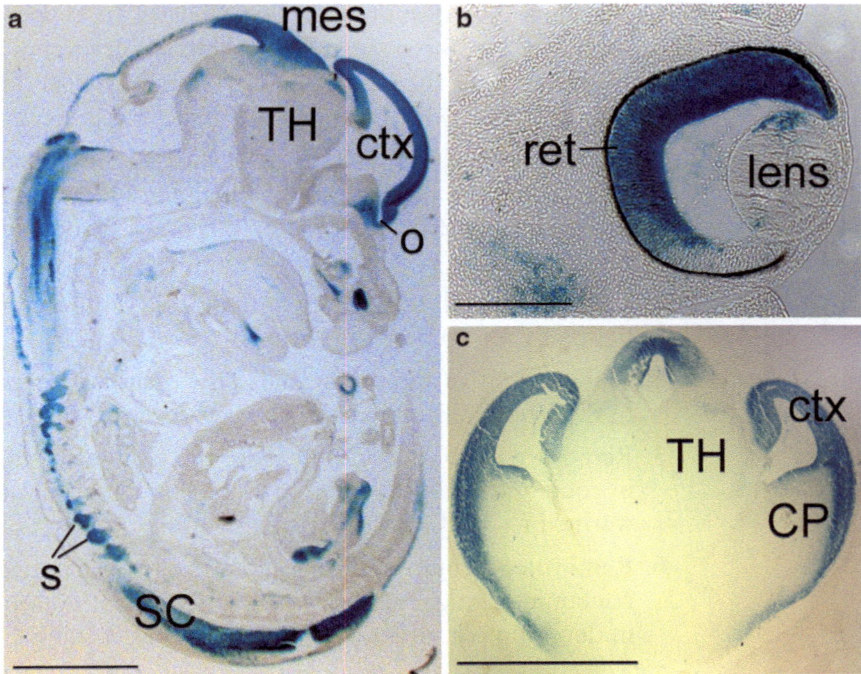

Fig. 2 Whole tissue staining of an E15 embryo. (**A**) Sagittal cross section of the embryo expressing an EphA5 receptor:LacZ fusion protein. (**B**) High magnification of the sectioned eye. Note the gradient in the retina and staining in the lens. (**C**) Coronal section through the brain. More precise expression details can be observed in high magnification images taken from sectioned tissue. *CP* caudate/putamen, *ctx* cortex, *lens* lens, *mes* mesencephalon, *o* olfactory lobe, *ret* retina, *s* spinal ganglia, *TH* thalamus. Scale bars = 1 mm (**A**, **C**), 200 μm (**B**)

3.3 Staining of Tissue Sections

1. For the analysis of older animals (E17, postnatal day (P)0, P6, Adult), dissect tissue and freeze immediately in Tissue-Tek O.C.T. Freezing Compound in embedding molds placed in powdered dry ice. Tissue blocks can be stored in an air-tight container at −80 °C.
2. Equilibrate tissue blocks to −16 to −20 °C inside the cryostat for 1 h before cutting.
3. Cut 10 μm cryosections onto plus-charged glass slides.
4. After drying, post-fix the slides for 2 min in fixation buffer (*see* **Note 7**).
5. Place slides in a slide mailer and follow Subheading 3.1, **steps 4–7**.
6. Dehydrate sections in a graded ethanol series, followed by two washes in xylene. Each step should be limited to 5 min due to the solubility of the X-gal stain in organic solvents. Slides can be coverslipped in Permount and imaged (*see* **Note 11**).

4 Notes

1. (a) Paraformaldehyde is toxic. Wear personal protection and work under a fume hood.

   (b) Fresh paraformaldehyde is recommended for optimal fixation; however, 4 % paraformaldehyde has routinely been frozen and stored at −20 °C. If using frozen paraformaldehyde, thaw completely in a water bath and bring the temperature up to 60 °C. If the temperature becomes too hot, the paraformaldehyde may denature and should be remade. Do not refreeze any unused portion.

   (c) Fixative waste must be collected in a labelled liquid waste storage container and disposed of properly.

   (d) Label equipment used to make paraformaldehyde and keep it separate from the other lab equipment.

   (e) Fixatives should be used ice cold.

2. Take care not to confuse Nonidet P-40 (octylphenoxypolyethoxyethanol) with another detergent (nonylphenoxypolyethoxyethanol) using the same acronym (NP-40). Unfortunately, Nonidet P-40 is currently out of production and no longer commercially available. Triton-X-100 and Igepal CA-630 are nearly chemically identical to Nonidet P-40 and should be acceptable alternatives for NP-40.

3. X-gal, or Bluo-gal, should be stored at −20 °C and protected from the light. In order to stay in solution, X-gal must be completely dissolved in DMF. Warming to room temperature quickens this process. The stock solution can be made fresh or stored in 1 mL aliquots at −20 °C. The solution should be clear and colorless. If the solution becomes yellow or pink, it should not be used. If the final concentration of DMF in the development buffer is too high and interferes with staining, making a 100× stock solution of X-gal may help.

4. Make twice the amount of equilibration buffer fresh before use and save half for the development buffer. Warm both buffers to 37 °C in a water bath ahead of use. Metal forceps should not be used to transfer tissue into or out of the equilibration or development buffer since the solutions may react with the iron in the metal and cause discoloring. The working concentration of potassium ferricyanide and ferrocyanide ranges from 5 to 35 mM; however, 5 mM solutions were found to be optimal.

5. Warm the X-gal stock to 37 °C and add to the pre-warmed equilibration buffer set aside for the development buffer. If the X-gal is added cold, a precipitate may form and cause solution to be unusable.

6. This protocol has been used for mouse embryos at these stages but earlier staged embryos and embryos from other species can also be used. Whole mount staining is unsuitable for tissue larger than mouse E15 embryos as the fixative and development with X-gal cannot permeate deeply or evenly to elicit reliable staining. If comparisons with older embryos need to be made, organs can be isolated or tissue can be cut into smaller pieces. Otherwise, large tissue, such as whole mouse liver or adult brain, is best stained by making fresh frozen cryostat sections and postfixing for 2 min (*see* Subheading 3.3).
7. The fixation time is crucial and should be determined empirically for specific types of tissue. Overfixation of the tissue will reduce β-galactosidase activity.

   If staining is to be used to compare intensity across different specimens, be sure to keep the fixation time constant. If direct comparisons are not critical, embryos staged E9 and younger do not require more than 15 min fixation time. Keep in mind that underfixation can leave a high background and poor tissue preservation. Testing a wild type tissue sample in parallel with the LacZ fusion-containing tissue will determine whether there is an issue. Increasing the pH of the PBS solution used to make reagents can help to decrease endogenous activity [17, 18].
8. Development time for the stain to appear is determined empirically. Be careful not to overdevelop the tissue as a greenish precipitate, Prussian green, may begin to form.
9. A very tiny pinch of sodium azide will help to prevent mold growth on tissue stored at 4 °C for several months in PBS. Staining in whole mount tissue can develop Prussian green precipitate from the residual development buffer within the tissue if left in PBS over several months so it is best to image promptly when the color looks the best.
10. Organic solvents can quench the staining; for this reason, it is important to be consistent in the treatment of the tissue specimens. If optimizing incubation times, keep steps in xylene or Histo-Clear as short as possible.
11. Although the staining is vulnerable in organic solvents, the permanence of the slides is greater than the whole mount tissue. Slides can be kept for at least a decade without any discernable loss of staining.

Acknowledgments

This work was supported by NIH grant RO1EY019012 and PO1HD023315 to R.Z.

References

1. Alam J, Cook JL (1990) Reporter genes: application to the study of mammalian gene transcription. Anal Biochem 188:245–254
2. Silhavy TJ, Beckwith JR (1985) Uses of lac fusions for the study of biological problems. Microbiol Rev 49:398–418
3. Conquet F (1995) Inactivation in vivo of metabotropic glutamate receptor 1 by specific chromosomal insertion of reporter gene lacZ. Neuropharmacology 34:865–870
4. Feldheim DA, Nakamoto M, Osterfield M, Gale NW, DeChiara TM, Rohatgi R, Yancopoulos GD, Flanagan JG (2004) Loss-of-function analysis of EphA receptors in retinotectal mapping. J Neurosci 24:2542–2550
5. Shuman HA, Silhavy TJ, Beckwith JR (1980) Labeling of proteins with beta-galactosidase by gene fusion. Identification of a cytoplasmic membrane component of the Escherichia coli maltose transport system. J Biol Chem 255:168–174
6. Cooper MA, Crockett DP, Nowakowski RS, Gale NW, Zhou R (2009) Distribution of EphA5 receptor protein in the developing and adult mouse nervous system. J Comp Neurol 514:310–328
7. Washburn CP, Cooper MA, Zhou R (2007) Expression of the tyrosine kinase receptor EphA5 and its ligand ephrin-A5 during mouse spinal cord development. Neurosci Bull 23: 249–255
8. Dembinska ME, Stanewsky R, Hall JC, Rosbash M (1997) Circadian cycling of a PERIOD-beta-galactosidase fusion protein in Drosophila: evidence for cyclical degradation. J Biol Rhythms 12:157–172
9. Ben-Arie N, Hassan BA, Bermingham NA, Malicki DM, Armstrong D, Matzuk M, Bellen HJ, Zoghbi HY (2000) Functional conservation of atonal and Math1 in the CNS and PNS. Development 127:1039–1048
10. Miquerol L, Gertsenstein M, Harpal K, Rossant J, Nagy A (1999) Multiple developmental roles of VEGF suggested by a LacZ-tagged allele. Dev Biol 212:307–322
11. Horwitz JP, Chua J, Curby RJ, Tomson AJ, Darooge MA, Fisher BE, Mauricio J, Klundt I (1964) Substrates for cytochemical demonstration of enzyme activity. I. Some substituted 3-indolyl-beta-d-glycopyranosides. J Med Chem 7:574–575
12. Cohen RB, Tsou KC, Rutenburg SH, Seligman AM (1952) The colorimetric estimation and histochemical demonstration of beta-d-galactosidase. J Biol Chem 195:239–249
13. Cotson S, Holt SJ (1958) Studies in enzyme cytochemistry. IV. Kinetics of aerial oxidation of indoxyl and some of its halogen derivatives. Proc R Soc Lond B Biol Sci 148: 506–519
14. Lojda Z (1970) Indigogenic methods for glycosidases. I. An improved method for beta-D-glucosidase and its application to localization studies on intestinal and renal enzymes. Histochemie 22:347–361
15. Danielian PS, Muccino D, Rowitch DH, Michael SK, McMahon AP (1998) Modification of gene activity in mouse embryos in utero by a tamoxifen-inducible form of Cre recombinase. Curr Biol 8:1323–1326
16. Greenbaum D, Colangelo C, Williams K, Gerstein M (2003) Comparing protein abundance and mRNA expression levels on a genomic scale. Genome Biol 4:117
17. Sanchez-Ramos J, Song S, Dailey M, Cardozo-Pelaez F, Hazzi C, Stedeford T, Willing A, Freeman TB, Saporta S, Zigova T, Sanberg PR, Snyder EY (2000) The X-gal caution in neural transplantation studies. Cell Transplant 9:657–667
18. Weiss DJ, Liggitt D, Clark JG (1999) Histochemical discrimination of endogenous mammalian beta-galactosidase activity from that resulting from lac-Z gene expression. Histochem J 31:231–236

Chapter 19

Chromatin Immunoprecipitation Assay of Brain Tissues Using Percoll Gradient-Purified Nuclei

Baojin Ding and Daniel L. Kilpatrick

Abstract

Protein–DNA interactions are critical to maintain genome stability, DNA replication, chromosome segregation and to regulate gene expression. Chromatin immunoprecipitation (ChIP) is a powerful technique to study these interactions within living neurons and nervous tissue. In particular, ChIP analysis of chromatin in which protein–DNA interactions are first fixed in situ provides a valuable approach to identify specific transcription factor–DNA interactions and their regulation in the developing nervous system. Here we describe a procedure utilizing Percoll gradient purification of nuclei from fresh brain tissue pre-fixed with formaldehyde for ChIP analysis. This purification protocol provides an enrichment of neuronal nuclei in high yield. We also illustrate the suitability of chromatin prepared from Percoll-purified brain nuclei for ChIP analysis of regulated transcription factor interactions with neuronal gene promoters.

Key words Brain tissue, Formaldehyde fixation, Nuclei, Percoll gradient purification, Sonication, Antibody, Chromatin immunoprecipitation

1 Introduction

During neuron development, numerous *trans*-factors are involved in specific gene regulation events through their direct or indirect interactions with chromatin DNA [1–4]. These include temporal changes in transcription factor occupancy of target promoters that regulate the timing of neuronal gene expression [5]. Understanding how various transcription factors interact with chromatin and how these interactions are regulated in vivo is a critical but significant challenge in developmental neuroscience. Chromatin immunoprecipitation (ChIP) is an extremely valuable technique for exploring these in vivo interactions and their importance for gene regulation [6, 7]. Native ChIP (nChIP) [8] is frequently used to study changes in histone and genomic DNA marks that are associated with nervous system development, plasticity and disease. However, to study specific transcription factor–chromatin interactions, it is necessary to first cross-link protein and DNA in situ (e.g., using formaldehyde)

Renping Zhou and Lin Mei (eds.), *Neural Development: Methods and Protocols*, Methods in Molecular Biology, vol. 1018,
DOI 10.1007/978-1-62703-444-9_19, © Springer Science+Business Media, LLC 2013

in order to physically preserve their interactions during chromatin isolation (xChIP) [9, 10]. The chromatin is then sheared by sonication to fragment sizes of 200–1,000 base pairs [7, 9] and protein–DNA complexes containing the factor(s) of interest are immunoprecipitated using a specific antibody, which is typically coupled to agarose, Sepharose, or magnetic beads. The immune complexes are washed under stringent conditions to remove nonspecifically bound chromatin. The precipitated chromatin is then eluted from beads, the cross-links are reversed, and the enriched target DNA sequences are purified and detected using PCR [4, 6, 7].

For a successful xChIP assay, a sufficient amount of high quality fixed chromatin is important. In this procedure, we describe a strategy to isolate nuclei from brain tissue using Percoll gradient centrifugation. This technique provides preparations of high quality and yield that are enriched in neuronal nuclei (e.g., 80–95 % NeuN(+) nuclei from developing mouse cerebellum) [5]. The recovered nuclei are also more readily usable for preparing chromatin relative to nuclei prepared with standard sucrose density gradient fractionation, which can be "sticky" and difficult to suspend. This protocol also should be scalable to smaller amounts of fixed brain tissue using appropriate reductions in nuclei gradient reagent volumes. We also demonstrate the suitability of these chromatin preparations for conventional xChIP analysis of transcription factor–DNA interactions on neuronal gene promoters.

2 Materials

All solutions are prepared with ultrapure water. Hazardous waste disposal should follow appropriate regulations and procedures.

2.1 Brain Tissue Disruption and Fixation Components

1. Phosphate Buffered Saline (PBS): 137 mM NaCl, 2.7 mM KCl, 8 mM Na_2HPO_4, 1.46 mM KH_2PO_4. Dissolve 8 g of NaCl, 0.2 g of KCl, 1.4 g of Na_2HPO_4, and 0.2 g of KH_2PO_4 into 800 ml water. Adjust pH to 7.4 and add distilled H_2O to 1 L. Sterilize by autoclaving.
2. Frosted microscope slides (Fisher Scientific, #12-552-3).
3. Formaldehyde 36.5 % (Sigma, #F8775).
4. 1.25 M glycine: Dissolve 9.4 g glycine powder (American Bioanalytical, #AB730) into 90 ml water and add water to final volume of 100 ml.
5. Protease inhibitors: Dissolve 1 EDTA-free Protease Inhibitor Cocktail Tablet (Roche, #11873580001) into 2 ml of nuclease-free water to make 25× stock, keep at −20 °C.
6. Tube Shaker or Rotator.

2.2 Percoll Gradient Purification Components

1. 2.5 M sucrose: Dissolve 85.6 g of sucrose in water and add water to final volume of 100 ml. Make tenfold dilution of 2.5 M sucrose with water to prepare 0.25 M sucrose solution.
2. 90 % Percoll: Add 5 ml of 2.5 M sucrose to 45 ml of Percoll (Sigma, #P1644). Mix well and keep at 4 °C.
3. 60 % Percoll and 10 % Percoll: Dilute 90 % Percoll with 0.25 M sucrose to the final concentration and keep at 4 °C.
4. Cell lysis buffer: 0.25 M sucrose, 50 mM Tris-HCl (pH 8.0), 5 mM $MgCl_2$, 25 mM KCl, 0.1 % NP-40, 1 mM dithiothreitol (DTT), 1×protease inhibitor (*see* **Note 1**).
5. Nuclei purification buffer: 0.25 M sucrose, 50 mM Tris-HCl (pH 8.0), 5 mM $MgCl_2$, 25 mM KCl, 1×protease inhibitor (*see* **Note 1**).
6. Homogenizers (various). Size depends on tissue weight and sample volume.
7. Nylon mesh (74 μm). Cut to size and place in a funnel.
8. Nuclei lysis buffer: 1 % SDS, 10 mM EDTA, 50 mM Tris–HCl (pH 8.0), 1×protease inhibitor (*see* **Note 1**).
9. Trypan blue solution (0.4 %, Sigma, #T8154).
10. 10 ml syringe (Becton Dickinson).
11. Hemocytometer counting slide (Hausser Scientific).
12. Ultracentrifuge and Beckman 4-place rotor GH3.8 (or equivalent).

2.3 Chromatin Shearing and Immunoprecipitation Components

1. Sonicator or Bioruptor.
2. Agarose gel electrophoresis apparatus.
3. ChIP dilution buffer: 0.01 % SDS, 1.1 % Triton X-100, 1.1 mM EDTA, 167 mM NaCl, 20 mM Tris–HCl (pH 8.0). Keep at 4 °C (*see* **Note 2**).
4. Salmon sperm DNA/Protein A-agarose 50 % slurry (Millipore, #16-157), keep at 4 °C (*see* **Note 3**).
5. Agitator, Fisher Scientific, Hematology/Chemistry mixer 346 (or equivalent).
6. Antibodies, specific and negative control (*see* **Note 4**).
7. Low-salt wash buffer: 0.1 % SDS, 1 % Triton X-100, 2 mM EDTA, 150 mM NaCl, 20 mM Tris–HCl (pH 8.0) (*see* **Note 5**).
8. High-salt wash buffer: 0.1 % SDS, 1 % Triton X-100, 2 mM EDTA, 500 mM NaCl, 20 mM Tris–HCl (pH 8.0) (*see* **Note 5**).
9. LiCl wash buffer: 0.25 M LiCl, 1 % NP-40, 1 % deoxycholate, 1 mM EDTA, 20 mM Tris–HCl (pH 8.0) (*see* **Note 5**).
10. Non-stick RNase-free microfuge tubes (1.5 ml) (Applied Biosystems, #AM12450).

11. TE wash buffer: 10 mM Tris–HCl (pH 8.0), 1 mM EDTA (*see* **Note 5**).
12. Elution buffer: 1 % SDS, 100 mM $NaHCO_3$ (*see* **Note 6**).
13. Eppendorf 5415C desktop centrifuge (or equivalent).

2.4 Cross-link Reversal and DNA Recovery Solutions and Reagents

1. 5 M NaCl: Dissolve 146.3 g of NaCl into 350 ml water, add water to a final volume of 500 ml. Autoclave and keep at room temperature.
2. Proteinase K (20 mg/ml): Dissolve 20 mg of Proteinase K powder (Roche, PCR grade, #03115879001) into 1 ml of water. Aliquot and store at −20 °C.
3. RNase A (10 mg/ml): Dissolve 20 mg of RNAse A (Sigma, #R4642) into 2 ml of 10 mM sodium acetate (pH 5.3). Boil for 10 min to inactivate DNase. Aliquot and store at −20 °C.
4. Chloroform–isoamyl alcohol (24:1): Mix chloroform (EM Science, #CX1055) with isoamyl alcohol (Fisher Scientific, #A393) in a ratio of 24:1. Keep at 4 °C.
5. Phenol–chloroform–isoamyl alcohol (25:24:1) DNA extraction solution: Mix phenol (American Bioanalytical, #AB01616) saturated with Tris buffer (pH 7.9) and chloroform–isoamyl alcohol in the ratio of 1:1. Keep at 4 °C.
6. Glycogen (Invitrogen, 20 μg/μl, #10811-010).
7. Nuclease-free water (American Bioanalytical, #AB02128).
8. Ethanol (Decon Labs, #2701). Make 75 % and 95 % ethanol (v/v) solutions using nuclease-free water.

3 Methods

All solutions are ice-cold except where specified. All steps are performed on ice or in a cold room except as specified.

3.1 Brain Tissue Disruption and Formaldehyde Fixation

1. Dissect mouse brain and cut out the cerebellum and/or other relevant region(s). Place the tissue directly into ~1 ml of PBS in a petri dish. Measure the wet weight of tissues and keep on ice (*see* **Notes 7** and **8**).
2. Mince tissues between the frosted ends of the slides into small pieces in ice-cold PBS. Transfer tissues to a clean 15 ml conical centrifuge tube. Rinse the slides and petri dish, and try to transfer as much tissue as possible to maximize yield (*see* **Note 9**).
3. Suspend minced tissue in room temperature PBS containing 1×protease inhibitors (*see* **Note 10**) and add formaldehyde to a final concentration of 1 % (280 μl of 36.5 % formaldehyde/10 ml PBS). Rotate samples at room temperature for 10 min to cross-link chromatin.

4. Add 1/9 volume of 1.25 M glycine and incubate 3 min at room temperature with rotation to stop the cross-link reaction.
5. Centrifuge the fixed tissue mince at 1,100 × *g* at 4 °C for 3 min. Remove the supernatant and wash the pellet twice with 10 ml of ice-cold PBS containing protease inhibitors. Centrifuge each time at 1,100 × *g* for 3 min.
6. Remove the supernatant and suspend the pellet into 2 ml of cell lysis buffer. Homogenize the tissue using ten up and down strokes with a 5 ml frosted glass–glass homogenizer. Repeat using another ten strokes with a Dounce homogenizer (A pestle).
7. Filter the homogenate through nylon mesh (74 μm) (*see* **Note 11**) and collect into a 15 ml conical tube.
8. Rinse the homogenizer and nylon mesh with cell lysis buffer and pool the rinse and homogenate (*see* **Note 12**). Add cell lysis buffer up to 10 ml and incubated samples on ice for 10–20 min to fully lyse cells (*see* **Note 13**).
9. Determine nuclei yield for the pre-Percoll fraction using a hemocytometer.

3.2 Percoll Gradient Fractionation

1. Centrifuge homogenate at 1,100 × *g* at 4 °C for 3 min.
2. Remove the supernatant carefully and suspend the pellet into 5 ml nuclei purification buffer. Homogenize the sample using five strokes with a Dounce homogenizer (A pestle).
3. Mix 5 ml of nuclei sample with 5 ml of 10 % Percoll and transfer to a 30 ml clear round bottom centrifuge tube.
4. Slowly underlay the nuclei/5 % Percoll mix with 10 ml of 60 % Percoll using a 10 ml syringe. Two separate layers should be clearly apparent (*see* **Note 14**).
5. Centrifuge at 16,000 × *g* for 1 h at 4 °C in a swinging bucket rotor (e.g., Beckman GH3.8 rotor in a Beckman GS-6 desktop centrifuge) with the brake off. Take care not to disturb the layer interface when handling tubes.
6. Carefully remove the nuclei-enriched interface between the 5 and 60 % Percoll layers using a 1 ml pipette and transfer to a 15 ml conical tube (*see* **Note 15**).
7. Bring the volume up to 10 ml with ice-cold PBS and centrifuge at 1,100 × *g* for 5 min at 4 °C to pellet nuclei (*see* **Note 16**).
8. Remove the supernatant while avoiding contact with the nuclei pellet. Wash the nuclei pellet three times by repeated suspension in 10 ml of ice-cold PBS and centrifugation at 1,100 × *g* for 5 min at 4 °C.

Table 1
Yields and percent recoveries for nuclei purified on Percoll gradients from fixed fresh brain tissues

Tissue[a]	Wet weight (gram)	Nuclei number pre-Percoll (million)	Nuclei number post-Percoll (million)	% Recovery
Adult whole brain (1)	0.54	136	98.8	72.6
P21 Cortex (2)	0.59	42.6	35.5	83.3
P7 Cortex (10)	1.1	185	146	78.9
P7 Cerebellum (10)	0.35	279	225	80.1

[a]Values in parentheses: number of mice harvested for each tissue

9. Following the last wash, remove a small aliquot to count nuclei and to verify their purity (i.e., relative absence of cellular debris) using trypan blue and a hemocytometer. Nuclei recoveries relative to pre-Percoll fractions are typically ~70–80 % (Table 1).
10. After the last wash, add sufficient Nuclei lysis buffer to adjust the chromatin concentration to 20×10^6 nuclei equivalents/ml. Fully suspend the fixed chromatin by gentle pipetting (*see* **Note 17**) and place 1 ml aliquots into 1.5 ml plastic tubes. Either shear chromatin immediately (see below) or place at −80 °C for long-term storage.
11. (Optional) Verify the purity of neuronal nuclei by immunostaining with anti-NeuN antibody.

3.3 Chromatin Shearing

1. Shear the chromatin DNA using a sonicator or Bioruptor (*see* **Note 18**). Maintain samples in an ice water bath during shearing to minimize heat production.
2. Centrifuge at $1,500 \times g$ for 10 min at 4 °C. Place the supernatant in a clean 1.5 ml tube and discard the pellet.
3. Verify the shearing efficiency by running 5–10 μl on a 1 % agarose gel. Optimal shearing yields an apparent fragment size of 600–1,000 bp for cross-linked chromatin and between 150 and 500 bp following cross-link reversal (*see* Fig. 1 and **Note 19**).
4. Place aliquots of sheared chromatin equivalent to $\sim 2 \times 10^6$ nuclei (100 μl) into 1.5 ml tubes. Samples are either processed immediately or frozen at −80 °C.

3.4 ChIP Assay

1. Dilute sheared chromatin tenfold using ChIP dilution buffer. We routinely employ 2×10^6 nuclei equivalents per ChIP reaction, although smaller amounts of chromatin are suitable [11].

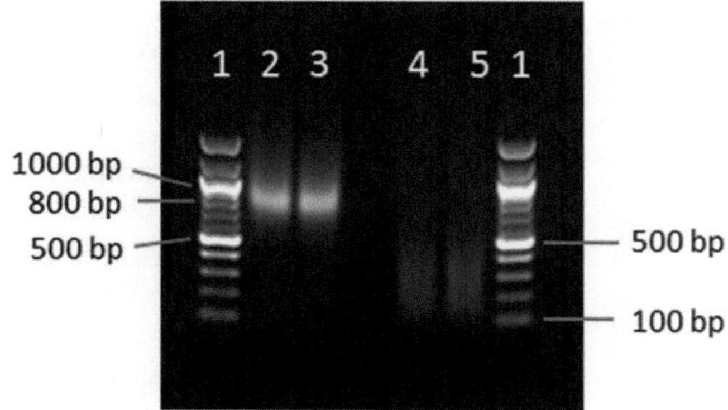

Fig. 1 Assessment of chromatin shearing after sonication with and without cross-link reversal. Samples were resolved on a 1 % agarose gel. Lanes: *1*, 100 bp DNA ladder; *2* and *3*, cross-linked chromatin from P7 and P21 mouse cerebellum, respectively; *4* and *5*, P7 and P21 cerebellar chromatin, respectively, after cross-link reversal

2. Add 50 μl of salmon sperm DNA/Protein A-agarose to each sample and incubate at 4 °C for 0.5–1 h with rotation (*see* **Note 20**).
3. Centrifuge at 1,000 × *g* for 1 min and transfer the supernatant to a new 1.5 ml non-stick RNase-free microfuge tube (*see* **Note 21**). Take care to not touch the bead pellet during supernatant removal. Discard the beads.
4. Remove an aliquot of pre-cleared chromatin and save as an Input sample and put aside. For example, take 50 μl from a total of 1,000 μl supernatant (=5 % of total Input).
5. Add a sufficient amount of relevant experimental or control antibody to each ChIP reaction tube (~950 μl supernatant) (*see* **Note 22**). Incubate samples at 4 °C overnight with rotation.
6. Add 40 μl of Protein A-agarose slurry to each sample. Incubate at 4 °C for 1–2 h with rotation to collect immune complexes.
7. Centrifuge at 1,000 × *g* for 1 min to pellet agarose beads and carefully remove the supernatant.
8. Wash beads with 1 ml of the following buffers: low-salt wash (1×), high-salt wash (1×), LiCl salt wash (1×), and TE (2×). For each wash, agitate samples in a cold room for 10 min, then centrifuge at 1,000 × *g* for 1 min to pellet beads.
9. Remove the wash supernatant (*see* **Note 23**) and add 260 μl of freshly prepared Elution buffer to beads. Vortex briefly and incubate samples for 15 min at room temperature with rotation. Centrifuge 1,000 × *g* for 1 min and transfer 250 μl of supernatant into a fresh 1.5 ml tube.

10. Repeat elution as in **step 9** and combine the supernatants (500 μl total sample volume).
11. Add 20 μl of 5 M NaCl and 3 μl of RNase A (10 mg/ml) to the 500 μl eluate as well as to the Input sample (*see* **Note 24**). Vortex and briefly centrifuge. Incubate samples at 65 °C for at least 4 h to reverse cross-links (*see* **Note 25**).
12. Add 10 μl of 0.5 M EDTA, 20 μl of 1 M Tris–HCl (pH 6.5), and 5 μl of proteinase K (20 mg/ml) to each sample and incubate at 50 °C for 1 h.
13. Purify immunoprecipitated DNA using phenol–chloroform extraction (*see* **Note 26**). Add 500 μl of phenol–chloroform–isoamyl alcohol (25:24:1) to each sample and vortex vigorously for 20 s. Centrifuge at 16,000 × *g* for 5 min at 4 °C. Carefully remove as much of the supernatant as possible without contacting the interface and place in a clean 1.5 ml tube.
14. Add 500 μl of chloroform to the supernatant and shake vigorously for 20 s (*see* **Note 27**). Centrifuge at 16,000 × *g* for 5 min at 4 °C. Transfer the supernatant to a clean 1.5 ml tube.
15. Add 1 μl of glycogen (20 mg/ml) and 1 ml of 100 % ethanol to each supernatant, and mix well by inversion several times. Place on ice or at −20 °C for 2 h to overnight. Centrifuge at 16,000 × *g* for 30 min at 4 °C to precipitate DNA.

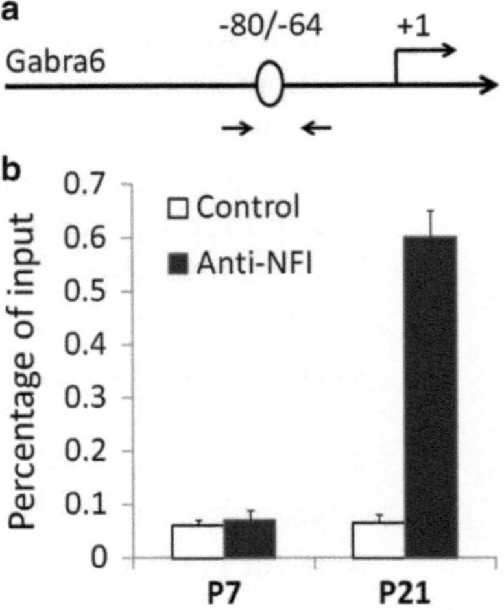

Fig. 2 Quantitative PCR analysis of ChIP samples. (**a**) Nuclear Factor One (NFI) binding site within the mouse Gabra6 proximal promoter region. *Arrows* indicate the region amplified in the qPCR reaction. (**b**) Relative NFI occupancy of the Gabra6 gene promoter showing a temporal increase in NFI occupancy in mouse cerebellum between P7 and P21. Anti-NFI: NFI antiserum, control: Pre-immune serum

16. Wash pellet with 75 % ethanol twice and 95 % ethanol once (*see* **Note 28**). Centrifuge at 16,000 × *g* for 5 min at 4 °C for each wash. Remove the liquid after the final wash and air-dry the pellet at room temperature for 5 min.
17. Dissolve the pellet in 50 μl of nuclease-free water or in 10 mM Tris-HCl (pH 8.0) (*see* **Note 29**) and store at −20 °C.
18. Analyze samples by real time PCR (Fig. 2).

4 Notes

1. Add dithiothreitol and protease inhibitors just before use.
2. Add protease inhibitors just before use.
3. Close the lid tightly once opened to avoid evaporation. Seal with paraffin.
4. ChIP grade antibodies are recommended. For the negative control antibody, use either generic purified antibody (e.g., IgG) or pre-immune serum from the appropriate host species to match the type of specific antibody used.
5. Pre-chill on ice before use. Store all wash buffers at 4 °C.
6. Prepare this freshly.
7. In this protocol, the wet weight of fresh tissues should be no more than 0.5 g for each gradient preparation. If >0.5 g of tissue is used, separate into multiple parallel gradients.
8. Excess PBS volume will reduce the mincing efficiency.
9. For small amounts of tissue (<0.1 g), a loose-fitting glass–glass homogenizer is recommended to fully retain the minced tissue and minimize loss of nuclei.
10. If the wet weight of tissue is between 0.1 and 0.5 g, suspend minced tissue in 10 ml of PBS, and if less than 0.1 g, suspend in 5 ml PBS (minimum volume).
11. Wet the nylon mesh with cell lysis buffer before filtering the tissue mince.
12. Try to collect all material at each step to maximize recovery.
13. Complete cell lysis can be verified microscopically using trypan blue staining.
14. A dye such as phenol red can be added to the 60 % Percoll, if necessary, to clearly observe the two layers.
15. Carefully remove the upper layer first, and then harvest the nuclei-enriched interface. Do not allow the gradients to sit unnecessarily to avoid gradient diffusion.

16. If >5 ml of nuclear interface volume is collected, nuclei may not efficiently sediment due to elevated Percoll concentrations. Dilute with additional PBS and split into multiple tubes for centrifugation, as required.
17. Avoid frothing or generation of bubbles during the pipetting step.
18. Optimal shearing conditions may vary for different sonicators and this should be determined beforehand. For a Sonics Model CV18 sonicator, we use parameter settings of 60 % of Amplitude; 10 s ON, 10 s OFF, with a total sonication time of 10 min for each 1 ml sample.
19. Note that the apparent DNA fragment size can vary depending on the agarose concentration used for gel resolution. To determine the size of DNA following cross-link reversal, incubate a small aliquot (20–50 μl) of sheared chromatin at 65 °C for 4 h before gel analysis.
20. This pre-clearing step is important to decrease non-specific binding to beads. We have found that 50 μl of Protein A-agarose slurry is enough for a 1 ml sample (~2×10^6 nuclei equivalents). If a different bead type is used, determine the optimal bead amount beforehand.
21. Non-stick RNase-free microfuge tubes are recommended in the following steps to decrease non-specific background signal.
22. Optimal amounts of specific antibody can be determined empirically beforehand. We typically use 5 μg of IgG-purified antibody or 5–10 μl of antiserum for each sample (2×10^6 nuclei equivalents). An appropriate non-specific antibody control should be used to determine background chromatin DNA binding (*see* **Note 4**).
23. After removal of the supernatant, re-centrifuge for 1 min at $1,000 \times g$ and remove as much residual liquid as possible using a pipette without disturbing the pellet.
24. First add 450 μl of ChIP dilution buffer to each 50 μl Input sample (500 μl final volume). Process the Input in parallel with ChIP samples in all subsequent steps.
25. This brings the final concentration of NaCl to 0.2 M, which is critical for subsequent ethanol precipitation of DNA. It is important to seal the tube lid tightly (e.g., using paraffin) to avoid sample loss during this digestion step due to the elevated temperature. Overnight incubation is recommended to completely reverse cross-links.
26. An alternative to phenol–chloroform extraction for DNA recovery is the use of DNA purification columns, especially for ChIP with low binding affinity. These also can reduce effects of inhibitors (e.g., organic solvents) on subsequent PCR analysis. Our experience is that sample recovery is lower for columns

Table 2
Recoveries of input DNA purified by phenol–chloroform extraction (P/C) or by column kits, based on qPCR analysis

DNA starting material (% input)	% Recovery for P/C extraction	% Recovery for QIAquick purification column
10 % of input	100 ± 7.8	67.4 ± 3.1
3 % of input	102 ± 11	54 ± 2.3
1 % of input	105 ± 5.4	21.2 ± 1.7

relative to phenol–chloroform and ethanol precipitation (Table 2).

27. We find that vigorous shaking during chloroform extraction is more efficient than vortex mixing.
28. Multiple washes with 75 % ethanol together with the final 95 % ethanol wash serve to minimize organic solvent inhibitory effects.
29. Nuclease-free water is preferable for DNA suspension since some buffer components may interfere with subsequent analyses, including qPCR. However, 10 mM Tris-HCl (pH 8.0) is better for long-term storage (>2 months).

References

1. Felsenfeld G, Groudine M (2003) Controlling the double helix. Nature 421(6921): 448–453
2. Schubeler D, Elgin SC (2005) Defining epigenetic states through chromatin and RNA. Nat Genet 37(9):917–918
3. Bertrand N, Castro DS, Guillemot F (2002) Proneural genes and the specification of neural cell types. Nat Rev Neurosci 3(7):517–530
4. Collas P (2010) The current state of chromatin immunoprecipitation. Mol Biotechnol 45(1):87–100
5. Wang W, Shin Y, Shi M, Kilpatrick DL (2011) Temporal control of a dendritogenesis-linked gene via REST-dependent regulation of nuclear factor I occupancy. Mol Biol Cell 22(6): 868–879
6. O'Neill LP, Turner BM (1996) Immunoprecipitation of chromatin. Methods Enzymol 274:189–197
7. Nelson JD, Denisenko O, Bomsztyk K (2006) Protocol for the fast chromatin immunoprecipitation (ChIP) method. Nat Protoc 1(1): 179–185
8. O'Neill LP, Turner BM (2003) Immunoprecipitation of native chromatin: NChIP. Methods 31(1):76–82
9. Kuo MH, Allis CD (1999) In vivo cross-linking and immunoprecipitation for studying dynamic Protein:DNA associations in a chromatin environment. Methods 19(3):425–433
10. Solomon MJ, Varshavsky A (1985) Formaldehyde-mediated DNA-protein crosslinking: a probe for in vivo chromatin structures. Proc Natl Acad Sci USA 82(19): 6470–6474
11. Dahl JA, Collas P (2008) MicroChIP–a rapid micro chromatin immunoprecipitation assay for small cell samples and biopsies. Nucleic Acids Res 36(3):e15

Chapter 20

In Vivo Dual Luciferase Reporter Assay with Chick Neural Tube In Ovo Electroporation System

Hui Wang and Michael P. Matise

Abstract

Luciferase reporter systems are widely employed to provide a quantitative readout of gene expression for studies of transcriptional regulation, translation efficiency, and cell signaling. The most common application of luciferase involves transient transfections into cells in vitro or in vivo. In both cases, the normal variability inherent in transfection approaches can introduce significant errors into the data that makes comparison between separate experiments problematic. The dual luciferase reporter assay system (DLR, Promega, WI, USA) is designed to control for this technical issue by using a co-transfection approach with two separate reporter proteins that emit at distinct wavelengths: one from firefly (*Photinus pyralis*) and the second from *Renilla* (*Renilla reniformis*). By normalizing experimental luciferase readings to an internal control transfected under the same conditions, these problems can be largely negated. Here, we describe a method for applying this technique to an in vivo system, the developing chick embryo neural tube. This system provides a physiologically relevant context for functional studies in a spatially and/or temporally controlled manner.

Key words Dual luciferase reporter assay, In ovo electroporation, Firefly luciferase, *Renilla* luciferase

1 Introduction

Reporter genes have become powerful tools in studying many aspects of cellular physiology. Two important characteristics of good reporter systems are (1) a minimal physiological effect on the transfected host cells/organism, and (2) biochemical properties that provide quantifiable stability and detection kinetics. Commonly used detection/measurement methods include indirect fluorescence, luminescence, and absorbance. Luciferase-based reporters are based on a class of enzymes that can catalyze an oxidation reaction of distinct luciferin substrates, producing photons as a by-product [1–3]. In addition to little obvious or immediate effect on host cells, the luciferase enzyme reaction is highly sensitive and extends over a wide range of linear responses [4]. Also the

Renping Zhou and Lin Mei (eds.), *Neural Development: Methods and Protocols*, Methods in Molecular Biology, vol. 1018, DOI 10.1007/978-1-62703-444-9_20, © Springer Science+Business Media, LLC 2013

short half-life of luciferase protein (approximately 3 h) in cells allows near real time monitoring of expression [5]. It is therefore widely used as a bioluminescent reporter gene, especially for studying promoter activities, via monitoring gene expression in a quantitatively controlled manner.

The Promega DLR system, a popular low-throughput kit, provides very sensitive dual-reporter assays using firefly and *Renilla* luciferases. Both enzymes provide immediate activity following translation since they are monomeric proteins and need no post-translational modifications [6, 7]. Due to distinct evolutionary origins, firefly and *Renilla* luciferases can catalyze the oxidization of different luciferin substrates, which prevents their cross-activation. Thus, the DLR assay system is developed to sequentially measure each luciferase activity by (1) injecting the firefly luciferase substrate, beetle luciferin, to generate the first round of luminescent signal, followed by (2) the simultaneous quenching of firefly luciferase luminescence during introduction of coelenterazine, the *Renilla* luciferase substrate, to produce a second round of luminescent signal. Typically, one of these luciferases will be under the control of an experimental DNA sequence of interest, while the other is under the control of a basic promoter to generate a baseline level of luciferase activity. This strategy controls for variables such as transfection efficiency, cell proliferation, or apoptosis. Thus, the distinct light-intensity assay permits normalization of experimental data and improved accuracy and reproducibility.

Besides the DLR assay system, many commercial luciferase reporter assay kits are available. For instance, Dual-Light system (Applied Biosystems, MA, USA) is another highly sensitive low-throughput dual-reporter assay kit that is designed for sequentially detecting firefly luciferase (experimental reporter) and beta-galactosidase (control) activities. It is important to note that both DLR and Dual-Light systems suffer from at least two limitations: (1) using one reporter as internal control, the dual-reporter assays can only measure the expression of single gene at a time; (2) a relatively inconvenient cell lysis procedure is required for both, which inhibits high-throughput analysis. With this in mind, multiple new techniques have been developed. First, Ohmiya's lab pioneered a multicolor luciferase assay system in which three different luciferase genes (emitting green, orange, and red luminescence) that are monitored simultaneously with distinct optical filters in the presence of a common substrate, beetle luciferin [8]. This triple-reporter system allows for the measurement of two reporter genes, providing an opportunity to study more complicated cellular activities such as cross-talk between distinct signaling cascades, comparison of the transcriptional activities of different target promoters, and promoter–promoter interactions. As for high-throughput analysis, the Dual-Glo luciferase assay system (Promega, WI, USA) is designed to simplify the cell lysis procedure by combining this step with the first reporter assay. Recently, several secreted bioluminescent or chemiluminescent reporter genes, including *Cypridina* (*Cypridina noctiluca*) luciferase, *Gaussia* (*Gaussia princeps*) luciferase, *Metridia* (*Metridia longa*) luciferase, and secreted Alkaline Phosphatase (SEAP), have been incorporated into some commercial dual secreted reporter kits (Thermo Scientific Pierce, IL, USA; Clonetech Laboratories, CA, USA; [9]). One obvious advantage of these secreted reporter assays is that gene expression can be monitored in real time within living cells. Accordingly, these kits are suitable for high-throughput analysis with multiple-well cultured cells as well.

Another issue is that most published procedures of reporter gene analysis are based on in-vitro assays with cell lines or primary cell cultures. However, in vivo studies undoubtedly provide the ideal, physiologically relevant environment to conduct reporter assays. For this reason, our lab has combined the DLR assay with in ovo chick neural tube electroporation system to characterize how key developmental transcriptional regulators function to control a *cis-regulatory module* (*CRM*) of *Nkx2.2*, a Sonic Hedgehog (Shh) signaling target gene expressed in a specific embryonic spinal cord progenitor domain [10]. The protocol presented here provides the details on how to manipulate DLR assay within the developing chick embryo system. This approach takes advantage of the easy availability and accessibility of chick embryos in ovo, and allows the approach to be combined with additional biomedical and pharmaceutical assays that together provide a powerful approach for studying gene regulation in vivo.

2 Materials

2.1 Animals

Specific pathogen free (spf) fertilized White Leghorn chicken eggs (Aichi Line) are available from suppliers such as Charles River (MA, USA).

2.2 Plasmids

Experimental vector uses pGL3 basic luciferase reporter with multiple cloning sites to insert the DNA sequence of interest (e.g., *Nkx2.2-CRM* and chicken beta-actin combined promoter). The phRL vector was used as an internal control (*see* **Note 1**). For each individual experiment, some factors (e.g., certain expression plasmids, synthesized RNAi oligomers, and small molecule drugs) can be applied to directly regulate pGL3 transcription/translation, or indirectly by modulating target signaling cascades. To optimize the transfection stability and sensitivity among different samples within one assay, we also used carrier plasmids or PUC18 vector to balance the overall DNA quantity for each set of transfections (discussed below). All plasmids are purified from Maxi-prep.

2.3 Reagents

1. Dual-Luciferase Reporter Assay System (100× assays, DLR, Promega, WI, USA): Passive Lysis Buffer (5×; PLB), Luciferase Assay Buffer II (LABII), Luciferase Assay Substrate (LAS), Stop&Glo Buffer (SGB), Stop&Glo Substrate (50×; SGS) (*see* **Note 2**).
2. Others: Leibovitz's L-15 Medium, distilled deionized water (ddH_2O), Plasmid Elution Buffer (10 mM Tris–Cl, pH 8.5), 4 % Trypan Blue Solution, Dulbecco's Phosphate-Buffered Saline 1× (DPBS).

2.4 Others

22G1½ needles, Borosilicate glass capillaries (WPI, 1.2 mm OD, 0.68 mm ID, 4 in. length), BD Vacutainer Blood Collection Set (Model 367251), cellophane tape, micro scissors, micro forceps, disposable transfer pipettes, petri dishes, a small brush, 15 and 1.5 mL eppendorf tubes, 5 mL syringe and a large gauge needle, pushpins, and an appropriate platform for holding the egg stably during electroporation.

3 Methods

3.1 In Ovo Electroporation (Details Described in Chapter 13)

1. Incubate eggs for 53 h to H&H stage 14–15 embryos to do electroporation (this is an ideal time window for the injection of plasmids for functional studies in the developing spinal cord) (*see* **Note 3**).
2. Design an appropriate combination of reporter, effector, and "balance" plasmid DNA mixture (*see* **Note 4**).
3. Carefully fill the neural tube central canal with the DNA mixture with a single injection before electroporation (*see* **Note 5**) and return to incubator for appropriate time.
4. Collect decapitated embryos at desired H&H stage (*see* **Note 6**).

3.2 Active Lysis of Cells (See Note 7)

1. Wash each embryo with prechilled DPBS in a large petri dish, transfer it into one 1.5 mL eppendorf tube, and carefully remove residual DPBS using vacuum suction.
2. Prepare fresh 1× PLB by diluting 5× PLB with ddH_2O. Add 200 μL 1× PLB for each embryo (*see* **Note 8**).
3. Homogenize embryos with 22G1½ needles on ice (*see* **Note 9**). Incubate on ice for 15 min.
4. Subject the cell lysate to two freeze–thaw cycles (dry ice/room temperature) (*see* **Note 10**).
5. Centrifuge cell lysate at 13,800 RCF for 30 s. Transfer the supernatant into a fresh 1.5 mL tube (*see* **Note 11**).

3.3 Luminescence Measurements (See Notes 12–14)

1. Thaw LARII and SGB aliquots at room temperature. Further dispense LARII into 100 μL aliquots. Prepare SGR (*see* **Note 15**).
2. Pre-warm TD-20/20n Single-Tube Luminometer (Tuner Biosystems, CA, USA). Program it with a 2-s pre-read delay followed by a 10-s measurement period for each assay (*see* **Note 16**).
3. Transfer 20 μL cell lysate into LARII and mix by gently pipetting three times (*see* **Note 17**). Put the tube in the luminometer to measure luminescence (*see* **Note 18**).
4. Once ready, take the tube out, transfer 100 μL SGR, and mix by briefly vortexing. Return it to the luminometer and start reading (*see* **Note 18**).
5. Remove the tube and move forward to the next sample (*see* **Note 19**).
6. When the assay is completed, the rest of the cell lysate can be stored at −80 °C for future use.

4 Notes

1. *Renilla* luciferase can interact with green fluorescent protein (GFP), which could influence its emitted light intensity by resonance energy transfer [11]. For this reason, it is obligatory to avoid using GFP expression plasmid for the DLR assay.
2. Upon receiving the DLR assay kit (100× assays): (1). Store 5× PLB at −20 °C and prepare fresh 1× PLB with ddH_2O just before use; (2). Prepare Luciferase Assay Reagent II (LARII) by dissolving LAS with 10 mL LABII, dispense it into 1 mL aliquots, and store at −80 °C; (3). Dispense SGB into 1 mL aliquots and store at −80 °C; (4). Store 50× SGS at −20 °C and prepare Stop & Glo Reagent (SGR) by mixing 20 μL 50× SGS with one 1 mL SGB aliquot just before use.
3. Since incubated chicken embryos are continuously developing organisms, this in vivo environment is much more complex than cell lines and primary-cultured cells, i.e., it is not very easy to achieve stable experimental outcomes with in ovo chicken system. To this end, it is important to select experimental animals of a similar size and equivalent HH stage.
4. First, the overall DNA concentration of the plasmid mixture is up to 2 μg/μL because highly concentrated DNA makes the solution too sticky to inject consistently into the neural tube, which results in variable electroporation efficiencies and inconsistent, unreliable luminescent readings. Second, to avoid *trans* effects between promoters, it is useful to add a small volume of internal control vector to generate a constitutive normalization signal at a relatively low-level. 10:1 to 50:1 is a recommended

ratio range for experimental reporter vector to internal control vector. To optimize the formula for plasmid mixture, several sets of preliminary co-transfections are needed.

5. Stable and efficient transfections are critical for in ovo luciferase quantitation, since multiple injections could cause the leakage of plasmid out of the neural tube lumen.
6. pGL3 and hRL vectors, as well as the effector plasmid, are all conventional expression vectors that can be stably expressed in early embryonic chick neural tube for approximately 48 h but degrade thereafter.
7. PLB is applicable for both active and passive lysis. Passive lysis is sufficient for cultured cells and can be performed by incubating at room temperature for 15 min. However, for tissue samples, active lysis is required since cells are still held together by complex cellular components. Therefore, a physical homogenization process is required followed by a few freeze–thaw cycles in order to prepare whole tissue for luciferase analysis.
8. Based on our experience, the overall amount of protein that can be purified from seven decapitated H&H stage 22 chicken embryos is equal to approximately one 10 cm petri dish of 293t cells at 95 % confluence. If embryos are collected earlier than E3.5, 200 μL of PLB therefore is sufficient to completely lyse one decapitated embryo. If desired, tissues can be store in PLB at −80 °C for weeks.
9. Tissue should not be homogenized too vigorously to avoid denaturing proteins through the introduction of heat generated by friction.
10. Notably, luciferase activity will gradually decay with more than three freeze–thaw cycles.
11. Clear cell lysate can be stored effectively at −80 °C for weeks.
12. This procedure is designed for a single-tube manual luminometer.
13. Though typically low, background subtraction is crucial for accurate reading outcomes, especially for monitoring low-level luciferase activity. Several preliminary measurements may be needed to detect system background. Light contamination from the environment and static electricity on the sample tube contribute to light artifacts that can affect accurate readings of luciferase activity. To control for this background, measurements from several sets of non-transfected control cell lysates are required. In addition to system background described above, background luminescence also arises from autoluminescence of coelenterazine (*Renilla* luciferin substrate) and residual firefly luciferase luminescence in the second round of *Renilla* luciferase reading. With appropriate precautions, autoluminescence

of coelenterazine is greatly prevented by PLB and SGR reagents and the first round firefly luciferase luminescence is quenched more than 1,00,000-fold, both of which should be reduced to an undetectable value. Thus, if carefully performed, and if *Renilla* activity level is not too low, these new generated background signals can be largely ignored.

14. Once cell lysate is added into the LARII, the entire reading procedure should be completed within 1 min.
15. Before making a luminescence measurement, LARII and SGR should be kept at ambient room temperature; exposure sunlight or bright bulbs should be avoided.
16. The luminometer quantifies a measurement of light intensity over a defined period.
17. Vigorously mixing will leave a little solution in the tube wall that may escape from being quenched by SGR and influence the following *Renilla* luminescent readout.
18. If the luminometer is not connected to a printer or computer, manually record the readout value.
19. To get precise results, each experiment (in ovo electroporation plus luciferase reading) needs to be replicated at least three times.

References

1. Baldwin TO (1996) Firefly luciferase: the structure is known, but the mystery remains. Structure 4(3):223–228
2. Fisher AJ, Thompson TB, Thoden JB, Baldwin TO, Rayment I (1996) The 1.5-A resolution crystal structure of bacterial luciferase in low salt conditions. J Biol Chem 271(36):21956–21968
3. Shimomura O (1985) Bioluminescence in the sea: photoprotein systems. Symp Soc Exp Biol 39:351–372
4. Naylor LH (1999) Reporter gene technology: the future looks bright. Biochem Pharmacol 58(5):749–757
5. Leclerc GM, Boockfor FR, Faught WJ, Frawley LS (2000) Development of a destabilized firefly luciferase enzyme for measurement of gene expression. BioTechniques 29(3):590–1, 4–6, 8 passim
6. Gould SJ, Subramani S (1988) Firefly luciferase as a tool in molecular and cell biology. Anal Biochem 175(1):5–13
7. Matthews JC, Hori K, Cormier MJ (1977) Purification and properties of Renilla reniformis luciferase. Biochemistry 16(1):85–91
8. Nakajima Y, Kimura T, Sugata K, Enomoto T, Asakawa A, Kubota H et al (2005) Multicolor luciferase assay system: one-step monitoring of multiple gene expressions with a single substrate. Biotechniques 38(6): 891–894
9. Wu C, Suzuki-Ogoh C, Ohmiya Y (2007) Dual-reporter assay using two secreted luciferase genes. BioTechniques 42(3):290, 2
10. Wang H, Lei Q, Oosterveen T, Ericson J, Matise MP (2011) Tcf/Lef repressors differentially regulate Shh-Gli target gene activation thresholds to generate progenitor patterning in the developing CNS. Development 138(17):3711–3721
11. Ward WW, Cormier MJ (1976) In vitro energy transfer in Renilla bioluminescence. J Phys Chem 80:2289–2291

Part III

Axon Pathways and Synapses

Chapter 21

Growth Cone Collapse Assay

Xin Yue, Alexander I. Son, and Renping Zhou

Abstract

Growth cone collapse is an easy and efficient test for detecting and characterizing axon guidance activities secreted or expressed by cells. It can also be used to dissect signaling pathways by axon growth inhibitors and to isolate therapeutic compounds that promote axon regeneration. Here, we describe a growth cone collapse assay protocol used to study signal transduction mechanisms of the repulsive axon guidance molecule ephrin-A5 in hippocampal neurons.

Key words Hippocampus, Neurons, Ephrins, F-actin, Lamellipodia, Filopodia

1 Introduction

The growth cone is a highly dynamic structure at the growing tip of axons and dendrites capable of sensing various attractive and repulsive cues that guide the migrating processes to their targets during development [1, 2]. Growth cones were first discovered by Ramón y Cajal more than a century ago and described as "a concentration of protoplasm of conical form" at the axonal tips in his fixed early chick embryonic spinal cord tissues [3]. In addition, Cajal correctly speculated that growth cones sensed guidance signals secreted by surrounding cells to reach their migratory destinations. Indeed, we now know that these structures respond quickly to guidance signals; attractive signals cause growth cones to migrate towards a given cue, while repulsive signals result in growth cones to turn away or collapse [2].

Growth cones are fan-shaped structures supported by a meshwork of filamentous actin (F-actin) in lamellipodia bundled into thick actin fibers to form filopodial protrusions [2]. Interspersed among the actin fibers are microtubules that regulate growth cone dynamics along with actin cytoskeleton [4]. Growth cone collapse was originally observed by Kapfhammer and Raper when they encountered signals located on inhibitory neurites *in vitro* [5].

Renping Zhou and Lin Mei (eds.), *Neural Development: Methods and Protocols*, Methods in Molecular Biology, vol. 1018, DOI 10.1007/978-1-62703-444-9_21, © Springer Science+Business Media, LLC 2013

Similarly, exposure of growth cones to repulsive axon guidance activities also led to the collapse of these structures [6].

The growth cone collapse assay has been important in isolating repellent axon guidance cues such as semaphorin 3A [7, 8]. In addition, due to its simplicity and ease of implementation, the assay has been used to dissect the molecular pathways that mediate the actions of repulsive guidance signals, such as ephrins [9–13], slits [14–17], and Nogo-A [18–21], and has also been used to identify pharmacological inhibitors that promote axon regeneration by counteracting axon growth inhibitors [22, 23].

The following protocol has been used to analyze signal transduction pathways underlying ephrin inhibition of rat embryonic hippocampal growth cones and is applicable to growth cone collapse assays using embryonic rat or mouse brain and spinal cord neurons. This protocol complements an excellent protocol published earlier by Kapfhammer et al. for chick embryonic neurons [24].

2 Materials

2.1 Hippocampal Neuron Culture Components

1. E18 pregnant Sprague Dawley rat.
2. Nunc Lab-Tek chamber slides (Fisher Scientific, Pittsburgh, PA).
3. Poly-D-lysine (Sigma-Aldrich, St. Louis, MO), stored at −80 °C, at stock concentration of 10 mg/mL. At the day of slide coating, add 200 μL of PDL to 4 mL of PBS and make the final concentration 0.5 μg/μL.
4. Laminin (Sigma-Aldrich, St. Louis, MO), stored at −80 °C, at stock concentration 1 μg/μL. At the day of coating, add 80 μL to 4 mL PBS and make the final concentration 20 μg/mL.
5. Sterile surgical scalpel blades (Fisher Scientific, Pittsburgh, PA).
6. Forceps (Fisher Scientific, Pittsburgh, PA).
7. 10 cm Corning Pyrex reusable petri dishes (Fischer Scientific, Pittsburgh, PA).

2.2 Neuron Culture Reagents

1. L-Glutamine, 200 mM (Life Technologies, Grand Island, NY).
2. B27 supplement, 50× (Life Technologies, Grand Island, NY).
3. Penicillin–streptomycin, 10,000 U–10,000 μg/mL (Sigma-Aldrich, St. Louis, MO).
4. Neurobasal medium (Life Technologies, Grand Island, NY) (*see* **Note 1**).
5. Neuron culture medium: To make neuron culture medium to 500 mL of Neurobasal medium, add the following components:

(a) 1.25 mL of 200 mM L-glutamine.
(b) 10 mL of 50× B27 supplement.
(c) And 5 mL of 10,000 U/mL of penicillin–streptomycin.
(d) Cover the media bottle with aluminum foil and store at 4 °C.

6. 2.5 % trypsin (Life Technologies, Grand Island, NY).
7. Fixation solution: 4 % paraformaldehyde and 0.25 % glutaraldehyde in a cacodylate buffer (0.1 M sodium cacodylate, 0.1 M sucrose, pH 7.4).

2.3 Ephrins

1. Ephrin-A5–Fc (R&D Systems, Minneapolis, MN).
2. AffiniPure rabbit anti-human IgG Fcγ (Jackson ImmunoResearch, West Grove, PA).
3. Texas Red® X-Phalloidin (Life Technologies, Grand Island, NY).

3 Methods

3.1 Coating Chamber Slides

1. Coat chamber slide with poly-D-lysine solution (0.5 μg/μL) in a 37 °C tissue culture incubator overnight.
2. Aspirate the poly-D-lysine solution and wash the chambers with PBS three times.
3. Coat the chambers with laminin by adding sufficient amounts of laminin solution (20 μg/mL) to cover the chamber bottom in a tissue culture incubator at 37 °C for 2 h.
4. Aspirate the laminin solution (DO NOT WASH WITH PBS). The chamber slide is now ready for neuron culture.

3.2 Preparation of Hippocampal Neuronal Explants (See Note 2)

1. Terminally anesthetize a pregnant rat on day 18 of pregnancy. Spray the skin of the rat with 70 % ethanol (*see* **Note 3**).
2. Make a surgical incision horizontally across the lower abdomen with scissors. Follow this with another incision vertically up the midline.
3. In a sterile workspace (laminar flow hood), remove the uterine sacs from the mother and place the sacs in a 10 cm petri dish filled with 10 mL of sterile PBS (*see* **Note 3**).
4. For each embryo sac, use two sharp forceps to pull open the sac and remove the embryo into a second clean petri dish filled with PBS until all embryos are collected.
5. Hold the embryo head with curved forceps. Use one tip of sharp Dumont forceps to punch a hole in the foramen magnum and gently tear open the skull. Use a second curved forceps to gently push the brain out of the skull.
6. Transfer the brain to a sterilized 10 cm glass dish containing 5 mL of PBS.

7. Make a diagonal cut between the cortex and mesencephalon using a scalpel.
8. Carefully remove the meninges using fine Dumont forceps and separate the hippocampus from the cortex.
9. Using a pair of sharp scalpels, separate the hippocampi from the entorhinal cortex.
10. Place the dissected hippocampi into Neurobasal medium and cut the tissue into very small pieces (*see* **Note 4**). Cut the explants by repeatedly crossing the points of two dissection knife blades and seed the tissue into slide chambers coated with poly-D-lysine and laminin.
11. Maintain explants at 37 °C in a humidified tissue culture incubator with 5 % CO_2 in Neurobasal medium supplemented with B27 and 2 mM L-glutamine.

3.3 Preparation of Dissociated Neurons (See Note 2)

1. Dissect the hippocampi from E18 rat brains in PBS and immediately transfer the tissues into 15 mL sterile conical tubes with 4.5 mL of Neurobasal medium.
2. Add 0.5 mL of 2.5 % trypsin for 15 min in a 37 °C water bath.
3. Dissociate neurons by trituration with Pasteur pipettes in the Neurobasal medium and plate the neurons onto chamber slides coated with poly-D-lysine and laminin.
4. While the hippocampal neurons settle to the bottom of the tube, pipette off the trypsin. Do not use a vacuum as neurons will be lost. Add 5 mL of media, mix, and let stand for 5 min. Repeat this step two more times to wash off any residual trypsin.
5. Add 5 mL of media and dissociate residual neuronal aggregates by pipetting up and down using a 1,000 μL micropipette.
6. Plate the dissociated neurons in coated culture wells at different densities (*see* **Note 5**).

3.4 Ephrin Cross-Linking

1. Cross-link ephrin with anti-Fc IgG at a ratio of 5:1 in micrograms. For example, 1 μg ephrin should be mixed with 0.2 μg of anti-IgG in a 40 μL volume of PBS (20 μL of 50 μg/mL ephrin stock mixed with 20 μL of 10 μg/mL anti-IgG stock).
2. Mix and incubate the mixture in a 37 °C water bath for 2 h.
3. Dilute the cross-linked ephrin at the appropriate working concentrations. For hippocampal growth cone collapse, 0.2 μg/mL is effective in inducing growth cone collapse after 15 min.

3.5 Treatment of Neuronal Cultures for Growth Cone Collapse Assay

1. Treat hippocampal cultures for 15 min with 0.2–2 μg/mL ephrin-A5–Fc that has been preclustered for 2 h with rabbit anti-human IgG Fcγ.
2. Treat control cultures with the Fc fragment alone preclustered with anti-IgG antibody.

3.6 Fixation and Staining

1. Fix neurons with fixation solution for 30 min at 37 °C after ephrin treatment.
2. Wash the fixed neurons with PBS three times.
3. Incubate with 0.3 % Triton® X-100 in PBS for 20 min at room temperature.
4. Wash with PBS three times.
5. Stain with Texas Red® X-Phalloidin for 20 min at room temperature, and cover the slides with aluminum foil to prevent photobleaching of the fluorescent dye.
6. Texas Red® X-Phalloidin has a stock concentration of 200 U/mL (equivalent to approximately 6.6 μM) in methanol. Dilute 5 μL of the methanol stock solution into 200 μL PBS for each chamber to be stained.
7. Wash with PBS three times.
8. Mount the slide with the 1:1 mix of PBS and glycerol, and seal the edges of the coverslip with nail polish (*see* **Note 6**).
9. Examine for growth cone morphology under a fluorescence microscope. Specimens prepared in this manner retain actin staining for at least 2–3 days when stored in the dark at 4 °C.
10. Growth cone collapse effects can be quantified by counting the percent of neurites with growth cones after treatment compared with controls.

4 Notes

1. Glutamate has been removed according to the Neurobasal medium data sheet: http://tools.invitrogen.com/content/sfs/manuals/3956%20Neurobasal.pdf
2. The growth cone collapse assay can be done using either neural explants or dissociated single neurons. Explants allow for better growth and survival, as axons are able to radiate from these structures with growth cones at the outreaching tips. In contrast, dissociated neuronal cultures are more optimal when gene expression manipulation is required, as it allows access to individual neurons using calcium- or lipofection-mediated transfection or through the use of viral vectors.

3. One major issue during brain dissection for neuron culture is contamination. There are a number of ways to control contamination. First, all tools must be sterilized before the experiments; ideally, duplicates of tools should be sterilized and available to account for instruments mistakenly touching nonsterile surfaces during the procedure. Second, anesthetized pregnant rats can be better sterilized by soaking the entire animal in a 500 mL beaker with 250 mL of 70 % ethanol for 5 min. Third, a fresh beaker with 70 % ethanol should be prepared in case tools need to be sterilized during the procedure. Fourth, adding penicillin–streptomycin to the medium helps prevent bacterial contamination in the cultured neurons; however, be aware that this does not prevent fungal growth.
4. When generating explant cultures, smaller explants give the best results. Neurites in the smaller explants are better separated, allowing for easier quantification. The best size is around a few dozen neurons.
5. For dissociated neurons, proper density allowing for the observation of distinct growth cones while also sustaining robust neuronal growth is necessary for the collection of accurate data. Cultures with too high density will generate growth cones overlapping with neurons, making proper quantification (such as percent of neurites with growth cones) difficult. In addition, high-density cultures often contain large amounts of cellular debris which may also interfere with observations since these errant fragments can be stained with the Phalloidin dye. In contrast, neurons do not survive well when plated with too few cells. Seeding neurons in multiple wells with varying densities is therefore prudent for choosing samples with the appropriate concentration of neurons for the assay.
6. Make sure the coverslip seal with nail polish is complete, since the glycerol mounting medium is sticky and may damage or dirty the microscope lens. Clean the slides of any residual glycerol on the slides with 70 % ethanol to allow for clearer imaging.

References

1. Gallo G, Letourneau PC (2004) Regulation of growth cone actin filaments by guidance cues. J Neurobiol 58:92–102
2. Dent EW, Gupton SL, Gertler FB (2011) The growth cone cytoskeleton in axon outgrowth and guidance. Cold Spring Harb Perspect Biol 3:pii: a001800
3. Cajal RY (1890) A quelle epoque apparaissent les expansions des cellule nerveuses de la moelle epinere du poulet. Anat Anzerger 5:609–613
4. Buck KB, Zheng JQ (2002) Growth cone turning induced by direct local modification of microtubule dynamics. J Neurosci 22: 9358–9367
5. Kapfhammer JP, Grunewald BE, Raper JA (1986) The selective inhibition of growth cone extension by specific neurites in culture. J Neurosci 6:2527–2534
6. Cox EC, Muller B, Bonhoeffer F (1990) Axonal guidance in the chick visual system:

posterior tectal membranes induce collapse of growth cones from the temporal retina. Neuron 4:31–37

7. Luo Y, Raible D, Raper JA (1993) Collapsin: a protein in brain that induces the collapse and paralysis of neuronal growth cones. Cell 75: 217–227
8. Luo Y et al (1995) A family of molecules related to collapsin in the embryonic chick nervous system. Neuron 14:1131–1140
9. Drescher U et al (1995) In vitro guidance of retinal ganglion cell axons by RAGS, a 25 kDa tectal protein related to ligands for Eph receptor tyrosine kinases. Cell 82:359–370
10. Meima L et al (1997) AL-1-induced growth cone collapse of rat cortical neurons is correlated with REK7 expression and rearrangement of the actin cytoskeleton. Eur J Neurosci 9:177–188
11. Meima L, Moran P, Matthews W, Caras IW (1997) Lerk2 (ephrin-B1) is a collapsing factor for a subset of cortical growth cones and acts by a mechanism different from AL-1 (ephrin-A5). Mol Cell Neurosci 9:314–328
12. Wahl S, Barth H, Ciossek T, Aktories K, Mueller BK (2000) Ephrin-A5 induces collapse of growth cones by activating Rho and Rho kinase. J Cell Biol 149:263–270
13. Yue X, Dreyfus C, Kong TA, Zhou R (2008) A subset of signal transduction pathways is required for hippocampal growth cone collapse induced by ephrin-A5. Dev Neurobiol 68:1269–1286
14. Nguyen Ba-Charvet KT et al (1999) Slit2-mediated chemorepulsion and collapse of developing forebrain axons. Neuron 22: 463–473
15. Roche FK, Marsick BM, Letourneau PC (2009) Protein synthesis in distal axons is not required for growth cone responses to guidance cues. J Neurosci 29:638–652
16. Piper M et al (2006) Signaling mechanisms underlying Slit2-induced collapse of Xenopus retinal growth cones. Neuron 49:215–228
17. Wong EV, Kerner JA, Jay DG (2004) Convergent and divergent signaling mechanisms of growth cone collapse by ephrinA5 and slit2. J Neurobiol 59:66–81
18. GrandPre T, Nakamura F, Vartanian T, Strittmatter SM (2000) Identification of the Nogo inhibitor of axon regeneration as a Reticulon protein. Nature 403:439–444
19. Hsieh SH, Ferraro GB, Fournier AE (2006) Myelin-associated inhibitors regulate cofilin phosphorylation and neuronal inhibition through LIM kinase and Slingshot phosphatase. J Neurosci 26:1006–1015
20. Joset A, Dodd DA, Halegoua S, Schwab ME (2010) Pincher-generated Nogo-A endosomes mediate growth cone collapse and retrograde signaling. J Cell Biol 188:271–285
21. Naska S, Lin DC, Miller FD, Kaplan DR (2010) p75NTR is an obligate signaling receptor required for cues that cause sympathetic neuron growth cone collapse. Mol Cell Neurosci 45:108–120
22. Gaub P et al (2010) HDAC inhibition promotes neuronal outgrowth and counteracts growth cone collapse through CBP/p300 and P/CAF-dependent p53 acetylation. Cell Death Differ 17:1392–1408
23. Montolio M et al (2009) A semaphorin 3A inhibitor blocks axonal chemorepulsion and enhances axon regeneration. Chem Biol 16: 691–701
24. Kapfhammer JP, Xu H, Raper JA (2007) The detection and quantification of growth cone collapsing activities. Nat Protoc 2:2005–2011

Chapter 22

The Stripe Assay: Studying Growth Preference and Axon Guidance on Binary Choice Substrates In Vitro

Markus Weschenfelder, Franco Weth, Bernd Knöll, and Martin Bastmeyer

Abstract

Stripe assays are frequently used for studying binary growth decisions of cells and axons towards surface-bound molecules in vitro. In particular in the fields of neurodevelopment and axon guidance, stripe assays have become a routine tool. Several variants of the stripe assay have been developed since its introduction by Bonhoeffer and colleagues in 1987 (Development 101:685–696, 1987). In all variants, however, the principle is the generation of a structured binary growth substrate, consisting of two sets of cues, arranged in alternating stripes. There are two major classes of stripe assays, mainly distinguished by the source material used for stripe pattern manufacturing: membrane stripe assays, where the stripe patterns are generated with membrane fractions isolated from tissue or cells, and stripe assays with purified proteins, also called modified stripe assays. In this chapter we describe in detail the classical membrane stripe assay, the commonly used modified stripe assay employing purified proteins, and a novel stripe assay for high-affinity interacting proteins, like receptor/ligand pairs.

Key words Substrate patterning, Explant culture, Axon guidance, Growth cone, Retinotectal, Ephrin/Eph

1 Introduction

The stripe assay was originally developed by F. Bonhoeffer and colleagues [1] to investigate axon guidance mechanisms in the chick retinotectal system in vitro. In this assay, the optic tectum was cut into topographically defined pieces (anterior vs. posterior), from which membrane fractions were prepared. Employing two special silicone filtering templates and a vacuum device, the membranes derived from the tectal cells were adhered to nucleopore filters. The resulting carpets consisting of alternating stripes of tectal membranes of different topographic origin were used as choice substrates for outgrowing retinal ganglion cell axons. Employing this assay, the presence of a graded distribution of repulsive axon guidance

Renping Zhou and Lin Mei (eds.), *Neural Development: Methods and Protocols*, Methods in Molecular Biology, vol. 1018, DOI 10.1007/978-1-62703-444-9_22, © Springer Science+Business Media, LLC 2013

cues on tectal membranes was discovered [2, 3]. These repulsive guidance cues were later on identified as ephrin-As ([4], *for review see refs.* 5–7). To focus on issues of cell/protein instead of cell/ tissue interactions, Vielmetter et al. modified the membrane stripe assay for use with purified proteins [8, 9]. For this assay, the striped pattern is produced using a silicone matrix with parallel microchannels of defined spacing and width imprinted in the surface. After attachment of the structured matrix face to a flat plastic or glass surface and thereby sealing the channels, a first protein solution is applied into the microchannels through an inlet slot on the backside of the matrix. The molecules adsorb from solution to cover the accessible surface, reiterating the microchannel pattern. After removing the matrix, another protein solution can be applied. While covering the whole stripe pattern, the second protein will only adhere to those parts of the surface not yet blocked by the first, i.e., the alternate stripes, as long as the two proteins have low affinity for each other. The resulting stripe carpet substrate can be used to study differential reactions of cells or neurites towards the two proteins of interest. In the field of axon guidance, in particular for the ephrin/ Eph system, bidirectional signaling through both partners of a receptor/ligand pair has recently come into focus ([10]; *for review see refs.* 11, 12). For molecules interacting with high affinity, like receptor/ligand pairs, Vielmetter's version of the stripe assay is no longer applicable, because the first protein, instead of blocking the binding of the second, will now act as an affinity matrix. Therefore, we developed the receptor/ligand stripe assay [13]. This method combines protein contact printing [14] with physisorption from the microfluidic channel system. First, the microchannel silicone matrix is used for contact printing of one protein, and subsequently, while the matrix remains attached to the surface, the other protein is applied by physisorption from the channels. Thus, the two molecular species remain spatially separated during surface tethering, preventing their interaction. This yields an alternating pattern of sharply separated receptor and ligand stripes. The method provides a simple but powerful tool to get insights into receptor/ligand bidirectional signaling and should be equally suitable for studies of other interacting proteins. All three methods of generating stripe carpets as choice substrates for cell and tissue culture (membrane stripe assay, Vielmetter's modified stripe assay for purified proteins, and the receptor/ligand stripe assay) will be detailed in this chapter.

2 Materials

As the materials required for the membrane stripe assay (protocol 1, *see* Subheading 3.1) differ significantly from those required for stripe assays with purified proteins (Vielmetter's modified stripe assay and receptor/ligand stripe assay, protocol 2, *see* Subheading 3.2), they are listed separately.

2.1 Equipment and Reagents for Protocol 1

1. Refrigerated microcentrifuge.
2. Ultracentrifuge (e.g., Beckman Optima TLX with TLS-55 swinging bucket rotor, Beckman Coulter, Brea, CA, USA).
3. Ultracentrifuge rotor tubes 1.5 ml (Beckman Coulter, Brea, CA, USA).
4. Spectrophotometer, 0.1 ml quartz cuvettes.
5. Laminar flow hood.
6. Cell/tissue culture incubator.
7. Equipment for suction apparatus: vacuum source (vacuum pump or laboratory bench vacuum), suction bottle, nonreturn valve fitting in bottle, tubing to connect bottle, plastic clips to close tubing, suction filter plate (can be purchased from the lab of Martin Bastmeyer, bastmeyer@kit.edu), and manometer (*see* Fig. 1a, b).
8. Cell scraper.
9. Syringes, 1 ml.
10. Needles 26 G, 0.45 μm.
11. Hamilton® syringe (Hamilton, Bonaduz, Switzerland).
12. White and blue silicone templates. White templates contain microchannels for generating the stripe pattern; blue templates contain a nylon mesh (*see* Fig. 1c, d). Both types of templates can be purchased from the lab of Martin Bastmeyer.
13. Forceps with flat, round tips.
14. Polycarbonate nucleopore filters, 0.1 μm pore size (Whatman GE Healthcare, Little Chalfont, Buckinghamshire, UK).
15. Spermidine × 3 HCl (Sigma Aldrich, St. Louis, USA).
16. Leupeptin (Sigma Aldrich, St. Louis, USA), dissolve 25 mg in 26.3 ml autoclaved H_2O, and store aliquots at −20 °C; final concentration 50 mM.
17. Aprotinin (Sigma Aldrich, St. Louis, USA), dissolve 25 mg in 4.2 ml autoclaved H_2O, and store aliquots at −20 °C; final concentration 200 U/ml.
18. Pepstatin (Sigma Aldrich, St. Louis, USA), dissolve 25 mg in 18 ml methanol, and store aliquots at −20 °C; final concentration 2 mM.
19. Natural mouse laminin (Invitrogen, Carlsbad, CA, USA).
20. Fluorescent microspheres, 0.5 μm, e.g., FluoSpheres® carboxylate modified, red fluorescent (Invitrogen, Carlsbad, CA, USA).
21. Hank's balanced salt solution (HBSS) without Ca^{2+} and Mg^{2+} (Invitrogen, Carlsbad, CA, USA).
22. Sterile phosphate buffered saline (PBS) without Ca^{2+} and Mg^{2+} (Invitrogen, Carlsbad, CA, USA), add 500 U/ml penicillin and 50 μg/ml streptomycin.

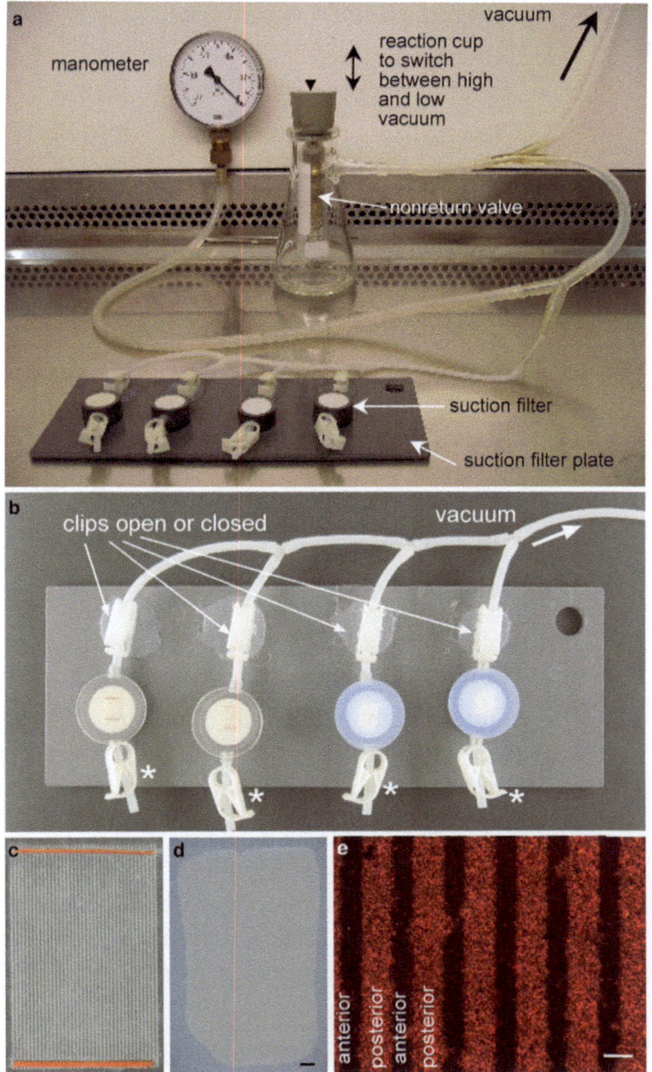

Fig. 1 Experimental setup for the membrane stripe assay. (**a**) Four suction filters mounted on a suction filter plate are connected to a vacuum pump via a nonreturn valve. Closing the hole (at *black triangle*) of the valve by inserting a reaction cup leads to a decrease of the pressure in the channel system (to −800 to −1,000 mbar; "high vacuum"). Removal of this cup increases the pressure (to −200 to −400 mbar; "low vacuum"). (**b**) Detailed view of the suction filter plate illustrating the mounting of the templates on the suction filters. The two *left* filters are covered with *white* templates to produce the first stripes. The two *right* filters are covered with *blue* templates to generate the set of second stripes. Clips marked with *white stars* have to be closed permanently. Clips on top are either closed (during changing nucleopore filters) or open (during suction operation). (**c**) Higher magnification of a *white* template showing the channel system. (**d**) Higher magnification of the *blue* template used to prepare the second stripes, depicting the innermost white mesh and the surrounding blue silicone base. (**e**) A typical pattern of alternating stripes derived from membrane fractions of the anterior and posterior chicken optic tectum, respectively. Posterior membranes are visualized with rhodamine-labeled beads. Scale bar: 100 μm

23. Homogenization Buffer (HB), 10 mM Tris–HCl, pH 7.4, 1 mM spermidine × 3 HCl, and 1.5 mM $CaCl_2 \times 2H_2O$.
24. 1.8 M sucrose in HB, sterile filtered (0.4 μm); store at −20 °C.
25. SDS, 2 % solution in H_2O.
26. Cell/tissue culture medium (e.g., DMEM).

2.2 Equipment and Reagents for Protocol 2

1. Refrigerated microcentrifuge.
2. Laminar flow hood.
3. Cell/tissue culture incubator.
4. N_2 stream.
5. Silicone matrices (can be purchased from the lab of Martin Bastmeyer).
6. Polystyrene Petri dishes, at least 60 mm in diameter.
7. Forceps with flat, round tips.
8. Sterile phosphate buffered saline (PBS) without Ca^{2+} and Mg^{2+}, 8 g/l NaCl, 0.2 g/l KCl, 1.15 g/l $Na_2HPO_4 \times 2H_2O$, 0.2 g/l KH_2PO_4; pH 7.4.
9. Sterile Hank's buffered salt solution (HBSS) without Ca^{2+} and Mg^{2+}, 8 g/l NaCl, 0.4 g/l KCl, 60 mg/l NaH_2PO_4, 60 mg/l $Na_2HPO_4 \times 2\ H_2O$, 0.35 g/l $NaHCO_3$, 1 g/l glucose, 7.76 g/l HEPES, 10 mg/l phenol red; adjust to pH 7.4 (*see* **Note 1**).
10. Ethanol 50 % and ethanol 99.8 %.
11. Purified proteins of interest. For the EphA3 and ephrin-A2 stripe assay described here, recombinant human ephrin-A2-Fc fusion protein and a recombinant chimera of the mouse EphA3-ectodomain with human Fc (EphA3-Fc; both R&D Systems, Minneapolis, MN, USA), anti-human-Fc-Alexa488, and anti-human-Fc-Alexa594 (both Invitrogen, Carlsbad, CA, USA) were used.
12. Natural mouse laminin (Invitrogen, Carlsbad, CA, USA).

3 Methods

3.1 Protocol 1: Membrane Stripe Assay

In the following we describe the procedure for the original version of the stripe assay employing crude membrane fractions. As the starting material for harvesting crude membrane fractions, two alternative sources can be used: tissue or a heterologous cell line. Diverse tissue sources (e.g., specific brain regions such as the chiasm [15], cortical areas [16], or hippocampus [17]) can be employed to analyze the activity of membrane-bound guidance cues expressed in a spatially differential manner (e.g., rostral vs. caudal or dorsal vs. ventral parts). In the original version of the stripe assay, guidance activity of the anterior vs. posterior chick E9 optic tectum was compared [2]. Alternatively, specific membrane-associated guidance cues under investigation can be overexpressed in cell lines (e.g., HEK293 cells in ref. 18). In either case, crude membrane fractions are harvested and employed in stripe assay experiments.

3.1.1 Preparations Before Starting Protocol 1

1. Prepare PBS with protease inhibitors (thereafter called PBS+) by adding 250 μl of leupeptin, 100 μl of aprotinin, and 10 μl of pepstatin stock solution to 10 ml PBS; store at 4 °C (prepare freshly for each experiment).
2. Prepare homogenization buffer with protease inhibitors (thereafter called HB+) by adding 125 μl of leupeptin, 50 μl of aprotinin, and 5 μl of pepstatin stock solution to 5 ml homogenization buffer; store at 4 °C (prepare freshly for each experiment).
3. Thaw 1.8 M sucrose stock and prepare 0.18 M solution by diluting in HB+; store on ice.
4. Boil blue and white templates for 5 min in autoclaved H_2O and store in autoclaved H_2O.
5. Cut nucleopore membrane into rectangular pieces (approximately 1.5–1.7 cm × 1.2–1.4 cm). Label rectangles by cutting left corner of the short side with shiny side facing downwards.
6. Boil precut nucleopore filters for 5 min in autoclaved H_2O and store in autoclaved H_2O.
7. For coating, cover the boiled nucleopore filters with 20 μg/ml laminin in HBSS and incubate for at least 1 h at 37 °C. Prior to use, wash with HBSS and store in PBS. Thaw laminin slowly on ice (approximately 1 h) to prevent aggregate formation.
8. Sterilize 1.5 ml ultracentrifuge rotor tubes under UV light for 10 min.

3.1.2 Preparation of Crude Membrane Fractions from Tissue

1. Dissect tissue of interest in HBSS and store tissue samples in ice-cold HB+ until all material has been collected. The amount of tissue used should range between approximately 40 and 60 mg. The optimal amount of starting material has to be determined according to the individual experimental design.
2. Wash tissue pieces with 500 μl of cold HB+.
3. Add 600 μl fresh HB+ after removal of the wash solution.
4. Homogenize tissue several times using a 1 ml syringe with a 26G needle. Avoid air bubble formation during homogenization. Homogenize slowly, i.e., allow 5–10 s for 600 μl homogenate to pass through the syringe. Keep homogenate on ice.

3.1.3 Preparation of Crude Membrane Fractions from Transfected HEK293 Cells

1. Prepare four 10 cm dishes with approximately $1.5–3 \times 10^6$ HEK293 cells/plate.
2. Transfect cells according to standard protocols and let them express the plasmid overnight.
3. Wash transfected cells twice with cold PBS.
4. Remove all excess PBS and add 250 μl PBS+ per dish.
5. Remove cells with a cell scraper.
6. Pool cells of all dishes with the same condition and transfer cell material to a reaction tube.
7. Centrifuge at 4 °C, 18,000 × *g* for 30 s.
8. Resuspend the pellet in 600 μl of ice-cold HB+.
9. Homogenize cells several times using a 1 ml syringe with a 26G needle. Avoid air bubble formation during homogenization. Homogenize slowly, i.e., allow 5–10 s for 600 μl homogenate to pass through the syringe. Keep homogenate on ice.

3.1.4 Sucrose Cushion Centrifugation

The following steps describe sucrose cushion centrifugation, which will remove nuclei and cytoplasmic proteins and retain the membrane fraction on top of a high-density sucrose cushion:

1. Pipette 350 μl of 1.8 M sucrose solution into a 1.5 ml rotor tube.
2. Slowly layer 150 μl of 0.18 M sucrose solution on top.
3. Load a maximum of 800 μl tissue or cell homogenate on top of the 0.18 M sucrose cushion.
4. Centrifuge at 52,000 × *g* for 10 min at 4 °C. Meanwhile prepare reaction cups with 1 ml PBS+ and store on ice.
5. After centrifugation, carefully remove the tubes from the rotor. Verify—by visual inspection—the presence of a white layer at an intermediate position in the rotor tube. This contains the crude membrane fractions (*see* **Note 2**).
6. Carefully transfer the white intermediate phase into the tube with 1 ml PBS+ with a Hamilton® syringe.
7. Centrifuge again to remove remaining sucrose for 7 min at 13,000 × *g* at 4 °C.
8. Discard the supernatant and resuspend the pellet in 1 ml cold PBS+ with the syringe with 26 G needle.

3.1.5 Determination and Adjustment of Membrane Fraction Concentration

1. Add 5 μl of each sample to 70 μl of 2 % SDS solution (prepare duplicates). Take 5 μl PBS+ plus 70 μl of 2 % SDS as reference. Determine OD at 220 nm and multiply by 15 to obtain the OD of the membrane stock suspension.

2. Adjust the membrane stock suspension to a calculated OD_{220nm} of 1.5 by diluting in cold PBS+ and store on ice (*see* **Note 3**). For one set of first stripes, 150 μl of membrane solution is required. For one set of second stripes, calculate a total of 300 μl, 150 μl that is required initially and an additional 150 μl to top up and prevent drying out of the nucleopore filters. As determined by Bradford assay, the adjusted membrane suspension approximately contains 60–70 μg of protein per ml.
3. Label the first stripe solution with fluorescent beads. For FluoSpheres® take a 1:1,000 dilution.

3.1.6 Preparation of Stripe Carpets

1. Set up the vacuum sucking device according to Fig. 1a.
2. Sterilize the suction filter plate by spraying with 70 % ethanol. Rinse individual suction filters with PBS.
3. Check the applied vacuum with all four clips closed: Toggling between "low" and "high" vacuum is achieved by insertion or removal of a 1.5 ml reaction cup into or from the hole of the bottle lid (*see* Fig. 1a). At "low" vacuum (no reaction cup in the hole of the bottle lid) the pressure should be around -200 to -400 mbar and at "high" vacuum (1.5 ml reaction cup inserted into the hole of the bottle lid) around -800 to -1,000 mbar.
4. Prepare 10 cm dishes with individual PBS+ drops, for intermediate storage of the nucleopore filters in **step 13**.
5. Cover each suction filter with one white template. Sterile blunt forceps should be used to handle templates and nucleopore filters. Rinse all templates with 1 ml of PBS by opening the clips. Red lines of the white template should be facing upwards (*see* Fig. 1b, c).
6. Close all clips (*see* Fig. 1b).
7. Add one nucleopore membrane on each of the four suction filters. The marked corner of the membrane should be oriented towards the upper left side (shiny surface faces bottom).
8. Pipette 150 μl of first stripe solution onto each of the four nucleopore filters.
9. Open clips. Apply high vacuum by closing the hole of the vacuum bottle with a reaction cup.
10. Start timer at -600 mbar and allow sucking for 90 s. During this period, filter regions above the template channels get clogged with membrane fragments, thereby forming the first stripes. Caution: Do not allow nucleopore membranes to dry out at any time.
11. Vent high vacuum by removing the reaction cup. Remove excess liquid.

12. Leave low vacuum until stripes get visible on the filter surface.
13. Immediately close clips. Transfer all four nucleopore membranes onto PBS+ drops (face up) in the Petri dish from **step 4**.
14. The stripe quality can be checked by fluorescence microscopy (*see* Fig. 1e).
15. Wash the white templates after use by opening clips and rinsing with 1 ml PBS.
16. Remove white templates.
17. Wash the suction filters three times with 1 ml of PBS by applying vacuum.
18. Add one blue template on each suction filter (in either orientation, *see* Fig. 1b, d).
19. Lift each blue template slightly at the edge and apply 1 ml PBS+ underneath. Blue templates must be positioned flatly on suction filters with no air bubbles underneath.
20. Retransfer nucleopore membranes (*see* **step 14**) onto suction filters, face up.
21. Immediately add 150 μl of second stripe solution and open clip. Leave low vacuum until all four clips are opened.
22. Apply high vacuum (insert reaction cup) and suck for 90 s once −600 mbar is reached. Do not allow nucleopore membranes to dry out at any time. Use the additional 150 μl to top up liquid if necessary. During this period, the flow will go through the as yet unclogged regions of the filter, generating the second stripe set.
23. Rinse nucleopore membranes with 1 ml of PBS+.
24. Close clip and vent the high vacuum (remove reaction cup).
25. Transfer processed nucleopore membranes to individual 35 mm Petri dishes and immediately add one drop of tissue culture medium onto the stripe carpets. Store dishes at 37 °C and use them the same day.

Timing: Preparing 10–12 carpets takes approximately 4–5 h.

For control experiments suggested for the membrane stripe assay, *see* **Note 4**.

3.2 Protocol 2: Modified Stripe Assay with Purified Proteins

In the following we present exemplary protocols for EphA3 and its interaction partner ephrin-A2. For any other set of proteins, you might have to adapt and optimize several parameters in this protocol, e.g., the sequence of application, concentrations, and incubation times (*see* **Note 5**). All steps have to be carried out under sterile conditions. For handling the matrices use flat, round tip forceps.

3.2.1 Preparations Before Starting Protocol 2

All annotated volumes are calculated per matrix.

1. Make up 150 µl sterile PBS containing 16 µg/ml ephrin-A2-Fc and 48 µg/ml goat anti-human-Fc-Alexa594. Incubate for 30 min at RT to cluster the ephrin-A2 protein with the anti-Fc antibody (*see* **Note 6**). Centrifuge at approximately 300 × *g* for 15 min at 4 °C; keep supernatant on ice.
2. Make up 150 µl of 30 µg/ml EphA3-Fc and 2 µg/ml anti-human-Fc-Alexa488 in sterile PBS (*see* **Note 7**); keep on ice.
3. Prepare 100 µl of 20 µg/ml laminin solution in HBSS; keep on ice.
4. Cleaning the silicone matrices. Rinse the matrix with 50 % ethanol and dry in an N_2 stream (*see* **Note 8**). Caution: Adjust the stream carefully. Excessive pressure may damage matrix microstructures.

   Adhere the matrix face to a clean plastic Petri dish and lift it off again. Repeat this procedure at least twice. If available, use a plasma-cleaned glass surface instead of a plastic Petri dish (*see* **Note 9**). Store cleaned matrices face down in a closed plastic Petri dish.

3.2.2 Modified Stripe Assay According to Vielmetter

This protocol describes the stripe assay introduced by Vielmetter et al. [8] for two noninteracting proteins (*see* **Note 10**), which was reviewed in detail by Knöll et al. [9]. The protein of interest (here EphA3-Fc) is applied to the surface via a silicone microchannel matrix and adsorbs from solution to the surface. The pattern is subsequently coated with another protein (here laminin).

To perform this type of stripe assay, start at **step 5** in Subheading 3.2.3.

3.2.3 Receptor/Ligand Stripe Assay

This assay, developed by Gebhardt et al. [13], combines protein contact printing with physisorption from the microfluidic channels. It is used for proteins that interact with high affinity, like receptor/ligand pairs. First, the microchannel silicone matrix is used for contact printing of one protein (here ephrin-A2-Fc) and afterwards—with the matrix still attached to the surface—the second protein (here EphA3-Fc) is applied into the channels. Thus, the interaction of the proteins is prevented by spatial separation during surface coating, resulting in an alternating pattern of sharply separated receptor and ligand protein stripes.

1. Place the matrix (*see* Fig. 2a′) in the bottom half of a 60 mm polystyrene Petri dish with the microchannel pattern facing upwards (*see* Fig. 2a).
2. Apply 150 µl ephrin-A2-Fc solution onto the channel field (*see* Fig. 2b, b′). Assure that the solution is spread out equally over the pattern. Be careful not to scratch the channels of the matrix with the pipette tip (*see* **Note 11**).

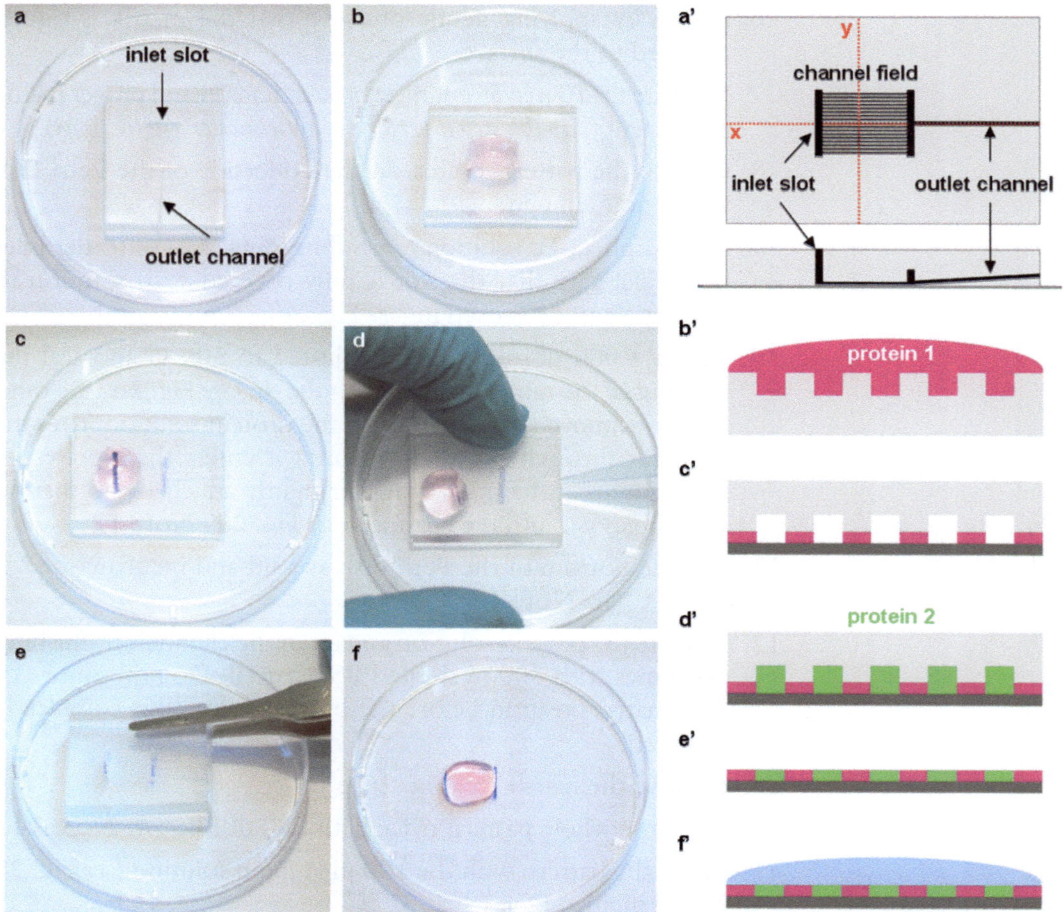

Fig. 2 Critical steps of the receptor/ligand stripe assay. Image of a silicone matrix used to generate the stripe pattern (**a**). Scheme of a top view and of a transverse section of a matrix along the indicated *x*-axis shown in *red* (**a′**). Pictures (**b**) to (**f**) and corresponding drawings of sections (**b′** to **f′**), along the indicated *y*-axis comprising five channels, depict the critical steps of the procedure. The first protein solution is applied onto the channel field (**b**) and adheres to the silicone surface (**b′**). After rinsing and drying the coated matrix, the first protein is printed onto the plastic surface (**c′**). The second protein solution is applied to the inlet slot (**c**) and sucked in (**d**). Now, the two interacting proteins are spatially separated during attachment to the same surface (**d′**). After incubation, protein adhesion, and flushing of the channels, the matrix is removed (**e**). The two proteins are now adhered in adjacent stripes on the same surface (**e′**) and can be covered by a growth-promoting ECM protein (**f, f′**)

3. Incubate for 1 h at 37 °C covered with the lid of the Petri dish.
4. Aspirate the protein solution, wash the matrix by dipping into sterile PBS five to ten times, and carefully dry in an N_2 stream. Make sure that the matrix—in particular the patterned face—is dry.
5. Place the matrix face down onto the inner side of the lid of the Petri dish (*see* Fig. 2c′). Position so that the outlet channel remains accessible with a pipette (*see* **step 10**). It is crucial not to relocate the matrix after the first touch, because the first attachment will define the printed pattern (*this remark can be ignored if performing Vielmetter's stripe assay*).

6. Assure by applying gentle pressure that the matrix is tightly attached to the surface.
7. Incubate for 10 min to optimize protein transfer and proper matrix attachment (*skip this step if performing Vielmetter's stripe assay*).
8. Outline the patterned area on the underside of the Petri dish with a pointed pen.
9. Apply EphA3-Fc solution (*see* Subheading 2.2) onto the inlet slot. Take care that the slot is fully covered with solution to prevent air to be sucked in (*see* Fig. 2c).
10. Take a 1,000 μl micropipette adjusted to 500 μl and seal the opening of the outlet channel with the tip (*see* Fig. 2d). Be careful not to move the matrix. Suck the protein solution from the inlet slot through the channels until it enters the pipette tip. Make sure that the matrix remains tightly attached and that no air bubbles get into the channels (*see* Fig. 2d′, **Note 12**).
11. Use the bottom of the Petri dish as a lid and incubate for 2 h at 37 °C (*see* **Note 13**).
12. Aspirate the protein solution and clean the channels by flushing two times with 300 μl sterile PBS. Crucial: Do not remove or alter matrix position before the protein solution is completely removed.
13. Remove the matrix (*see* Fig. 2e, e′).
14. Rinse the whole pattern twice with 300 μl PBS (*see* **Note 14**).
15. Cover the pattern with the 100 μl laminin solution (*see* Fig. 2f, f′) and incubate for 1 h at 37 °C.
16. Aspirate the laminin solution and wash two times with 300 μl PBS. The pattern can now be checked under the fluorescence microscope if labels have been used (*see* Fig. 3a and **Note 15**). Covered with PBS or HBSS, patterned surfaces can be stored at 4 °C at least 2 days before use (*see* **Note 16**).
17. Before application of cells or tissue explants, replace salt solution by the culture medium used subsequently. Incubate at least 15 min in a 37 °C incubator to equilibrate. Apply the cells or explants onto the pattern (*see* Subheading 3.3).

3.3 Plating of Explants or Cells on Stripe Carpets

Many types of neurons and nonneuronal cells have been studied with stripe assay experiments, including retinal ganglion cells [1, 2, 4, 8, 10, 13, 15, 18–21], sensory neurons of the vomeronasal epithelium [22] and from dorsal root ganglia [23], spinal motor neurons [24, 25], hippocampal [26, 27] and cortical [28] pyramidal cells, and cortical interneurons [29] as well as neural crest cells [30] and oligodendrocytes [31].

Advisable plating densities as well as the culture conditions depend on the origin of the cells studied and the scientific question

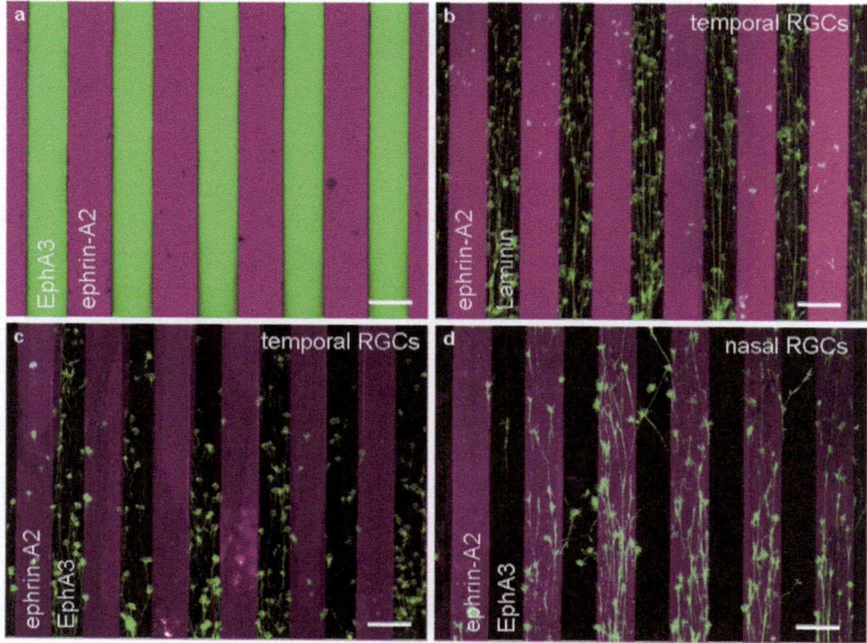

Fig. 3 Exemplary stripe patterns and stripe assay results. (**a**) Double-striped pattern generated with 30 μg/ml EphA3-Fc anti-Fc-Alexa488 antibody mix (physisorbed from solution, *green*) and 16 μg/ml ephrin-A2-Fc clustered with anti-Fc-Alexa594 antibody (printed, shown in *magenta*). The image was taken immediately after removing the matrix, with a ×10 objective. Note inevitable slight inhomogeneities of the printed ephrin-A2 pattern. (**b**) Typical result of an ephrin-A stripe assay (Vielmetter's type) with temporal chicken retinal ganglion cell axons (RGCs) avoiding the ephrin-A2 stripes (shown in *magenta*) and growing on laminin. (**c**, **d**) In a receptor/ligand stripe assay with ephrin-A2 (shown in *magenta*) and EphA3 (*not imaged*), the temporal RGC axons avoid ephrin-A2 and grow on EphA3, while nasal RGC axons avoid EphA3 and grow on ephrin-A2. Axons are stained for actin with Phalloidin-Alexa488. Scale bar: 100 μm

addressed in the experiment. Extensive axon fasciculation observed in dense neuronal cultures might interfere with axon guidance decisions with regard to the protein stripes. Exemplary images of fluorescently stained retinal axons on stripe patterns are shown in Fig. 3b–d.

4 Notes

1. PBS and HBSS can be used interchangeably.
2. If no or a fuzzy white intermediate phase is visible on the sucrose cushion, the amount of starting material is either too low or too high. Generally, membranes derived from transfected cell cultures tend to result in less sharp intermediate phases.
3. Aliquots of membrane fractions can be stored at −80 °C or in liquid nitrogen up to 1 month. However, membrane fractions lose some activity upon storage. Membranes may be stored in a 1:1 mixture of glycerol:PBS, which might preserve activity

somewhat better. Upon thawing, the membranes should at least not be frozen again.

4. Controls to perform for every set of membrane fractions and tissue sources:

   To avoid artifacts due to the sequence of stripe application (e.g., differences in membrane density between first and second stripes or contamination of first stripes with material from the second stripe), the order of allocating membrane fractions to first and second stripes should be reversed.

   Prepare striped carpets with both stripes containing the same type of membrane fragments. This should result in random outgrowth.

   To corroborate that a suggested activity is associated with a particular membrane fraction, heat inactivation can be employed. For posterior tectal membranes, incubation at 63 °C for 8 min was sufficient to eliminate all guidance activity [2]. More vigorous heat inactivation could lead to artificial activities induced by denatured protein.

   To investigate whether a particular receptor/ligand interaction is involved, mask the corresponding interaction by adding purified ligands or receptors to the medium [10, 32].

5. Parameters to check for every set of purified proteins:

   (a) *Functionality*: The mere presence of the protein on the surface need not reflect its biological functionality. Examples and mechanism of protein surface interactions interfering with protein functionality were lucidly reviewed by Butler in 2000 [33]. It has been shown that in many cases, adsorption to a surface can induce conformational changes, altered enzymatic activity, protein clustering, and even molecular unfolding. Fluorophore-labeled affinity probes (e.g., labeled ephrin-A2 for surface-bound EphA3) may provide more information about the functional state of the surface-bound protein than antibody staining.

   (b) *Protein printing vs. physisorption*: Not every protein is functional after printing (e.g., EphA3 cannot be printed functionally on a surface, while ephrin-A2 can be printed or physisorbed from the channels with comparable results). Check which combination of printing and adhesion from solution is the best for your proteins.

   (c) *Surface*: The chosen surface influences the amount and functionality of the adhered protein. At least for polystyrene and silicone, adsorption often involves conformational changes as proteins unfold to permit hydrophobic bonds with the solid phase [33]. At the same time these hydrophobic interactions enable stable protein binding. Pre-coating the surfaces with a linking protein like poly-L-lysine can increase the probability of retaining biological functionality of the protein of interest by moving it away from the hydrophobic plastic surface. Furthermore, polystyrene surfaces are not created equal. Some companies plasma treat their products to increase protein adsorption (e.g., Nunc, Dynatech), while others do not. It is not predictable which surface is the most suitable for your protein. For these reasons, it is advisable to test glass (hydrophilic) as well as different types of polystyrene Petri dishes (hydrophobic).

(d) *Concentration*: The protein concentration is very critical. It has to be adjusted individually, because the ratio of biologically functional surface-bound protein to the amount of applied protein is not equal for all proteins and surfaces. For IgG antibodies, a concentration of 5 μg/ml has been reported to be sufficient for the formation of a continuous protein monolayer on polystyrene [33]. On saturated surfaces, excess protein attaches on top of the support-bound initial layer. These secondary layers are less stable, but the probability of retaining biological functionality is increased. Thus, in some cases supersaturation of the surface may be useful. A concentration range from 1 to 50 μg/ml is reasonable in most cases.

(e) *Adhesion time*: At 37 °C, at physiological ionic strength and neutral pH, there is usually no further protein adsorption to the surface after 4 h, independent of protein concentration. Thus, it is unnecessary to use adhesion times longer than 4 h. Printed proteins need very short transfer times, as printing is diffusion independent. Increasing the time to 10 min (as described in this protocol) does not increase the amount of transferred protein but allows for more homogenous attachment of the silicone matrix to the surface.

(f) *Requirement of additional growth-permissive proteins*: The use of growth-permissive proteins (e.g., laminin) applied onto the whole pattern is necessary for many cells but adds another level of complexity to the system. The physics of adsorption—as mentioned for the stripe pattern proteins above—of course also determines the amount of biologically functional permissive protein. Different binding properties to the three potential attachment sites—free surface, protein 1, and protein 2—can result in different states of biological activity of the permissive protein in the two subsets of stripes. To rule out misinterpretations of growth decisions of axons due to such effects, controls using functionally inactivated or masked guidance proteins are advisable.

6. Most proteins need no clustering. Receptor tyrosine kinases (RTK), like EphA3, however, need to dimerize in the cell membrane for intracellular signaling. Clustered (at least dimerized) exogenous ligands induce this functional dimerization of the RTK in the membrane of treated cells. For EphA activation, an ephrin-A2 to antibody molar ratio of 1:1 has been established. Another reason for using conjugated antibodies at this point is the labeling of the protein for tracking it throughout the experiment (*see* **Note** 7).
7. If the used protein is to be labeled only, but not clustered, a protein-to-antibody molar ratio of 10:1 is sufficient. Any conjugated antibody can be used. For unspecific labeling of the stripes, the antibody even does not need to bind to the investigated guidance protein.
8. The N_2 stream, in addition to removing adherent liquid, cleans the channels from dust and other particles. Therefore, mere passive drying is not sufficient.
9. If fluorescent labels (e.g., fluorophore-labeled secondary antibodies) have been used, they are also transferred to the plastic surface at this step. This helps to monitor the cleaning success. Check the surface under a fluorescence microscope. Repeat cleaning until there is no fluorescent signal remaining.
10. "Noninteracting" in this context means that there are no specific interactions, like receptor/ligand interaction or cluster formation. Of course, there will be weak nonspecific interactions between most sets of proteins.
11. The hydrophobic surface of the silicone matrix makes this quite difficult. Use as much protein solution as needed to cover the pattern properly.
12. An alternative method to apply the protein solution is by using a Hamilton® syringe. Inject the protein solution directly via the outlet channel instead of using the inlet slot. By this means, smaller volumes of protein solution can be used (20–50 μl). Be careful not to damage matrix with the needle.
13. For other proteins up to 4 h at 37 °C might be optimal.
14. If the protein was successfully transferred to the surface, the PBS droplet will be square shaped because of the improved wettability in the region of the printed protein pattern. The protein stripes get visible during removal of the liquid.
15. The printed stripes tend to be somewhat more inhomogeneous than the stripes generated by absorption from solution.
16. The absence of exchangeable protein in a buffered salt solution as compared to culture media reduces the release of adsorbed protein to 1–2 % per day [33]. Routinely we use the patterned substrates within a week, if stored under buffered salt solution.

Acknowledgments

This work was supported by the German Research Foundation, DFG (grant BA1034/14-3 to M.B. and F.W.). B.K. is supported by the DFG, Schram-Foundation, Gottschalk-Foundation, and Non-Profit Hertie Foundation.

The authors thank Andrea Wizenmann for helpful comments on the manuscript.

References

1. Walter J, Kern-Veits B, Huf J et al (1987) Recognition of position-specific properties of tectal cell membranes by retinal axons in vitro. Development 101:685–696
2. Walter J, Henke-Fahle S, Bonhoeffer F (1987) Avoidance of posterior tectal membranes by temporal retinal axons. Development 101:909–913
3. Bonhoeffer F, Huf J (1982) In vitro experiments on axon guidance demonstrating an anterior-posterior gradient on the tectum. EMBO J 1:427–431
4. Drescher U, Kremoser C, Handwerker C et al (1995) In vitro guidance of retinal ganglion cell axons by RAGS, a 25 kDa tectal protein related to ligands for Eph receptor tyrosine kinases. Cell 82:359–370
5. Knöll B, Drescher U (2002) Ephrin-As as receptors in topographic projections. Trends Neurosci 25:145–149
6. McLaughlin T, O'Leary DD (2005) Molecular gradients and development of retinotopic maps. Annu Rev Neurosci 28:327–355
7. Suetterlin P, Marler KM, Drescher U (2012) Axonal ephrinA/EphA interactions and the emergence of order in topographic projections. Semin Cell Dev Biol 23:1–6
8. Vielmetter J, Stolze B, Bonhoeffer F et al (1990) In vitro assay to test differential substrate affinities of growing axons and migratory cells. Exp Brain Res 81:283–287
9. Knöll B, Weinl C, Nordheim A et al (2007) Stripe assay to examine axonal guidance and cell migration. Nat Protoc 2:1216–1224
10. Rashid T, Upton AL, Blentic A et al (2005) Opposing gradients of ephrin-As and EphA7 in the superior colliculus are essential for topographic mapping in the mammalian visual system. Neuron 47:57–69
11. Egea J, Klein R (2007) Bidirectional Eph-ephrin signaling during axon guidance. Trends Cell Biol 7:230–238
12. Feldheim DA, O'Leary DD (2010) Visual map development: bidirectional signaling, bifunctional guidance molecules, and competition. Cold Spring Harb Perspect Biol 2(11):a001768
13. Gebhardt C, Bastmeyer M, Weth F (2012) Balancing of ephrin/Eph forward and reverse signaling as the driving force of adaptive topographic mapping. Development 139:335–345
14. von Philipsborn AC, Lang S, Löschinger J et al (2006) Growth cone navigation in substrate-bound ephrin gradients. Development 133:2487–2495
15. Wizenmann A, Thanos S, von Boxberg Y et al (1993) Differential reaction of crossing and non-crossing rat retinal axons on cell membrane preparations from the chiasm midline: an in vitro study. Development 117:725–735
16. Hübener M, Götz M, Klostermann S et al (1995) Guidance of thalamocortical axons by growth-promoting molecules in developing rat cerebral cortex. Eur J Neurosci 7:1963–1972
17. Stein E, Savaskan NE, Ninnemann O et al (1999) A role for the Eph ligand ephrin-A3 in entorhino-hippocampal axon targeting. J Neurosci 19:8885–8893
18. Monschau B, Kremoser C, Ohta K et al (1997) Shared and distinct functions of RAGS and ELF-1 in guiding retinal axons. EMBO J 16:1258–1267
19. Simon DK, O'Leary DD (1992) Responses of retinal axons in vivo and in vitro to position-encoding molecules in the embryonic superior colliculus. Neuron 9:977–989
20. Nakamoto M, Cheng HJ, Friedman GC et al (1996) Topographically specific effects of ELF-1 on retinal axon guidance in vitro and retinal axon mapping in vivo. Cell 86:755–766
21. Mann F, Ray S, Harris WA et al (2002) Topographic mapping in dorsoventral axis of the Xenopus retinotectal system depends on signaling through ephrin-B ligands. Neuron 35:461–473

22. Knöll B, Zarbalis K, Wurst W et al (2001) A role for the EphA family in the topographic targeting of vomeronasal axons. Development 128:895–906
23. Snow DM, Lemmon V, Carrino DA et al (1990) Sulfated proteoglycans in astroglial barriers inhibit neurite outgrowth in vitro. Exp Neurol 109:111–130
24. Kao TJ, Kania A (2011) Ephrin-mediated cis-attenuation of Eph receptor signaling is essential for spinal motor axon guidance. Neuron 71:76–91
25. Bonanomi D, Chivatakarn O, Bai G et al (2012) Ret is a multifunctional coreceptor that integrates diffusible- and contact-axon guidance signals. Cell 148:568–582
26. Forster E, Tielsch A, Saum B et al (2002) Reelin, disabled 1, and beta 1 integrins are required for the formation of the radial glial scaffold in the hippocampus. Proc Natl Acad Sci USA 99:13178–13183
27. Knöll B, Kretz O, Fiedler C et al (2006) Serum response factor controls neuronal circuit assembly in the hippocampus. Nat Neurosci 9:195–204
28. Bagnard D, Lohrum M, Uziel D et al (1998) Semaphorins act as attractive and repulsive guidance signals during the development of cortical projections. Development 125: 5043–5053
29. Zimmer G, Schanuel SM, Bürger S et al (2010) Chondroitin sulfate acts in concert with semaphorin 3A to guide tangential migration of cortical interneurons in the ventral telencephalon. Cereb Cortex 20:2411–2422
30. Wang HU, Anderson DJ (1997) Eph family transmembrane ligands can mediate repulsive guidance of trunk neural crest migration and motor axon outgrowth. Neuron 18:383–396
31. Cohen RI, Rottkamp DM, Maric D et al (2003) A role for semaphorins and neuropilins in oligodendrocyte guidance. J Neurochem 85:1262–1278
32. Ciossek T, Monschau B, Kremoser C et al (1998) Eph receptor-ligand interactions are necessary for guidance of retinal ganglion cell axons in vitro. Eur J Neurosci 10:1574–1580
33. Butler JE (2000) Solid supports in enzyme-linked immunosorbent assay and other solid-phase immunoassays. Methods 22:4–23

Chapter 23

Microcontact Printing of Substrate-Bound Protein Patterns for Cell and Tissue Culture

Martin Fritz and Martin Bastmeyer

Abstract

Patterned distributions of signalling molecules play fundamental roles during embryonic development. Several attempts have been made to reproduce these patterns in vitro. In order to study substrate-bound or membrane proteins, microcontact printing (μCP) is a suitable method for tethering molecules on various surfaces. Here, we describe three μCP variants to produce patterns down to feature sizes of about 300 nm, which are highly variable with respect to shape, protein spacing, and density. Briefly, the desired pattern is etched into a silicon master, which is then used as a master for the printing process. Each variant offers certain advantages and the method of choice depends on the desired protein and the biological question.

Key words Axon guidance, Ephrin/Eph, Growth cone, Gradient, Soft lithography, Micropatterning

1 Introduction

Microcontact printing (μCP) is a versatile method to fabricate protein-patterned surfaces. The method is highly flexible with respect to geometries, feature sizes, and concentrations of the printed protein pattern. μCP was originally introduced by Whiteside and colleagues [1–3] for patterning self-assembled monolayers of alkanethiols on gold-coated surfaces and was further modified, for example, by the group of E. Delamarche [4, 5]. Using this technique, processes like neuronal differentiation, neurite outgrowth, or synaptogenesis have been studied. For example, rat hippocampal neurons differentially extend axons and dendrites along printed patterns of L1 and N-cadherin [6], demonstrating new aspects of neuron polarity and axon specification. Additionally, μCP has been successfully used to study synaptogenic molecules [7] and axon bifurcation [8]. The studies of Shi et al. [6] and Chiang et al. [8] demonstrate, furthermore, that μCP can be used to generate not only single-protein surfaces but also surface patterns with different proteins in close proximity.

Renping Zhou and Lin Mei (eds.), *Neural Development: Methods and Protocols*, Methods in Molecular Biology, vol. 1018,
DOI 10.1007/978-1-62703-444-9_23, © Springer Science+Business Media, LLC 2013

During the development of the nervous system, axon outgrowth is controlled by axon guidance molecules. These proteins either have a patterned distribution forming attractive pathways and repulsive boundaries or are distributed in a graded fashion. Molecular gradients play important roles during embryonic development by providing positional and directional information. They are essential for correct pathfinding and guidance of neuronal growth cones and are therefore responsible for proper development of the complex structure of the brain [9].

Several attempts have been made to produce gradients of axon guidance molecules in vitro [10]. To study diffusible molecules, in vitro assays based on soluble gradients should be applicable, whereas substrate-bound or membrane-anchored guidance molecules should be presented as substrate-bound gradients in vitro.

Generation of surface-bound gradients continues to be an important issue. Starting in 1992, the Bonhoeffer group has developed methods to generate striped gradients of membrane fragments [11, 12]. However, the composition of these membranes is ill defined and, therefore, reproducibility is limited.

μCP is a useful method to obtain stable and highly reproducible gradients, where gradient slope, spacing, and density of the protein patterns can be precisely controlled [13]. In case of the axon guidance molecule ephrin-A5, these gradients lead to differential responses of temporal and nasal RGC growth cones [14].

Here, we provide detailed protocols for three different variants of μCP. They all have in common that spatially precise patterns of subcellular resolution can be produced (down to feature sizes of about 300 nm). Using these methods, a variety of proteins can be printed in an almost unlimited number of structural variations.

We focus on μCP with a reusable silicone stamp (Subheading 3.3), the lift-off technique (Subheading 3.4), and the gold-thiol chemistry (Subheading 3.5). Each variant offers certain advantages and the method of choice depends on the desired protein and the biological question.

2 Materials

2.1 General Materials for All Microcontact Printing Methods

1. Polydimethylsiloxane, PDMS, Sylgard 184 (Dow Corning, Midland, MI, USA).
2. Ethanol, EtOH, >99.8 % p.a.
3. Acetone > 99.5 % p.a.
4. Sulfuric acid, H_2SO_4, >95 %.
5. Hydrogen peroxide, H_2O_2, ≥30 %.
6. Piranha solution, H_2SO_4 (> 95 %), H_2O_2 (≥30 %), mixed at 1:1 ratio (*see* **Note 1**).

7. Ammonia solution, $NH_3 \geq 25$ % p.a.
8. Ultrapure water, ddH_2O, sterile.
9. Nitrogen gas.
10. Phosphate buffered saline, PBS, without Ca^{2+} and Mg^{2+}, 8 g/l NaCl, 0.2 g/l KCl, 1.15 g/l $Na_2HPO_4 \times 2H_2O$, 0.2 g/l KH_2PO_4; adjust to pH 7.4 (*see* **Note 2**).
11. HANK's balanced salt solution without Ca^{2+} and Mg^{2+}, 8 g/l NaCl, 0,4 g/l KCl, 60 mg/l NaH_2PO_4, 60 mg/l $Na_2HPO_4 \times 2H_2O$, 0,35 g/l $NaHCO_3$, 1 g/l glucose, 7.76 g/l HEPES, 10 mg/l phenol red; adjust to pH 7.4 (*see* **Note 2**).
12. Solution of the protein to be patterned: Protein should be dissolved in 1× PBS (*see* **Note 3**).
13. Glass coverslips (*see* **Note 4**).
14. Clean forceps.
15. Heating plate.
16. Silicon masters containing the desired surface relief structure (*see* **Note 5**) (Fig. 1a).
17. Oxygen plasma cleaner (*see* **Note 6**).
18. Incubator (37 °C).
19. Equipment for preparation of primary cells or performing cell culture in general.

2.2 Additional Materials for Microcontact Printing with a Reusable PDMS Stamp

1. Octadecyltrichlorosilane (OTS), 1 mM solution in heptane (Sigma-Aldrich, St. Louis, MO, USA).
2. *n*-Heptane > 99 % p.a.
3. Scotch tape.
4. Solution to fill in the space between stamped areas: For example, growth substrates such as extracellular matrix (ECM) proteins (20 μg/ml laminin, 10 μg/ml fibronectin) or passivating proteins (1 % BSA) can be used. Protein solutions should be prepared in PBS.

2.3 Additional Materials for the Lift-Off Technique

1. Vacuum pump.
2. Fine point marker.
3. Solution to fill in the space between stamped areas: *see* Subheading 2.2, **item 4**.

2.4 Additional Materials for Gold-Thiol Chemistry

1. Gold-coated coverslips (*see* **Note** 7).
2. Octadecyl mercaptan (ODM), 1.5 mM solution in EtOH (Sigma-Aldrich, St. Louis, MO, USA).
3. Ethlyeneglycol-3O-mercaptan (EG3O-Me), 1.5 mM solution in EtOH (Prochimia, Sopot, Poland).

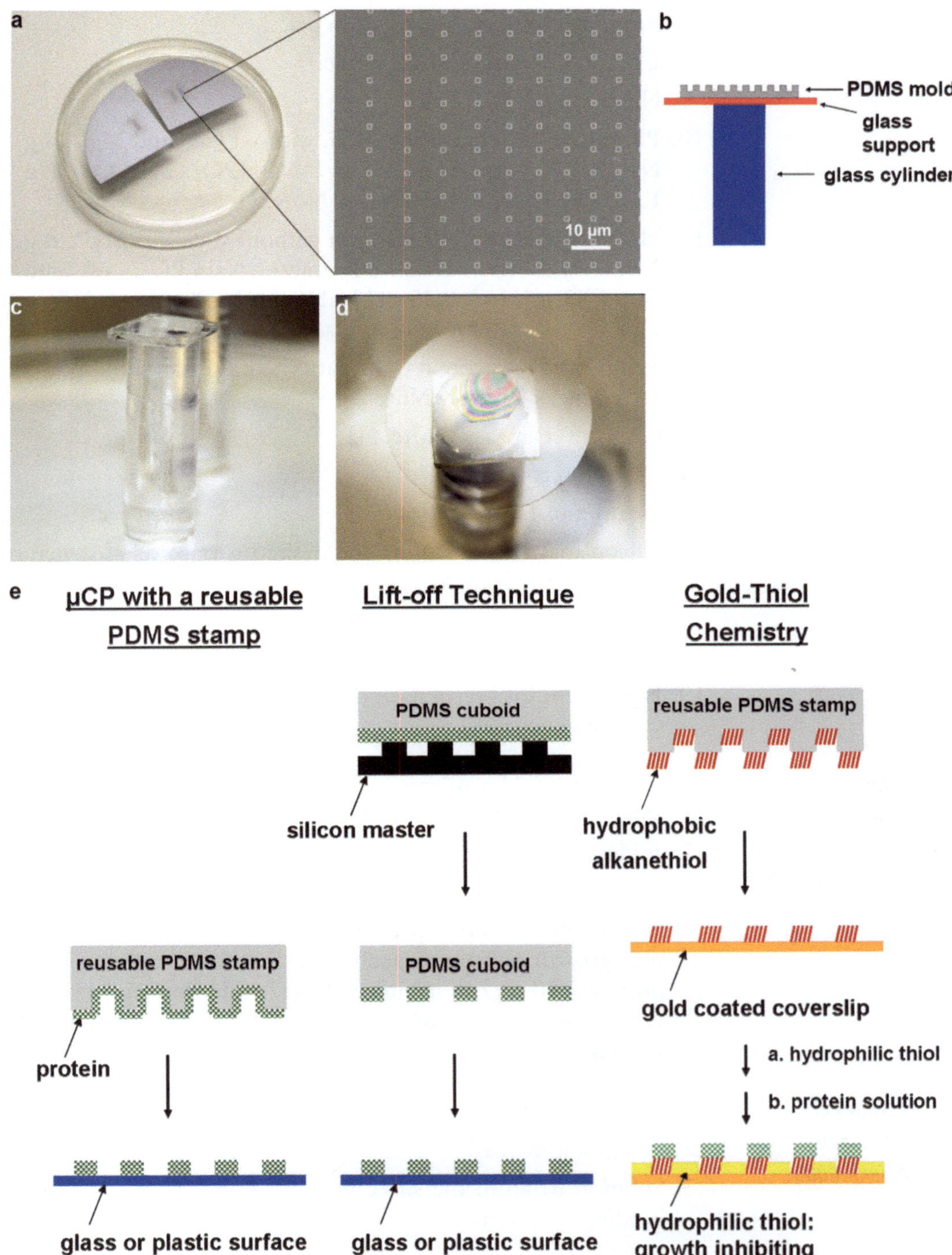

Fig. 1 Microcontact printing with a reusable PDMS stamp. (**a**) Two silicon masters necessary to mold reusable PDMS stamps (*left*). An exemplary dot structure imaged with a scanning electron microscope is shown to the *right*. (**b**) Schematic drawing of a reusable PDMS stamp with a glass cylinder for better handling, a glass support which prevents collapse of the stamp during the printing process and the PDMS mold of the master structure. (**c**) Image of a reusable stamp. (**d**) Successful adherence of a PDMS stamp to a glass coverslip indicated by the apparent Newton rings. (**e**) Comparison of the three different μCP variants, described in Subheadings 3.3–3.5

3 Methods

3.1 Cleaning of Silicon Masters

1. Place silicon masters (Fig. 1a) face up into a glass dish with piranha solution by using clean forceps (*see* **Note 8**).
2. Incubate masters in piranha solution at 80 °C on a heating plate until gas bubbles become visible (~8–10 min).
3. Remove masters using forceps and rinse with ddH_2O several times.
4. Dry masters carefully under a stream of nitrogen.
5. Repeat **steps 1–4**, until all visible contaminations are removed.
6. Incubate silicon masters in a solution of NH_3, H_2O_2, and ddH_2O (1:1:5) at 80 °C until gas bubbles become visible (~8–10 min).
7. Take masters out of the solution using forceps and rinse with ddH_2O several times.
8. Dry the silicon master carefully under a stream of nitrogen.

3.2 Cleaning of Coverslips

1. Wash coverslips for 15 min in a solution containing EtOH:ddH_2O (1:1) in an ultrasonic water bath.
2. Incubate coverslips for 6 h or overnight in a solution containing EtOH:acetone (1:1).
3. Wash coverslips 3 × 30 min in EtOH.
4. Dry coverslips in a dust-free place.
5. Bake coverslips for 6 h or overnight at 180 °C (*see* **Note 9**).

3.3 Microcontact Printing with a Reusable PDMS Stamp

With this method, protein patterns down to structure sizes of about 1–3 μm can be stamped onto various surfaces. In brief, a PDMS mold of the microstructured silicon master is fabricated and is later covered with the protein solution. This PDMS mold (Fig. 1b, c) is then used as a stamp to transfer the protein to the surface [1, 4, 5] (Fig. 1e). With this method, multiple stamps can be generated from one silicon master, used in parallel, and many patterns can be generated in a short time. In addition, expensive silicon masters are not necessary for the routine procedure. On the other hand, stamping with a reusable PDMS mold is restricted with respect to pattern sizes. Only patterns that do not lead to a collapse of the stamp can be printed successfully.

Detailed Procedure

1. Silanizing the silicon master: To mold a PDMS stamp, surfaces of the masters have to be rendered highly hydrophobic. Submerge the master in a solution containing 1 mM OTS in heptane for 10 min under an atmosphere of nitrogen (*see* **Note 10**).
2. Dry the silicon master under a stream of nitrogen.

3. Mix the PDMS according to manufacturer instructions and overlay the etched structure with ~50 μl of liquid PDMS. Air bubbles should be avoided! (*See* **Note 11**).
4. Place a piece of clean thin glass (~1 mm) on top of the PDMS.
5. Fix the glass using metal weights and incubate the silicon master overnight at 60 °C for polymerization.
6. Separate the polymerized PDMS mold from the master. For better handling, attach a glass cylinder to the PDMS mold using two-component adhesive (Fig. 1b, c).
7. Place the reusable PDMS stamp face up into a petri dish and cover the patterned surface with the protein solution. Incubate for 30 min to 2 h at 37 °C (incubation times may vary, depending on the protein).
8. Discard the protein solution and wash the PDMS surface twice with ddH_2O.
9. Dry the stamp under a stream of nitrogen (*see* **Note 12**).
10. Place the stamp onto a clean glass coverslip. Ideally, the stamp will adhere to the surface immediately. If not, slight pressure can be applied to the coverslip using forceps. Successful adherence is indicated by Newton rings (Fig. 1d). Mark the position of the pattern on the backside of the coverslip using a fine point marker under a stereo microscope.
11. Remove the stamp from the glass coverslip (*see* **Note 13**).
12. Clean the stamp with a small piece of scotch tape to remove any particles.
13. Wash the stamp by covering the patterned surface with EtOH. In case of persistent contaminations, incubate the stamp in EtOH solution in an ultrasonic bath for 10 min.
14. Dry the stamp under a stream of nitrogen (*see* **Note 14**).

3.4 Lift-Off Technique

The lift-off technique offers higher flexibility with respect to feature size. A PDMS cuboid is homogeneously covered with protein solution and is placed directly onto the silicon master. The non-etched (protruding) areas of the master will remove the protein from the PDMS surface. The PDMS cuboid is then used as a stamp to transfer the remaining protein onto a glass or plastic surface (Fig. 1e). This technique bypasses the limitations due to stamp collapse leading to smaller pattern sizes down to ~0.3 μm [4, 5]. A disadvantage of the lift-off technique is that expensive silicon masters are always necessary in the routine procedure which requires the cleaning of the master after each stamping process.

Detailed Procedure

1. Mix the PDMS according to manufacturer instructions and pour into a clean and dust-free petri dish. The height of PDMS should be 5–7 mm (*see* **Note 15**).
2. Place the petri dish with the liquid PDMS into a vacuum chamber and apply a partial vacuum of ~500 mTorr to remove air bubbles. Continue with **step 3** if no more air bubbles are visible.
3. Incubate overnight at 60 °C (*see* **Note 16**).
4. Cut the PDMS in small cuboids using a scalpel with a clean blade. Edges of the cuboid should be clean and straight. The PDMS surface facing the bottom of the petri dish will later be covered with the protein solution. Cutting should be done in a dust-free environment.
5. Place the cuboids in another petri dish, with the stamp side up (*see* **Note 16**).
6. Cover the cuboids with the protein solution and incubate for 30 min to 2 h at 37 °C (incubation times may vary, depending on the protein). For 1 cm cuboids, about 250 μl of solution is needed (*see* **Note 17**).
7. Grasp the cuboid with forceps and discard the protein solution (*see* **Notes 18** and **19**).
8. Wash the cuboid twice by submerging into ddH_2O.
9. Dry the stamp under a stream of nitrogen.
10. Place the protein-covered surface of the PDMS cuboid onto the patterned structure of the silicon master. The etched area should be completely covered by the stamp. The stamp should adhere to the surface immediately. Do not relocate the stamp after contact with the etched area.
11. Mark the position of the pattern on the PDMS cuboid with a fine point marker under the stereo microscope.
12. Remove the PDMS cuboid with forceps from the silicon master and place onto a glass coverslip. The protein-covered area of the cuboid should be in contact with the glass surface. The cuboid should adhere to the surface immediately.
13. Transfer the marks of the PDMS cuboid to the coverslip using a fine point marker.
14. Remove the cuboid from the glass coverslip (*see* **Notes 13** and **19**).
15. Grasp the silicon master with forceps without contacting the etched pattern and wash with EtOH:ddH_2O (1:1) and EtOH > 99.8 % p.a.

16. Dry the master under a stream of nitrogen.
17. Clean the silicon master using oxygen plasma for 30 s to 1 min to remove organic contaminants (*see* **Note 20**).

3.5 Gold-Thiol Chemistry

The methods described so far use direct protein stamping to generate microstructured patterns. However, some proteins lose their bioactivity during this process even if they are successfully transferred to the surface. Patterns of such proteins might be produced by using gold-thiol chemistry [1, 15]. This technique utilizes the fact that sulfur establishes semi-covalent bonds to gold. In a first step, a hydrophobic, methyl-terminated alkanethiol, like ODM [16], is stamped in the desired pattern on gold-coated coverslips. Unstamped areas are then passivated by a second, hydrophilic thiol like EG3O-Me [17, 18]. When this thiol pattern is incubated with a protein solution, proteins preferentially adsorb to the hydrophobic thiol [4, 15] (Fig. 1e). With this method, proteins adsorb from solution to the surface without the hazard of conformational change. However, unstamped areas are completely passivated by the second thiol and therefore cannot be covered with a second protein, for example, a growth-supportive substrate.

Detailed Procedure

1.–6. *See* Chapter 3.3.

7. Cover the stamp with a solution of a hydrophobic thiol, for example, 1.5 mM ODM in EtOH, and incubate for 30 s to 1 min at room temperature.
8. Discard the hydrophobic thiol solution and dry the stamp under a stream of nitrogen.
9. Place the stamp onto a gold-coated coverslip. Ideally the stamp should adhere to the surface immediately. If not, slight pressure can be applied to the backside of the coverslip using forceps. Successful adherence is indicated by Newton rings (*see* **Note 21**).
10. Remove the stamp from the gold coverslip.
11. Flush the coverslip with EtOH and dry it under a stream of nitrogen.
12. Overlay the stamped surface with a solution of a second, hydrophilic thiol, for example, 1.5 mM EG3O-Me in EtOH, and incubate for 30 min at room temperature.
13. Flush the coverslips with EtOH and dry it under a stream of nitrogen (*see* **Note 22**).
14. Overlay the patterned surface with the desired protein solution. Conditions for incubation may vary, depending on the protein, for example, 1 h at 4 °C in case of 10 μg/ml fibronectin.
15. Wash the surfaces twice in PBS.

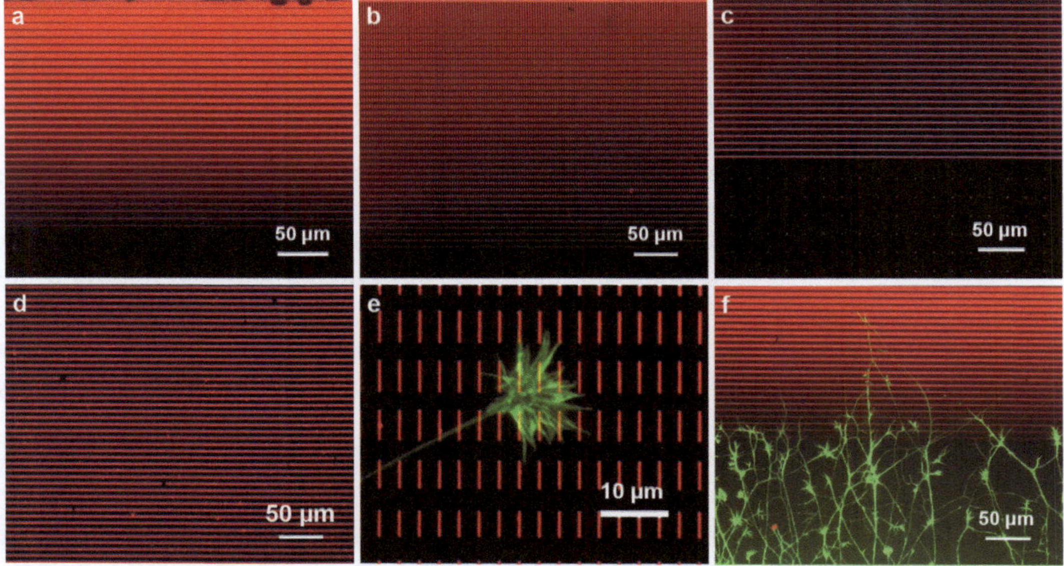

Fig. 2 Fluorescent protein patterns and retinal ganglion cell growth cones confronted with patterns. (**a**–**d**) Examples of ephrin-A5 patterns printed in gradients with variable steepness (**a**, **b**) or as stripes with variable width (**c**, **d**). (**e**) Growth cone of a temporal retinal axon growing in a pattern printed with a low ephrin-A5 concentration (2 μg/ml). (**f**) Several growth cones of temporal retinal axons stopping in front of a pattern printed with a high ephrin-A5 concentration (8 μg/ml). Preclustered ephrin-A5 is shown in *red*, phalloidin-stained axons in *green*

16. Clean the stamp with a small piece of scotch tape to remove any particles.
17. Wash the stamp by covering the patterned surface with EtOH. In case of persistent contaminations, incubate the stamp in EtOH solution and place it in an ultrasonic bath for 10 min.
18. Dry the stamp under a stream of nitrogen (*see* **Note 14**).

3.6 Cell Culture on Printed Patterns

1. Cover the printed protein pattern (from Subheading 3.3 to Subheading 3.4) with 150–200 μl of an appropriate growth substrate solution (either 20 μg/ml laminin or 10 μg/ml fibronectin can be used) or a solution to passivate the unstamped areas and incubate at 37 °C for 1 h (*see* **Note 23**).
2. Remove the solution and wash the patterned surfaces three times in PBS.
3. Seed the cells onto the patterned surface and incubate at appropriate temperatures in suitable growth media. In Fig. 2e, f, growth cones of retinal ganglion cells, either navigating in microcontact-printed ephrin-A5 patterns or stopping in front of repulsive ephrin-A5 patterns, are shown, as an example.

4 Notes

1. Piranha solution removes organic contaminations because of its high oxidizing potential. Its acidity as well as the oxidizing capacity makes piranha solution very hazardous. Solutions should be prepared by adding H_2SO_4 into previously supplied H_2O_2. Mixing the components leads to a highly exothermic reaction and may result in explosion and skin injuries if not handled with extreme care.

   Advice for disposal: Dilute piranha solution in an excess of H_2O and consider local standards for hazardous waste disposal.

2. In washing steps, PBS and HANK's balanced salt solution are interchangeable.

3. We have successfully stamped several commercially available proteins, including ephrin-A5 (8 μg/ml, R&D systems, Minneapolis, MN, USA) (Fig. 2), ephrin-A2 (8 μg/ml, R&D systems, Minneapolis, MN, USA), vitronectin (50 μg/ml, Sigma-Aldrich, St. Louis, MO, USA), or laminin (20 μg/ml, Invitrogen, Carlsbad, CA, USA). Protein concentrations necessary to obtain bioactive surfaces may vary and should be optimized, dependent on the protein. Some proteins (e.g., EphA3 and fibronectin) lose their activity during the direct stamping process, even if they are successfully transferred to the surface.

4. Protein patterns can also be stamped on plastic petri dishes. However, surface properties of plastic dishes may vary, depending on manufacturers. Therefore, it is highly recommended to test their applicability for μCP.

5. Silicon masters are fabricated by lithography methods. Standard photolithography can be used for feature sizes larger than 1–2 μm and electron beam lithography for structures down to a few nm. Reusable stamps require geometric patterns that should not collapse during printing. They should have an aspect ratio not exceeding five (ratio of structure height to width) [19]. Our masters were fabricated using a positive tone resist (PMMA) and a lift-off process for chromium (25 nm) to generate a mask for ion etching [20]. Structures were etched 650 nm deep into the silicon surface. Silicon masters can be stored at room temperature in a dust-free environment.

6. In general, plasma is a special state of a gas in which atoms are excited to higher energy states and are partially or fully ionized. It is created by using high-frequency voltages applied to oxygen. The activated high-energy components of the plasma are able to break most organic bonds and are therefore effective in removing organic contaminations [21]. We use the

plasma system 100E (PVA Techniques Plasma AG) with a pressure of 0.8 Torr and an RF power of 200 W to clean silicon masters.

7. Gold-coated coverslips are commercially available (i.e., Platypus Technologies, Madison, WI, USA or Sigma-Aldrich, St. Louis, MO, USA) or can be self-made. We use the Tectra Mini-coater (tectra GmbH) operated at a pressure of $\leq 1 \times 10^{-6}$ mbar. First, a titanium layer of 3–5 nm, serving as adhesion mediator, is deposited onto the coverslips, followed by a gold layer of 15–20 nm. Gold coverslips can be stored for up to 1 week in a petri dish.
8. Handle the silicon master with care and avoid contact with the etched surface of the master. Silicon masters are very brittle and can break easily.
9. Cleaned coverslips can be stored for several months in a dust-free environment.
10. OTS is one of many hydrophobic silanes useful for proper master silanization [22]. It exhibits many dangerous properties: It is explosive when mixed with water or oxidizing substances and has to be handled under an atmosphere of inert gas, such as nitrogen. Skin contact should be avoided. To test for hydrophobicity of the master, pipette a drop of water onto the silanized master. If the contact angle is large and the drop becomes spherically shaped, the silanization was successful.
11. We use Sylgard 184 PDMS supplied by Dow Corning [1, 5, 23]. PDMS has to be degassed before using in stamp fabrication. Place a petri dish with liquid PDMS into a vacuum chamber and apply a partial vacuum of ~500 mTorr to remove air bubbles.
12. It is recommended to perform **steps 8–11** quickly and carefully. Stamps should not be dry for an unnecessarily long period.
13. Dry patterns of many printed proteins can be stored at 4 °C for several days. Ideal storage conditions may vary depending on the protein.
14. Reusable PDMS stamps should be stored at room temperature in a dust-free environment.
15. If more PDMS is poured into the petri dish, cuboids will become too high. This leads to a loss of accuracy in marking the pattern of the master on the PDMS cuboid. If less PDMS is poured into the petri dish, it will become difficult to pick cuboids with forceps.

    The stiffness of polymerized PDMS varies, depending on the ratio of its components and the cross-linking conditions. Stamps should neither be too soft nor too brittle.

16. Petri dishes of polymerized PDMS and PDMS cuboids can be stored in a dust-free environment for several months.
17. Before using the PDMS cuboids, it is highly recommended to check carefully for dust particles on the surface. Clean the cuboids with EtOH, if required.
18. Protein solutions can be reused several times. However, the activity of the protein might be gradually lost.
19. **Steps 7–14** should be performed quickly. Stamps should not be dry for an unnecessarily long period.
20. Before treating silicon masters with the oxygen plasma, check carefully for PDMS contaminations. PDMS is vitrified under plasma conditions. Masters would be irreversibly destroyed in this case.
21. Because of high diffusion rates, especially of short-chained alkanethiols [5, 24], the stamping process should be performed quickly. Otherwise, surface background will increase dramatically. Using alkanethiols with longer alkyl chains will improve the process.
22. Dry gold coverslips containing the stamped thiol patterns can be stored for up to 3 days in a dust-free environment at room temperature.
23. In case of working with gold-thiol chemistry, unstamped areas are already passivated by the second hydrophilic thiol. Here, no growth substrate can be applied in these areas.

Acknowledgments

This work was supported by the German Research Foundation, DFG (grant BA 1034/14-3). The authors thank Franco Weth for helpful comments on the manuscript.

References

1. Kumar A, Whitesides GM (1993) Features of gold having micrometer to centimeter dimensions can be formed through a combination of stamping with an elastomeric stamp and an alkanethiol "ink" followed by chemical etching. Appl Phys Lett 63:2002–2004
2. Jackman RJ, Wilbur JL, Whitesides GM (1995) Fabrication of submicrometer features on curved substrates by microcontact printing. Science 269:664–666
3. Mrksich M, Chen CS, Xia Y et al (1996) Controlling cell attachment on contoured surfaces with self-assembled monolayers of alkanethiolates on gold. Proc Natl Acad Sci USA 93:10775–10778
4. Bernard A, Renault JP, Michel B et al (2000) Microcontact printing of proteins. Adv Mater 12:1067–1070
5. Michel B, Bernard A, Bietsch A et al (2001) Printing meets lithography: soft approaches to high-resolution patterning. J Res Dev 45: 697–719
6. Shi P, Shen K, Kam LC (2007) Local presentation of L1 and N-cadherin in multicomponent, microscale patterns differentially direct neuron function in vitro. Dev Neurobiol 67:1765–1776

7. Cornish T, Branch DW, Wheeler BC et al (2002) Microcontact printing: a versatile technique for the study of synaptogenic molecules. Mol Cell Neurosci 20:140–153
8. Chiang L, Poole K, Oliveira BE et al (2011) Laminin-332 coordinates mechanotransduction and growth cone bifurcation in sensory neurons. Nat Neurosci 14:993–1000
9. McLaughlin T, O'Leary D (2005) Molecular gradients and development of retinotopic maps. Annu Rev Neurosci 28:327–355
10. Keenan TM, Folch A (2008) Biomolecular gradients in cell culture systems. Lab Chip 8:34–57
11. Baier H, Bonhoeffer F (1992) Axon guidance by gradients of a target-derived component. Science 255:472–475
12. Rosentreter SM, Davenport RW, Löschinger J et al (1998) Response of retinal ganglion cell axons to striped linear gradients of repellent guidance molecules. J Neurobiol 37:541–562
13. von Philipsborn AC, Lang S, Bernard A et al (2006) Microcontact printing of axon guidance molecules for generation of graded patterns. Nat Protoc 1:1322–1328
14. von Philipsborn AC, Lang S, Löschinger J et al (2006) Growth cone navigation in substrate-bound ephrin gradients. Development 133:2487–2495
15. Prime KL, Whitesides GM (1991) Self-assembled organic monolayers: model systems for studying adsorption of proteins at surfaces. Science 252:1164–1167
16. Prime KL, Whitesides GM (1993) Adsorption of proteins onto surfaces containing end-attached oligo(ethyleneoxide): a model system using self-assembled monolayers. J Am Chem Soc 115:10714–10721
17. Morhard F, Pipper J, Dahint R et al (2000) Immobilization of antibodies in micropatterns for cell detection by optical diffraction. Sens Actuators B Chem 70:232–242
18. Kanda V, Kariuki JK, Harrison DJ et al (2004) Label-free reading of microarray-based immunoassays with surface plasmon resonance imaging. Anal Chem 76:7257–7262
19. Bietsch A, Michel B (2000) Conformal contact and pattern stability of stamps used for soft lithography. J Appl Phys 88:4310–4318
20. David C, Hambach D (1999) Line width control using a defocused low voltage electron beam. Microelectron Eng 46:219–222
21. O'Kane DF, Mittal KL (1974) Plasma cleaning of metal surfaces. J Vac Sci Technol 11:567–569
22. Tien J, Xia Y, Whitesides GM (1998) Microcontact printing of SAMs. Thin Films 24:227–250
23. Xia Y, Whitesides GM (1998) Soft lithography. Angew Chem Int Ed 37:550–575
24. Balmer TE, Schmid H, Stutz R et al (2005) Diffusion of alkanethiols in PDMS and its implications on microcontact printing (μCP). Langmuir 21:622–632

Chapter 24

Semiautomated Analysis of Dendrite Morphology in Cell Culture

Eric S. Sweet, Chris L. Langhammer, Melinda K. Kutzing, and Bonnie L. Firestein

Abstract

Quantifying dendrite morphology is a method for determining the effect of biochemical pathways and extracellular agents on neuronal development and differentiation. Quantification can be performed using Sholl analysis, dendrite counting, and length quantification. These procedures can be performed on dendrite-forming cell lines or primary neurons grown in culture. In this protocol, we describe the use of a set of computer programs to assist in quantifying many aspects of dendrite morphology, including changes in total and localized arbor complexity.

Key words Neuron, Dendrite, Axon, Sholl, Morphology, Tracing, Computer assisted

1 Introduction

Precise dendrite patterning is important for correct neuronal function. Normal communication, signal processing, and action potential propagation are all dependent on correct dendrite morphology [1–4]. Alterations in the dendrite arbor have been associated with diseases such as Alzheimer's disease, epilepsy, and Rett syndrome [5–10]. Understanding how factors affect the dendrite arbor is crucial to understanding how neurons function in both healthy and disease states.

There are several ways to quantify dendrites in growing cells. Sholl analysis is a measure of dendrite complexity that is typically performed on the whole dendritic arbor of a cell. Concentric, equidistant circles are drawn radiating from the cell body. Marks are noted on the circles where dendrites pass through the circles, and a graph can be drawn to show dendrite number at specific distances

Renping Zhou and Lin Mei (eds.), *Neural Development: Methods and Protocols*, Methods in Molecular Biology, vol. 1018, DOI 10.1007/978-1-62703-444-9_24, © Springer Science+Business Media, LLC 2013

from the cell body. Measurements of dendrite type can be made by counting the number of dendrites that extend directly from the cell body (primary), dendrites that branch from a primary (secondary), and dendrites that branch off of any other type of branch (tertiary, quaternary, etc.). Length is typically measured from the cell body to the tip of the dendrite. Sholl analysis, dendrite number, or dendrite length measurements performed in the normal fashion, by hand, must be done separately and require repeat analysis to capture to all of the information present in a neuron image. The use of the Bonfire program allows capture of all data discussed above while only requiring one set of manipulations to be done on the images. Bonfire also separates the data for specific subregions of the dendrite, such as length of secondary dendrites. The resulting data make very complex analysis of dendrite branching possible, and many changes that alter the arbor can do so without changing the total arbor measurements [11].

2 Materials

2.1 Images

1. 1.8-bit images TIFF images (*see* **Note 1**).
2. Invert color scheme of images to black dendrites on white background for ease of tracing (*see* **Note 2**).

2.2 Software

The Bonfire program requires the use of several open source programs to perform the quantitation in a semiautomated fashion:

1. ImageJ—http://rsbweb.nih.gov/ij/.
2. NeuronJ plugin for ImageJ—http://www.imagescience.org/meijering/software/neuronj/.
3. NeuronStudio—research.mssm.edu/cnic/tools-ns/html.
4. Bonfire Program—http://lifesci.rutgers.edu/~firestein.
5. MATLAB—www.mathworks.com/products/MATLAB.

2.3 Organized File Structure

The Bonfire program requires very specific information and folder layouts to work properly. Please follow directions precisely (Fig. 1):

1. Experiment folder (containing multiple conditions folders and Bonfire folder).
2. Condition folder (containing all images for analysis of that condition) (*see* **Note 3**).
3. Bonfire folder (containing Bonfire MATLAB scripts).

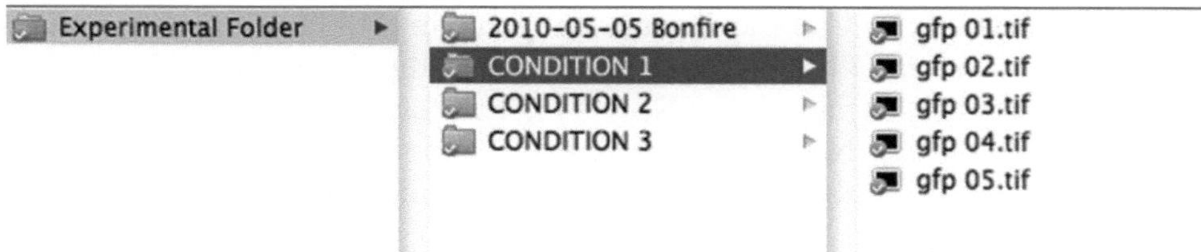

Fig. 1 Initial folder structure. Initial folder structure as called for in Subheading 2.3

3 Methods

3.1 Perform Computer-Assisted Tracing of Dendrites Using NeuronJ

1. Load the NeuronJ plugin into ImageJ (*see* **Note 4**).
2. Select the "Set parameters" button and choose "Dark" as the dendrite appearance.
3. Open the image to be traced using the "Open" button in NeuronJ (*see* **Note 5**).
4. Adjust the brightness/contrast and zoom as needed to properly visualize dendrites during tracing (*see* **Note 6**).
5. Select the "Add traces" button on the NeuronJ toolbar and trace the perimeter of the cell body.
6. Select the "Label tracings" button and then select the cell body trace. Label the tracing with tracing ID, "N1," and type, "Type 06." This is the only trace that will be labeled.
7. Trace all the dendrites using the "Add traces" button. It is recommended that you start at the cell body and trace outward. These traces should not be labeled in any way (*see* **Note 7**).
8. Save the tracing by selecting the "Save traces" button and save the trace (*.ndf) in the same directory where the image is stored in the "Condition folder."
9. Select the "Export tracings button" and export a tab-delineated text file with a separate file for each tracing. Let NeuronJ name these files for you (*see* **Note 8**).
10. Select the "Measure tracings button" and select "Display tracing measurements" (*see* **Note 9**). Run this command and save the file as "*_info" where * is your exact image file name. If your computer attaches the extension ".xls," then delete it (*see* **Note 10**).
11. Complete this process for each neuron in all the conditions to be analyzed. All of the images and generated files for a single condition should be contained in that "Condition folder."

3.2 Convert Tracings into an Arbor Connection Map Using Bonfire

1. Use the Bonfire program in MATLAB and run the script "bonfire_load" to reorganize the file structure of the "Condition folder" (*see* **Note 11**). This folder contains the image files and NeuronJ-generated files above. Each "Condition folder" should now contain a folder for each cell (*see* **Note 12**).
2. Use the Bonfire program in MATLAB and run the script "bonfire_ndf2swc" to create the arbor connection map from the trace file (*see* **Note 13**).
3. Select the condition folder that you wish to convert and allow the program to convert all of the .ndf files to .swc files. This script will convert all of the files in the condition folder but needs to be run once for each "Condition folder."

3.3 Correct Errors in the Arbor Connection Map Using NeuroStudio and Bonfire

1. Open the image file associated with the arbor connection map to be modified in NeuroStudio by selecting File→Open and choosing the appropriate image.
2. Select Run→Settings and change the X, Y, and Z boxes in the "Voxel size" window to 1.
3. Select File→Import SWC to open the appropriate SWC file and overlay it on the image opened in **step 1**.
4. Use the tools in NeuroStudio to correct the branches and end-points of the SWC file (*see* **Note 14**).
5. Follow these rules to ensure proper reading of the arbor map: (1) Each branch point (yellow dots) may only have two daughter branches and one mother; (2) there should only be one cell body (red dot) (*see* **Note 15**).
6. Save the dendrites in the same folder as the image and SWC file by selecting File→Save neurites. Use default name and do not rename.
7. Using Bonfire in MATLAB, run the script "bonfire_trace_check." This script will check all of the arbor maps in a single conditions folder. Select the condition you wish to analyze and repeat analysis for each condition (*see* **Note 16**).

3.4 Collect and Analyze Data from Arbor Maps Using Bonfire

1. Using Bonfire in MATLAB, run the script "bonfire." This script will analyze one condition folder. Select the condition folder you wish to analyze and select run (*see* **Note 17**).
2. The script "bonfire_results" will generate graphs that can give you an idea of the results of the experiment, but no statistics are possible from these graphs so it is only recommended for use as an indicator.
3. Repeat for each condition.
4. Using Bonfire in MATLAB, run the script "bonfire_export" to export the results to excel for further analysis (*see* **Note 18**).

3.5 Data Assembly in Excel

This section contains recommendations for analysis raw data after the Bonfire program has compiled it. These methods are recommendations only and other scientifically valid methods may be used:

1. Bonfire generates a single book in excel for each condition of an experiment. Each book contains multiple sheets with data on specific regions of the arbor as well as total arbor differences.
2. It is recommended that you compile statistics for each condition within the excel book created by Bonfire. These condition statistics are then copied to a new book that combines all conditions of an experiment. This allows for easier graphing and exporting to other statistical analysis software.

3.6 Further Data Analysis

1. Statistical power for branching analysis is generated by the number of neurons that are analyzed. Try to image >15 images for each condition, in each of three trials, since the variability in neurite number from cell to cell can be significant. It is not recommended that you use the number of dendrites as the "*N*" for you experiment.
2. Bonfire generates a large amount of data that may or may not be relevant to the question that you are trying to answer. However, the large amount of data are generated can draw out interesting phenomena. For example, a reduction in secondary branching may be compensated for by an increase in tertiary branches.
3. When analyzing the data generated by the Sholl analysis, it can be useful to consider portions of the curve separately instead of the curve as a whole. For example, a total Sholl graph could be broken into distance groups, such as 1–100 μm and 100–200 μm, or a method such as two-way ANOVA, can be used to compare the effect of condition and distance on the branching behavior (Fig. 2). Deciding where to place the edges of the bins with the grouping method can be somewhat arbitrary, and caution should be taken with the two-way ANOVA since it is not an appropriate test for a continuous variable. Although not formally correct, experience with two-way ANOVA shows that the analysis yields appropriate results. This analysis method does not consider the branches at a certain distance from the soma to be continuous since areas of significance tend to cluster together in large groups.

3.7 Graphical Presentation

1. Dendrite number, length, and tip number are typically presented as a simple column graph.
2. Sholl analysis is typically presented as an *X*–*Y* coordinate graph with "# of Intersections" as the *Y*-axis and "Distance from soma" as the *X*-axis (Fig. 2).

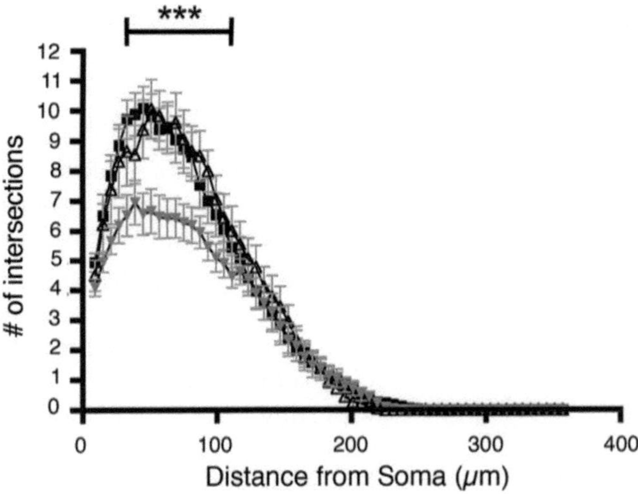

Fig. 2 Example graph of Sholl analysis data. Significance in this graph was determined by using two-way ANOVA and indicated *p* value given to the group of distances that showed same significant difference from control. This data were generated from *N* = 45 neurons

4 Notes

1. Images used in the development of protocol were taken at 200× measuring 512 × 640 pixels with a μm/pixel ratio of 1.5/1. This μm/pixel number is hard coded into the program but can be changed by replacing the value of the variable "pix_conv" with the desired value. The program is size independent, but if consistent trouble is encountered later, a last resort may be to try adjusting your files to match these conditions.
2. Black dendrites on a white background are normally easier to trace. You may use white dendrites on black background but be sure to select the "Light" dendrite tracing in NeuronJ.
3. This folder should not contain any other folders when you start the analysis. Only the images for analysis of that condition should be included.
4. The NeuronJ plugin for ImageJ must be installed in ImageJ before use in this protocol.
5. You must use the open button or the image will load using ImageJ and not the NeuronJ plugin.
6. When adjusting brightness and contrast, make sure that you are able to see the feature you are examining without overexposing the edges and making accurate tracing difficult.
7. While not required that you stop at each branch point and start a new trace, doing so can help if unknown errors occur in later steps. Trace whatever neurites are visible with a 2× zoom; the

program will eliminate all neurites below a certain size as those are considered filopodia and not neurites.

8. It is important that the files all be named based off of the image name. NeuronJ will do so and add important numbering data to the name.
9. If you only wish to examine total dendrite length, then you can select "display the group measurements" and manually make the μm to pixel conversion. This, however, is not recommended because many of the changes that occur in dendrite arbor patterning are often seen only in more complex dendrite relationships. You may also capture specific information about type of dendrites using NeuronJ by labeling each dendrite properly, but this is extremely time intensive and very error prone.
10. For Mac users, bring up the file info menu to delete the extension; otherwise, some versions of the OS will not completely delete the extension.
11. To run a script in MATLAB, you must first select the "Bonfire folder" as your master folder. The Bonfire folder is located in your "Experiment folder." After loading the correct master folder, type the name of the script as written into the MATLAB command window and follow the prompts and protocol. You only need to select this once unless you restart/close MATLAB.
12. If the load command does not work, double check the file structure and check that the extensions have been deleted of the "*.info" file.
13. The SWC file is a file that is readable in NeuroStudio. This will allow you to view all the connections and correct any errors in the automated conversion from the trace file to the SWC file.
14. It is recommended that you become familiar with the dendrite tool in NeuroStudio before you begin analysis.
15. As you connect various dendrites to the cell body and change the connections between branch points, the cell body should remain red. The rest of the red dots should change to the appropriate color. If at any time in the process your cell body changes color to a different color, it is recommended that you start the correction process again with that arbor map.
16. If there are errors, reopen the SWC file with NeuroStudio, correct the errors, and re-save the SWC file. Rerun "bonfire_trace_check" again until all errors are gone for all conditions.
17. This script creates a *.mat file which can only be read by MATLAB and must be exported to excel for analysis.
18. This will export the data from one condition and must be repeated for each folder. Bonfire does not perform statistical calculations.

Acknowledgments

This work was supported by NSF grants IBN-0919747 and IBN-0548543, March of Dimes Foundation grant 1-FY08-464, and NJ Governor's Council for Medical Research and Treatment of Autism grant 10-406-SCH-E-0.

References

1. Peng Y-R, He S, Marie H, Zeng S-Y, Ma J, Tan Z-J, Lee SY, Malenka RC, Yu X (2009) Coordinated changes in dendritic arborization and synaptic strength during neural circuit development. Neuron 61:71–84
2. Vetter P, Roth A, Häusser M (2001) Propagation of action potentials in dendrites depends on dendritic morphology. J Neurophysiol 85:926–937
3. van Elburg RAJ, van Ooyen A (2010) Impact of dendritic size and dendritic topology on burst firing in pyramidal cells. PLoS Comput Biol 6:e1000781
4. Schaefer AT, Larkum ME, Sakmann B, Roth A (2003) Coincidence detection in pyramidal neurons is tuned by their dendritic branching pattern. J Neurophysiol 89:3143–3154
5. Yamada M, Wada Y, Tsukagoshi H, Otomo E, Hayakawa M (1988) A quantitative Golgi study of basal dendrites of hippocampal CA1 pyramidal cells in senile dementia of Alzheimer type. J Neurol Neurosurg Psychiatry 51:1088–1090
6. Moolman DL, Vitolo OV, Vonsattel J-PG, Shelanski ML (2004) Dendrite and dendritic spine alterations in Alzheimer models. J Neurocytol 33:377–387
7. Dickstein DL, Brautigam H, Stockton SD, Schmeidler J, Hof PR (2010) Changes in dendritic complexity and spine morphology in transgenic mice expressing human wild-type tau. Brain Struct Funct 214:161–179
8. Jentarra GM, Olfers SL, Rice SG, Srivastava N, Homanics GE, Blue M, Naidu S, Narayanan V (2010) Abnormalities of cell packing density and dendritic complexity in the MeCP2 A140V mouse model of Rett syndrome/X-linked mental retardation. BMC Neurosci 11:19
9. Larimore JL, Chapleau CA, Kudo S, Theibert A, Percy AK, Pozzo-Miller L (2009) Bdnf overexpression in hippocampal neurons prevents dendritic atrophy caused by Rett-associated MECP2 mutations. Neurobiol Dis 34:199–211
10. Zoghbi HY (2003) Postnatal neurodevelopmental disorders: meeting at the synapse? Science 302:826–830
11. Langhammer CG, Previtera ML, Sweet ES, Sran SS, Chen M, Firestein BL (2010) Automated Sholl analysis of digitized neuronal morphology at multiple scales: whole cell Sholl analysis versus Sholl analysis of arbor subregions. Cytometry A 77:1160–1168

Chapter 25

Monitoring Synaptic Plasticity by Imaging AMPA Receptor Content and Dynamics on Dendritic Spines

Hiroshi Makino and Bo Li

Abstract

Time-lapse imaging techniques are widely used to monitor dendritic spine dynamics, a measurement of synaptic plasticity. However, it is challenging to follow the dynamics of spines over an extended period in vivo during development or in deep brain structures that are beyond the reach of traditional microscopes. Here, we describe an AMPA receptor-based optical approach to monitor recent history of synaptic plasticity. This method allows the identification of spines that have recently acquired synaptic AMPA receptors in a single imaging session, so that synaptic plasticity that occurs in vivo in a variety of conditions can be simply imaged in an ex vivo preparation.

Key words Two-photon imaging, Dendritic spine, Synaptic plasticity, AMPA receptor trafficking, Super Ecliptic pHluorin, Fluorescence recovery after photobleaching

1 Introduction

Optical methods based on two-photon laser-scanning microscopy (TPLSM) are widely used to monitor neuronal structure and function [1]. Using time-lapse imaging techniques, for example, one can follow the number, size, and shape of the dendritic spines of a particular neuron over time, both in vitro and in vivo [2–10]. These measurements provide important information about plastic changes in the structure and strength of excitatory synapses, given that most excitatory synapses are placed on dendritic spines [11] and that spine size correlates with synaptic strength [12, 13]. It is challenging, however, to follow the dynamics of the same dendritic spines in vivo over an extended period during early development or in deep brain structures such as hippocampus. It is also impossible to infer the recent plastic history of a given spine based on its size, because spine size, although it correlates with synaptic strength, is a consequence of plasticity integrated over a longer period of time.

Renping Zhou and Lin Mei (eds.), *Neural Development: Methods and Protocols*, Methods in Molecular Biology, vol. 1018, DOI 10.1007/978-1-62703-444-9_25,

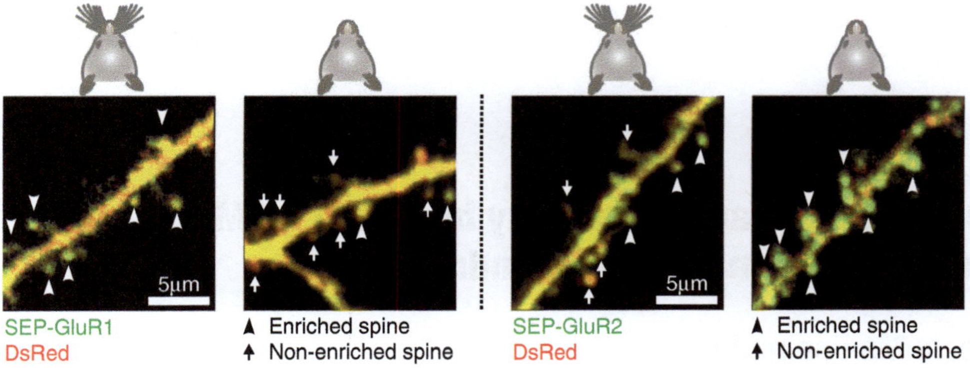

Fig. 1 Spine enrichment of AMPA receptors as an indicator for experience-driven plasticity or sensory deprivation-induced synaptic up-scaling. *Left panels* show examples of SEP-GluR1 and DsRed-expressing neurons in whisker-intact and whisker-trimmed animals. Whisking experience drives GluR1 into synapses and therefore increases SEP-GluR1 spine enrichment. *Right panels* are examples of SEP-GluR2- and DsRed-expressing neurons in whisker-intact and whisker-trimmed animals. Sensory deprivation by whisker trimming causes the insertion of GluR2 into synapses and therefore increases SEP-GluR2 spine enrichment. *Arrowheads* indicate AMPA receptor-enriched spines, whereas *arrows* indicate non-enriched spines (*Reproduced from Makino and Malinow, 2011 with permission from Elsevier*)

It would be of tremendous help if one can determine, in a snapshot, whether a given dendritic spine has recently undergone plasticity. To address this issue, Makino and Malinow took advantage of the findings that AMPA receptor trafficking into synapses underlies synaptic plasticity [14, 15] and developed an AMPA receptor-based optical approach to identify recently potentiated synapses [16, 17]. In this method, AMPA receptor subunit GluR1 or GluR2 (also referred to as GluA1 and GluA2), which mediates the activity-driven synaptic plasticity [14] or homeostatic synaptic scaling [18], respectively, is tagged with a pH-sensitive form of green fluorescent protein (Super Ecliptic pHluorin, SEP), so that the receptors on the cell surface are fluorescent, whereas those retained in the intracellular organelles are not detectable because their acidic environment quenches the SEP fluorescence [2]. In a series of experiments, Makino and Malinow [16, 17] demonstrate that enrichment of SEP-tagged AMPA receptors on the surface of dendritic spines, but not the size of spines, correlates with tagged AMPA receptors recently accumulated in the synapse, estimated by measuring the fluorescence recovery after photobleaching (FRAP). FRAP can be used to estimate synaptic receptor content because AMPA receptors display reduced mobility once they are incorporated into synapses, thereby causing a reduction in FRAP [16, 17, 19–21]. Thus, by measuring the enrichment of SEP-tagged AMPA receptors on the surface of dendritic spines, one can determine which spines have recently acquired new synaptic AMPA receptors (Fig. 1). This method makes it possible to assess, in a reduced ex vivo preparation, synaptic plasticity that occurs in vivo either during early development or in adult in various brain structures.

2 Materials

2.1 DNA Constructs

To implement this method, a Cre/loxP-mediated inducible expression system, where the transcription of genes of interest is regulated by a floxed stop cassette [22], is used. This includes several DNA constructs: (1) pCALNL-SEP-GluR1 (or pCALNL-SEP-GluR2; see below), which harbors a floxed stop cassette followed by the cDNA for SEP-GluR1 (or SEP-GluR2) driven by a CAG promoter; (2) pCALNL-DsRed, which harbors a floxed stop cassette followed by the cDNA for DsRed, a red cytoplasmic marker (to visualize cell morphology), driven by a CAG promoter; and (3) the 4-OHT-dependent Cre recombinase-expressing plasmid, pCAG-ER^{T2}-CreER^{T2}. All constructs can be obtained from Addgene (*see* **Notes 1** and **2**).

2.2 Imaging Setup

Although we use the following imaging setup, an equivalent system can also be used: a two-photon laser-scanning microscope (Prairie) equipped with a 40× 0.8 NA objective lens and a 1.4 NA oil condenser (Olympus), a Ti:sapphire laser (Chameleon Ultra II, Coherent), a set of dichroic mirrors and filters (Chroma) for the separation of green and red fluorescence signals, and photomultiplier tubes (PMTs) for the collection of photons.

2.3 Materials for In Utero Electroporation

1. Tweezer electrodes (Harvard Apparatus 450165).
2. Electroporator (Harvard Apparatus 450052).
3. DNA plasmids (see below).
4. Fast green (Sigma-Aldrich F7252).

3 Methods

3.1 In Utero Electroporation

1. Preparation of cDNAs: Mix the following plasmids at a ratio of 20:4:1 (a:b:c) with Fast green. These plasmids are amplified with the endotoxin-free Maxi kit (QIAGEN 12362) (*see* **Notes 3** and **4**).
   (a) pCALNL-SEP-GluR1 (or pCALNL-SEP-GluR2): ~5 μg/μl
   (b) pCALNL-DsRed: ~5 μg/μl
   (c) pCAG-ER^{T2}CreER^{T2}: ~5 μg/μl
2. Procedure: Neurons are transfected with these plasmids following an in utero electroporation procedure. Specific modifications of the procedure are needed depending on brain areas or cell types of interest [23–25]. For example, to transfect cortical layer 2/3 progenitor cells, E15–16 timed pregnant C57BL/6J mice (Charles River) can be used. Under anesthesia with an isoflurane-oxygen mixture (Lei Medical), the uterus

containing embryos is taken out. Approximately 0.5 μl of DNA solution containing Fast green is pressure injected through a beveled-glass capillary tube into the right lateral ventricle of each embryo. The head of each embryo is placed between tweezer electrodes (Harvard Apparatus 450165) with the anode contacting the right hemisphere (for targeting cortical regions). Electroporation is achieved with five square pulses (duration = 50 ms, frequency = 1 Hz, voltage = 25 V) using an electroporator (Harvard Apparatus 450052) (*see* **Note 5**). More detailed procedures for in utero electroporation can be found in refs. 23–25.

3.2 Cre Recombinase Activation by 4-OHT

The expression of SEP-GluR1 (or SEP-GluR2) and DsRed is induced by systemic application of 4-hydroxy-tamoxifen (4-OHT, Sigma-Aldrich H7904) to activate Cre recombinase 2 days before the imaging experiments. 4-OHT is dissolved in ethanol at a concentration of 20 mg/ml and diluted with 9 vol. of corn oil (Sigma-Aldrich C8267). Diluted 4-OHT (2 mg/ml) is intraperitoneally injected into each mouse. The volume of injection depends on the weight of animals. For example, for mice at postnatal day 11 (~6 g), 100 μl of the diluted 4-OHT per animal is sufficient.

3.3 Preparation of Acute Brain Slices for Imaging

A general method for making acute brain slices can be followed. Slices (350 μm thick) are cut with a vibratome (Leica VT1000 S) in gassed (95 % O_2 and 5 % CO_2) ice-cold solution containing 25 mM $NaHCO_3$, 1.25 mM NaH_2PO_4, 2.5 mM KCl, 0.5 mM $CaCl_2$, 7 mM $MgCl_2$, 25 mM d-glucose, 110 mM choline chloride, 11.4 mM sodium ascorbate, and 3.1 mM sodium pyruvate. Slices are then incubated in artificial cerebrospinal fluid (ACSF) containing 118 mM NaCl, 2.5 mM KCl, 26 mM $NaHCO_3$, 1.2 mM NaH_2PO_4, 11 mM d-glucose, 4 mM $MgCl_2$, and 4 mM $CaCl_2$ at 35 °C for 30 min and then at room temperature until being used.

3.4 Imaging

1. Z stack or single plane images of neurons expressing SEP-GluR1 (or SEP-GluR2) and DsRed are acquired on a Prairie TPLSM (or an equivalent two-photon system) using a mode-locked laser to 910 nm in constant perfusion of ACSF at 30 °C. Both epifluorescence (collected by a 40× 0.8 NA objective lens) and trans-fluorescence (collected by a 1.4 NA oil condenser) photons are collected and summed by PMTs coupled to the objective lens and condenser, respectively, to maximize the collection and usage of photons.
2. The enrichment of SEP-GluR1 (or SEP-GluR2) on spine surface is measured as the spine SEP signal normalized for spine area and for neuronal expression level of the SEP-tagged protein. Specifically, SEP and DsRed fluorescence in spines and dendrites

is measured as integrated green and red fluorescence, respectively, after subtraction of background and bleed-through between the two channels. To measure the density of spine surface AMPA receptors as an enrichment value, spine SEP fluorescence is normalized to

$$(4\times\pi)^{1/3}\times(3\times R_{\text{Spine}})^{2/3},$$

where R_{Spine} represents spine DsRed fluorescence (i.e., spine volume is converted to spine area assuming that spine heads are spherical). To compare across different cells, these values are then normalized to the fluorescence signal of common dendritic regions. Thus, spine enrichment values are calculated as

$$\frac{\left\{G_{\text{Spine}}/\left[(4\times\pi)^{1/3}\times(3\times R_{\text{Spine}})^{2/3}\right]\right\}}{\left\{G_{\text{Dendrite}}/\left[(4\times\pi)^{1/3}\times(3\times R_{\text{Dendrite}})^{2/3}\right]\right\}},$$

where G_{Spine} and G_{Dendrite} represent spine and dendrite SEP fluorescence, respectively, and R_{Dendrite} dendrite DsRed fluorescence.

3. To verify that the enrichment of SEP-GluR1 (or SEP-GluR2) on spine surface correlates with the fraction of immobile (and therefore synaptic) receptors, fluorescence recovery of spine SEP is measured at +25 and +30 min after photobleaching and compared to baseline fluorescence obtained at −10 and −5 min prior to photobleaching. SEP fluorescence on individual spines can be photobleached by scanning a region of interest covering a spine at a single z section ~50 times with higher intensity of the laser power (at 910 nm). Immobility of AMPA receptors is calculated as immobility = 1 − fluorescence recovery.

4 Notes

1. The advantages of using the inducible Cre-dependent expression system are the following: (1) it allows the sparse labeling of neurons, facilitating the isolation of microstructures from background during imaging. Typically less than 1 % of layer 2/3 cortical neurons express both SEP-GluR1 (or SEP-GluR2) and DsRed with the amount of 4-OHT mentioned in Methods; and (2) it allows the acute expression of SEP-GluR1 (or SEP-GluR2) at a much later time point after in utero electroporation. This allows the detection of plasticity events that occur only after the induction process (i.e., 4-OHT injection), thereby avoiding the accumulation of SEP-GluR1 (or SEP-GluR2) in the synapse due to the integration of plasticity over long time.

2. This approach allows the study of synaptic plasticity that occurs in a variety of situations, such as during development or in deep brain structures, in one imaging session because the imaging procedure can be simply performed in acute brain slices (for synaptic plasticity induced in vivo) or even in cultures (for synaptic plasticity induced in vitro).
3. The choice of SEP-GluR1 or SEP-GluR2 depends on the question of interest. For example, if the task is to monitor activity- or experience-dependent synaptic plasticity, then SEP-GluR1 should be used [16, 17]. If the task is to monitor synaptic scaling caused by inactivity, then SEP-GluR2 should be used [17, 18].
4. For the formation of homomeric GluR2, SEP-GluR2(R586Q) is expressed. Heteromeric AMPA receptors can also be used by coexpressing untagged-GluR2(edited) with either SEP-GluR1 or SEP-GluR3 at a 1:1 molor ratio [17].
5. The age of the embryo is an important determinant of which cortical layers to target. For example, if one wants to target deeper cortical layers (e.g., layer 5), electroporation needs to be performed at earlier embryonic stages [24]. The angle and polarity of the tweezer electrodes should also be changed depending on the brain structure of interest. For example, one can target the hippocampal CA1 region by flipping the polarity of the tweezer electrodes.

Acknowledgements

We thank R. Malinow for critical reading of the manuscript. This work was supported by NIH, the Dana Foundation, NARSAD (B.L.), and the Uehara Memorial Foundation (H.M.).

References

1. Yuste R, Konnerth A (2005) Imaging in neuroscience and development: a laboratory manual. Cold Spring Harbor, NY Laboratory Press
2. Kopec CD, Li B, Wei W, Boehm J, Malinow R (2006) Glutamate receptor exocytosis and spine enlargement during chemically induced long-term potentiation. J Neurosci 26:2000–2009
3. Kopec CD, Real E, Kessels HW, Malinow R (2007) GluR1 links structural and functional plasticity at excitatory synapses. J Neurosci 27:13706–13718
4. Holtmaat AJ et al (2005) Transient and persistent dendritic spines in the neocortex in vivo. Neuron 45:279–291
5. Holtmaat A, Wilbrecht L, Knott GW, Welker E, Svoboda K (2006) Experience-dependent and cell-type-specific spine growth in the neocortex. Nature 441:979–983
6. Matsuzaki M, Honkura N, Ellis-Davies GC, Kasai H (2004) Structural basis of long-term potentiation in single dendritic spines. Nature 429:761–766
7. Yang G, Pan F, Gan WB (2009) Stably maintained dendritic spines are associated with lifelong memories. Nature 462:920–924
8. Xu T et al (2009) Rapid formation and selective stabilization of synapses for enduring motor memories. Nature 462:915–919

9. Lai CS, Franke TF, Gan WB (2012) Opposite effects of fear conditioning and extinction on dendritic spine remodelling. Nature 483:87–91
10. Fu M, Yu X, Lu J, Zuo Y (2012) Repetitive motor learning induces coordinated formation of clustered dendritic spines in vivo. Nature 483:92–95
11. Harris KM, Kater SB (1994) Dendritic spines: cellular specializations imparting both stability and flexibility to synaptic function. Annu Rev Neurosci 17:341–371
12. Harris KM, Stevens JK (1989) Dendritic spines of CA 1 pyramidal cells in the rat hippocampus: serial electron microscopy with reference to their biophysical characteristics. J Neurosci 9:2982–2997
13. Takumi Y, Ramirez-Leon V, Laake P, Rinvik E, Ottersen OP (1999) Different modes of expression of AMPA and NMDA receptors in hippocampal synapses. Nat Neurosci 2:618–624
14. Malinow R, Malenka RC (2002) AMPA receptor trafficking and synaptic plasticity. Annu Rev Neurosci 25:103–126
15. Kessels HW, Malinow R (2009) Synaptic AMPA receptor plasticity and behavior. Neuron 61:340–350
16. Makino H, Malinow R (2009) AMPA receptor incorporation into synapses during LTP: the role of lateral movement and exocytosis. Neuron 64:381–390
17. Makino H, Malinow R (2011) Compartmentalized versus global synaptic plasticity on dendrites controlled by experience. Neuron 72:1001–1011
18. Gainey MA, Hurvitz-Wolff JR, Lambo ME, Turrigiano GG (2009) Synaptic scaling requires the GluR2 subunit of the AMPA receptor. J Neurosci 29:6479–6489
19. Borgdorff AJ, Choquet D (2002) Regulation of AMPA receptor lateral movements. Nature 417:649–653
20. Ehlers MD, Heine M, Groc L, Lee MC, Choquet D (2007) Diffusional trapping of GluR1 AMPA receptors by input-specific synaptic activity. Neuron 54:447–460
21. Triller A, Choquet D (2005) Surface trafficking of receptors between synaptic and extrasynaptic membranes: and yet they do move! Trends Neurosci 28:133–139
22. Matsuda T, Cepko CL (2007) Controlled expression of transgenes introduced by in vivo electroporation. Proc Natl Acad Sci USA 104: 1027–1032
23. Saito T (2006) In vivo electroporation in the embryonic mouse central nervous system. Nat Protoc 1:1552–1558
24. Hatanaka Y, Hisanaga S, Heizmann CW, Murakami F (2004) Distinct migratory behavior of early- and late-born neurons derived from the cortical ventricular zone. J Comp Neurol 479:1–14
25. Borrell V, Yoshimura Y, Callaway EM (2005) Targeted gene delivery to telencephalic inhibitory neurons by directional in utero electroporation. J Neurosci Methods 143: 151–158

Chapter 26

Morphological Analysis of Neuromuscular Junctions by Immunofluorescent Staining of Whole-Mount Mouse Diaphragms

Haitao Wu and Lin Mei

Abstract

Immunofluorescence or IF is a technique allowing the visualization of a specific protein or antigen in cells or tissues by binding a specific antibody chemically conjugated with a fluorescence dye. Immunofluorescent staining is widely used in life science research, particularly for neuroscience. Here, we describe the immunofluorescent staining of whole-mount neonatal mouse diaphragms to study the morphological patterns of the neuromuscular junction (NMJ) by using of presynaptic neuronal marker-neurofilament (NF) and synaptophysin antibodies; postsynaptic acetylcholine receptors (AChRs) were labeled with Alexa Fluor 594-conjugated α-bungarotoxin (α-BTX). Immunofluorescence-stained diaphragms were examined under a confocal microscope.

Key words Immunofluorescence, Whole mount, Mouse diaphragms, Neuromuscular junction, AChRs, α-BTX, Light microscopy

1 Introduction

Immunofluorescence is a sensitive method to determine the distribution patterns of interested proteins. It has been widely used both in life science research and clinical diagnostics. This technique was developed based on pioneering work by Coons and Kaplan [1] and later by Mary Osborne [2]. Immunofluorescent staining can be used on both fresh and fixed samples. In immunofluorescence techniques, antibodies that are chemically conjugated to various fluorescent dyes such as fluorescein isothiocyanate (FITC), tetramethyl rhodamine isothiocyanate (TRITC), or the Alexa Fluor dye series—a series of superior fluorescent dyes that span the visible spectrum—represent a major breakthrough in the development of fluorescent labeling reagents. These labeled antibodies bind directly (direct IF) or indirectly (indirect IF) to interested proteins which allows for

Renping Zhou and Lin Mei (eds.), *Neural Development: Methods and Protocols*, Methods in Molecular Biology, vol. 1018, DOI 10.1007/978-1-62703-444-9_26,

antigen detection through fluorescence techniques. Direct IF uses a single antibody that is chemically linked to a fluorophore. The antibody recognizes the target molecule and binds to it, and the fluorophore it carries can be detected via microscopy. Indirect IF uses two antibodies—the unlabeled primary antibody, which specifically binds the target protein, and fluorescent dye-conjugated secondary antibody, which recognizes the primary antibody and binds to it. Multiple secondary antibodies can bind a single primary antibody. This provides signal amplification by increasing the number of fluorophore molecules per antigen [3].

Neuroscience is one of the fields in which immunofluorescence labeling and confocal laser scanning have been widely used. It provided a powerful tool to detect neuronal-specific proteins associated with individual cells on cytological preparation (immunocytochemistry) or within a tissue biopsy (immunohistochemistry) such as brain, spinal cord, and diaphragm. Here we introduce a modified immunofluorescence technique: whole-mount staining of the neonatal mouse diaphragms to study the morphological patterns of the neuromuscular junction (NMJ)—a type of peripheral synapse formed between motoneurons and skeletal muscle fibers [4–6]. Due to its large size and the experimental accessibility, NMJ has contributed greatly to the understanding of the general principles of synaptogenesis and to the development of potential therapeutic strategies for muscular disorders [4].

Previous studies of developing mammalian NMJs emphasized the precise apposition of nerve terminals to AChR-rich postsynaptic sites [4, 5, 7]. Thus, we examined phrenic nerve innervations and postsynaptic AChR clustering on the diaphragm—a typical area to study the development and function of the NMJs [4, 8–11]. In the mouse diaphragm muscle, we used a two-color fluorescent immunofluorescence technique to simultaneously visualize innervating axons and presynaptic nerve terminals and motor end plates. Laser-scanning confocal microscopy was then used to optically scan the double-labeled diaphragm muscle and create z-stack images of the NMJs.

2 Materials

Prepare all solutions using ultrapure water (Millipore) and analytical grade reagents unless otherwise indicated. All reagents and buffer were prepared and stored at room temperature unless otherwise indicated. Diligently follow all waste disposal regulations when disposing waste materials. All mice were housed in a room with a 12 h light/dark cycle with ad libitum access to food and water. Experiments with animals were approved by institutional animal care and use committee (IACUC).

2.1 Mouse Anesthesia Reagent and Surgical Instruments

1. Isothesia (1-chloro-2,2,2-trifluoroethyl difluoromethyl ether) (isoflurane, U.S.P.; Butler Schein Animal Health) 100 mL (*see* **Note 1**).
2. Straight and curved Moria ultra fine forceps and sharp scissors (F.S.T). Two or more for each.
3. Olympus SZX10® research stereo microscope (1×).

2.2 Mouse Diaphragm Fixative Components

1. 0.2 M phosphate buffer (PB), pH 7.4: Weigh 21.8 g Na_2HPO_4 and 6.4 g NaH_2PO_4 and transfer to the cylinder. Add water to a volume of 900 mL. Mix and adjust pH with HCl (*see* **Note 2**). Make up to 1 L with water.
2. 0.1 M phosphate buffer (PB), pH 7.4: Add 500 mL 0.2 M phosphate buffer (PB), pH 7.4 to 500 mL distilled water.
3. 4 % paraformaldehyde (w/v) in 0.1 M phosphate buffer: Add 40 g paraformaldehyde in 0.1 M phosphate buffer, pH 7.4 (*see* **Note 3**). Heat to 60–65 °C while stirring. Add a few drops of 1 N NaOH until solution clear. Continue to stir to dissolve. Cool the solution and filter them out before storage at 4 °C.

2.3 "Immunofluorescent" Staining Buffers

1. 0.1 M glycine in 1× phosphate buffered saline (PBS) (Hyclone): Weigh 0.375 g glycine and transfer to the conical tube and add 1× PBS to a final volume of 50 mL. Store at 4 °C.
2. 10 % Triton X-100 in 1× PBS (Hyclone): Add 10 mL Triton X-100 (Sigma-Aldrich) to 90 mL 1× PBS and vortex. Store at 4 °C.
3. Blocking buffer (2.5 % BSA and 5 % goat serum in 0.5 % Triton X-100/PBS): Weigh 2.5 g BSA and add together with 5 mL goat serum (Sigma-Aldrich) and 5 mL 10 % Triton X-100/PBS to 90 mL 1× PBS. Mix to a volume of 100 mL. Store at 4 °C.
4. Washing buffer (0.5 % Triton X-100/PBS): Add 5 mL 10 % Triton X-100/PBS to 95 mL 1× PBS and vortex. Store at 4 °C.

2.4 Antibodies and Reagents

Majority of chemicals were purchased from Sigma-Aldrich Company unless otherwise indicated. Alexa Fluor 594-conjugated α-bungarotoxin (α-BTX) (Invitrogen, Carlsbad, CA) (B-13423; 1:3,000), aliquot and store at −20 °C; neurofilament (Millipore, Billerica, MA) (AB1991; 1:1,000), store at −20 °C; synaptophysin (Dako, Carpinteria, CA) (A0010; 1:2,000), store at 4 °C; Alexa Fluor 488 goat anti-rabbit IgG (Invitrogen, Carlsbad, CA) (A-11034, 1:1,000), store at 4 °C. Mounting medium: 10 mL (H-1000, VECTASHIELD® Vector Laboratories). Store at 4 °C in the dark (*see* **Note 4**).

2.5 Imaging Instrument and Software

Zeiss confocal laser-scanning microscopy (LSM 510 META 3.2) equipped with a 30 mW argon krypton laser (488, 568 nm) or Ar laser (458, 477, 488, 514 nm, 30 mW); HeNe laser (543 nm,

1 mW); and HeNe laser (633 nm, 5 mW) and filter sets suitable for the detection of fluorescein and Alexa Fluor dye series. Confocal images were acquired and analyzed with LSM5 Image Examiner (Carl Zeiss, Inc.).

3 Methods

Carry out all procedures at room temperature unless otherwise specified.

3.1 Mouse Diaphragm Dissection and Fixation

1. Neonatal (P0) mice were anesthetized by dropping them into a container containing cotton ball with isoflurane in the hood.
2. Anesthetized neonatal mice were sacrificed by decapitation with sharp scissors.
3. Cutting the end of the tail of the mouse with scissors (Part D) for genotyping if necessary (Fig. 1).
4. Cut through thoracic cavity along line A with sharp scissors (Fig. 1) and remove lung and heart tissues under a stereo microscope.
5. Cut through abdominal cavity along line B (Fig. 1) and remove liver and guts with sharp scissors without destroying the diaphragm (*see* **Note 5**).
6. Save part C between two cuts (Fig. 1) in a 2 mL vial with cold 4 % PFA fixative and incubate at room temperature for 30 min. Store at 4 °C for 2 more days until staining.

3.2 Whole-Mount Staining of Mouse Diaphragms

1. Intact dissection of the diaphragm from the tissue fixed in 4 % PFA and rinsed with 1× phosphate buffered saline (PBS) (pH 7.3) for 30 min three times at room temperature.
2. Diaphragms were incubated with 0.1 M glycine in 1× PBS for 1 h (*see* **Note 6**) and followed by rinsing with 0.5 % Triton X-100 in PBS for 30 min; repeat rinsing three times.
3. Diaphragms were blocked in the blocking buffer (2.5 % BSA and 5 % goat serum in 0.5 % Triton X-100/PBS) for 3–4 h at room temperature or overnight at 4 °C.
4. Diaphragms were incubated with primary antibodies (neurofilament, AB1991, 1:1,000; and synaptophysin, A0010, 1:2,000) in the blocking buffer overnight at 4 °C.
5. Rinse three times for 1 h each with 0.5 % Triton X-100 in PBS.
6. Diaphragms were incubated with Alexa Fluor 488 goat anti-rabbit antibody (A-11034, 1:1,000) and Alexa Fluor 594-conjugated α-bungarotoxin (α-BTX) (B-13423, 1:3,000) for 4 h at room temperature or overnight at 4 °C.

7. Rinse three times for 1 h each with 0.5 % Triton X-100 in PBS.
8. Rinse once with 1× PBS and flat mounted in mounting medium (*see* **Note** 7). Store at 4 °C until confocal microscopy scanning.

3.3 Confocal Laser-Scanning Microscopy

1. We examine immunofluorescent staining specimens using a Zeiss LSM 510 META 3.2 laser-scanning confocal microscopy (*see* **Note 8**).
2. We first start the LSM 5 program. After the start of the "Expert Mode" or the "Routine Mode," instrument initialization is performed and can be monitored in the initialization window and interrupted with a click on the "Cancel" button, if required.
3. We routinely use 10× objective to acquire and collect the laser-scanning images. To get the images with higher resolution and magnification, 25× or 40× oil immersion objectives also can be used.
4. We used a laser power between 1 and 30 %, and standard setup for imaging Alexa Fluor 488 and Alexa Fluor 594, which uses 522/32 and 617 nm emission filters, respectively.
5. The pinhole was kept at around 100 μm (1 Airy unit). Laser power was monitored periodically to ensure consistency between experiments. Virtually no cross talk was observed between channels (green, 505–530 nm band-pass filter; red, 585 nm long-pass filter).

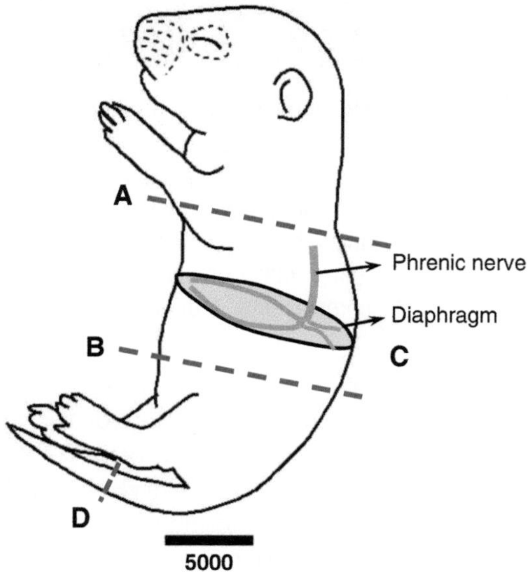

Fig. 1 Neonatal mice diaphragm dissection and fixation. Part (*A*) shows the line scissors cut through the thoracic cavity. Part (*B*) shows the line scissors cut through the abdomen. Part (*C*) including the diaphragm was fixed in 4 % paraformaldehyde for fluorescence staining. Part (*D*) was used for genotyping is necessary

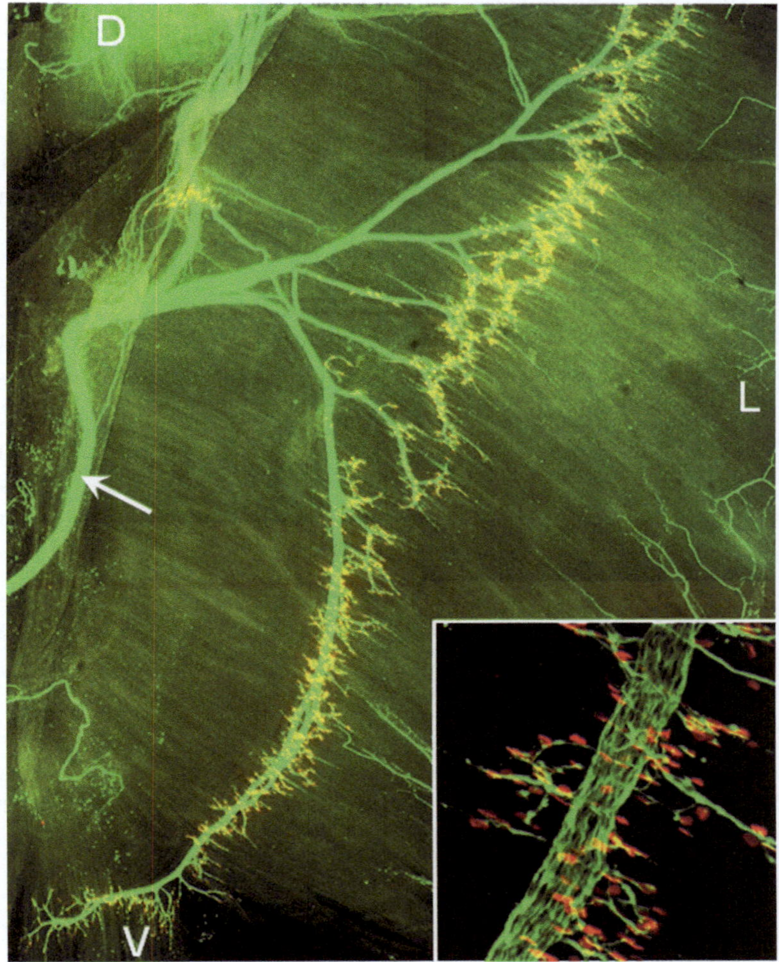

Fig. 2 Whole-mount staining illustrates the NMJ pattern in neonatal diaphragm. P0 diaphragms were stained whole mount with α-BTX (*red*) to label AChR clusters and with NF/synaptophysin antibodies (NF/Syn) to label motor nerve terminals, which was visualized by Alexa Fluor 488-conjugated goat anti-rabbit antibody (*green*). Shown were the *left*, ventral areas of hemidiaphragms. *Insets*, representative images of the NMJs in the central region of the diaphragm. *Arrow*, phrenic nerve. *D* dorsal, *L* lateral, *V* ventral

6. We routinely acquire the images using z-stack mode. Briefly, after adjustment of the focus to find the approximate center slice and perform color adjustment, continue focusing to determine where should be the first slide of the z-stack, and then continue to adjust the focus to the last slice of the z-stack. Z serial images were collected and finally collapsed into a single image. The images were transferred to Adobe Photoshop for display and analysis (Fig. 2).

3.4 Quantitative Analysis of Images

Image analysis and quantification procedures were performed with LSM5 Image Examiner (Carl Zeiss, Inc.) (*see* **Note 8**).

1. Scan a confocal z-stack of the selected view of the NMJ staining samples. Go for saturated signals in all two or three channels but avoid overexposure. The resolution of 1,024 × 1,024 pixels was highly recommended.
2. Choose the "Open" tab on the main menu, and open the z-stacked images.
3. Generation maximum projections of all three z-stacks.
4. For better contrast, select the "Contr" tab and adjust to the best results.
5. To measure the size of single AChR clusters and the end plate bandwidth, the red channel was selected by manipulating the "channel" table. The "Ch1-T1" (red) channel was open only to illustrate the staining of postsynaptic AChR clusters (*see* **Note 9**).
6. Single AChR cluster was selected by using of the selected tools within "Overlay" panel to measure the size (area, diameter, or circumference, etc.) of the single AChR clusters. In addition, the total number of AChR clusters in each view field can be determined either by manually counting (preferentially using the projections) or manual segmentation of individual synapses.
7. To measure the end plate bandwidth, a polygon was marked to include most peripheral AChR clusters and the myotube length contained in the polygon was measured using the "Ruler" tab listed in the "Overlay" panel.
8. To measure the length of the axon branches, the "Irregular lines" toolbar listed in the "Overlay" panel was selected to trace the projection of the axons. The value will be shown after selection of the "Ruler" tab.

4 Notes

1. Isoflurane may be used without a vaporizer. The liquid anesthetic is applied to gauze or cotton which is placed into a container for induction (bell jar) or in a conical tube for maintenance of anesthesia. This reagent should be stored and performed on a downdraft table or in a non-recirculating (fume) hood or special biosafety cabinet with a carbon filter.
2. Concentrated HCl (12 N) can be used at first to narrow the gap from the starting pH to required pH. From then on it would be better to use a series of HCl (e.g., 6 N and 1 N) with lower ionic strengths to avoid a sudden drop in pH below the required pH.
3. Wear a mask when weighing paraformaldehyde powder. To avoid exposing paraformaldehyde to coworkers, cover the weigh boat containing the weighted paraformaldehyde with another weigh boat (same size to the original weight boat con-

taining the weighed paraformaldehyde) when transporting it to the fume hood. Transfer the weighed paraformaldehyde to the cylinder inside the fume hood and mix on a hot-plate magnetic-stirrer device placed inside the hood.

4. VECTASHIELD® mounting medium is a unique, stable formula for preserving fluorescence, which prevents rapid photobleaching of fluorescent reagents such as fluorescein, rhodamine, Alexa Fluor 488, and Alexa Fluor 594. Samples mounted in VECTASHIELD® mounting medium retain fluorescence during prolonged storage at 4 °C.
5. In some case, to avoid destroying or stretching the diaphragm, some liver tissue can be maintained together with the diaphragm for fixation. These residual liver tissues later on can be easily dissected from the diaphragm after fixation.
6. 0.1 M glycine in 1× PBS incubation can quench the residual paraformaldehyde (PFA) cross-linking activity.
7. Putting few drops of 1× PBS on the slides to make the diaphragm samples in suspension on the slide without folds. Apply vacuum to remove PBS completely and dispense one to two drops of mounting medium on the surface of each diaphragm. The clean coverslip is carefully lowered onto the drop of mounting medium in such way to prevent the formation of bubbles. Small pieces of filter paper are placed around the edge of the coverslip to absorb excess mounting medium. Seal the edge of the coverslip with clear nail polish and let it dry.
8. The LSM 510 is a laser hazard class 3 B instrument and is marked as such. This moderate-risk class embraces medium-power lasers. Confocal microscopy users must take care not to be exposed to the radiation of such lasers. In particular, never look into the laser beam.
9. The images with high quality were used for quantitative analysis, and only images of fully identical imaging settings were compared.

Acknowledgements

This work was supported by NIH (L.M. and W.C.X.) and MDA (L.M.) grants.

References

1. Coons AH, Kaplan MH (1950) Localization of antigen in tissue cells; improvements in a method for the detection of antigen by means of fluorescent antibody. J Exp Med 91(1):1–13
2. Weber K, Bibring T, Osborn M (1975) Specific visualization of tubulin-containing structures in tissue culture cells by immunofluorescence. Cytoplasmic microtubules, vinblastine-induced paracrystals, and mitotic figures. Exp Cell Res 95(1):111–120
3. Fritschy J-M, Härtig W (2001) Immunofluorescence. eLS

4. Wu H, Xiong WC, Mei L (2010) To build a synapse: signaling pathways in neuromuscular junction assembly. Development 137(7): 1017–1033
5. Sanes JR, Lichtman JW (1999) Development of the vertebrate neuromuscular junction. Annu Rev Neurosci 22:389–442
6. Sanes JR, Lichtman JW (2001) Induction, assembly, maturation and maintenance of a postsynaptic apparatus. Nat Rev Neurosci 2(11):791–805
7. Burden SJ (1998) The formation of neuromuscular synapse. Genes Dev 12:133–148
8. Li XM et al (2008) Retrograde regulation of motoneuron differentiation by muscle beta-catenin. Nat Neurosci 11(3):262–268
9. Lin W et al (2000) Aberrant development of motor axons and neuromuscular synapses in erbB2-deficient mice. Proc Natl Acad Sci USA 97(3):1299–1304
10. Lin W et al (2001) Distinct roles of nerve and muscle in postsynaptic differentiation of the neuromuscular synapse. Nature 410(6832): 1057–1064
11. Yang X et al (2001) Patterning of muscle acetylcholine receptor gene expression in the absence of motor innervation. Neuron 30(2):399–410

Part IV

Morphological Analyses of the Developing Nervous System

Chapter 27

Routine Histology Techniques for the Developing and Adult Central Nervous System

W. Geoffrey McAuliffe

Abstract

The preparation of tissue for histological study is a multi-step process in which potential loss of quality and the introduction of artifacts can occur during each step. Knowledge of the process and the potential pitfalls at each step will serve the investigator well. Here I describe the most basic histologic techniques, worked out so long ago they are often absent from current literature.

Key words Central nervous system, Formaldehyde, Perfusion fixation, Immersion fixation, Light microscopy, Nissl, Staining

1 Introduction

The preparation of tissue for microscopic observation entails (1) fixation of the tissue, (2) the cutting of sections, and (3) staining those sections to provide adequate contrast. One of the best ways to become competent in this, or any new technique, is to become an "apprentice" to a laboratory thoroughly versed in the technique in question. A few days or even a few hours observing an expert will be invaluable. Alternatively, appropriate collaborations could be established. Ideally this is done during the design of the experiment so that new knowledge can be incorporated and potential problems dealt with early in the investigation. Since it is not possible to cover all of the possible intricacies of histological technique in a single chapter, a general text [1] is valuable to have in the laboratory.

2 Materials

Many commercial vendors offer ready-to-use fixatives, processing solutions and stains for sale. While these are usually of good quality and do reduce laboratory preparation time, their composition is

Renping Zhou and Lin Mei (eds.), *Neural Development: Methods and Protocols*, Methods in Molecular Biology, vol. 1018, DOI 10.1007/978-1-62703-444-9_27,

invariably secret. If the manufacturer decides to change the composition for economic or other reasons, the end user has no way of knowing this. Putting one's valuable experiment at the mercy of unknowns is not to be recommended, especially since most histological reagents are simple and economical to prepare.

2.1 Fixation

It is an axiom that good histology begins with good fixation. Decisions about fixation (choice of fixative and duration) will be dictated by the goals of the experiment. It is best to consult the literature during experimental design to avoid errors that waste time and materials. Do recognize that the goals and scope of the study may expand depending on findings.

The purpose of fixation is to stop autolysis, immobilize cellular components and antigens, and to prepare the tissue for subsequent processing. To this end the animal is usually perfused with or tissue immersed in buffered solutions of formaldehyde [1–5]. Formaldehyde nomenclature can be confusing to the newcomer. Formaldehyde is a gas that is bubbled through water until saturation, about 37–40 %. A small amount of methanol is added to commercial solutions for stability. This solution is usually diluted with 9 parts of water then buffered to make "formalin" or "10 % formalin" which contains about 4 % formaldehyde. Some investigators prefer to make a 4 % solution fresh from paraformaldehyde polymer to avoid the small amount of methanol present in commercial solutions. For routine work this is unnecessary. In either case the solution is buffered with either phosphate or calcium salts to pH 7.2–7.4. Two common formulas are as follows:

Formaldehyde, 37–40 %	100 ml
Distilled water	900 ml
Sodium phosphate monobasic, $NaH_2PO_4\ H_2O$ when dissolved add	4.0 g
Sodium phosphate dibasic, Na_2HPO_4	6.5 g

20 g of calcium acetate, $Ca(C_2H_3O_2)_2 \times H_2O$, may be substituted for the two phosphate salts above to improve the retention of phospholipids. Note that calcium salts cannot be added to phosphate buffers due to the immediate precipitation of calcium phosphate.

With either fixative prepare some buffer using the recipe above but substituting water for the formaldehyde. This will be needed for pre-perfusion washout and/or rinsing the tissue after fixation.

Some investigators add up to 8 % sucrose to both the fixative solution and the buffer for beneficial osmotic effects, especially if the fixative is to be delivered by perfusion. In this case the sucrose may or may not be removed in the post-fixation washings, depending on how the tissue will be subsequently processed.

Various concentrations of graded ethanols will be needed for dehydration if the tissue is to be embedded in paraffin and for making permanent slides. To prepare a 50 % concentration, put 50 ml of 95 % ethanol in a graduated cylinder and fill to the 95 ml mark with distilled water. Starting with 70 ml of 95 % ethanol and filling to the 95 ml mark will give a 70 % concentration. Thus, one can easily make any desired concentration. This is much more economical than using 100 % ethanol and the final concentration is close enough for our purposes.

2.2 Sectioning

A sliding microtome (Leica, Thermo Scientific) equipped with a freezing specimen stage is very useful for cutting frozen sections. The older Spencer AO 860 model is an excellent sliding microtome if available. Cutting frozen sections on a sliding microtome is probably the easiest method for a novice to master. The rotary microtome (Leica, Thermo Scientific) is the most common type of microtome for cutting sections of tissue embedded in paraffin. A cryostat (Leica, Microm, Sakura, Thermo Scientific) is a rotary microtome inside a freezing cabinet for cutting sections of frozen tissue. A vibrating blade microtome (Vibratome) is used to cut thicker sections of fixed or unfixed material that has not been frozen or embedded.

A sharp knife is essential for cutting good sections. Microtome knives come in a variety of lengths; 125 mm is the most convenient length for a rotary microtome. It is possible to use longer knives with a sliding microtome, but great care must be taken to avoid injury. In recent years disposable knives have become popular as they avoid the inconvenience of keeping multiple knives, since dull knives need to be sent out for periodic re-sharpening. Disposable knives are also an advantage when cutting embryos since cartilage and bone dull the knife quickly. Disposable knives fit into a holder which may be designed for a particular type of disposable knife or microtome. The manufacturer of the microtome should be consulted regarding the choice of a blade holder.

Microscope slides and cover glasses (coverslips) are essential supplies. The only real decision about slides is whether to purchase slides coated with substances to insure section adhesion or to buy plain slides and coat them yourself. In the latter instance, it may be necessary to wash the slides prior to coating; "pre-cleaned" slides may not be very clean. Plain slides can be coated with a gelatin solution, "subbed," as follows. Dissolve 2 g of gelatin in 1,000 ml of distilled water at 50–60 °C. A magnetic stirrer is helpful. Cool and add 0.25 g of chromium potassium sulfate. Store at 4–6 °C. Make fresh every few weeks. Even if contamination is not visible to the naked eye, it will show up under the microscope. Clean glass slides are dipped in the solution several times, air-dried in a vertical position, and stored in a dust-free box. Slides are good indefinitely. If harsh conditions or very long staining times will be required, *see* ref. 8 for coating slides with 3-aminopropyltriethoxy-silane.

Cover glasses of #1or #1½ thickness are satisfactory. Cover glasses are mounted with a drop of DPX, a synthetic mounting medium dissolved in xylene and available from the usual laboratory supply houses.

2.3 Staining

Use only dyes certified by the Biological Stain Commission (BSC) for the preparation of staining solutions. The BSC tests and certifies each batch sent to them by the manufacturer and they or the manufacturer affixes a certification label to each bottle of stain. Most stains are also identified by a color index (CI) number. Since most stains are simple percentage solutions in water or buffer, it is most economical to prepare staining solutions in the lab rather than buying commercial preparations of uncertain composition.

Cresyl violet or thionine (CI 52000) are the most common stains for nuclei and Nissl substance in the CNS. Preparation is easy: dissolve 0.1 g of cresyl violet acetate or thionine in 100 ml of distilled water. Add 0.5 ml of 10 % acetic acid or 100 μl of glacial acetic acid. The pH needs to be about 4 to insure specificity for nuclei and Nissl substance. Keeps for several months.

The Azure-Eosin stain [1, 3] gives a coloration similar to hematoxylin and eosin but is easier to prepare and apply. Make stock solutions of 0.1 % aqueous Azure A (CI 52005) and 0.1 % aqueous Eosin B (CI 45400) in distilled water. These keep for several months. For the stain mix 16 ml of Azure A stock solution with 16 ml Eosin B stock solution. Add 6.8 ml 02 M acetic acid, 1.2 ml 0.2 M sodium acetate, 20 ml acetone, and 100 ml distilled water. Optimal pH for formalin-fixed tissue is about 4. The staining solution must be mixed fresh each day and discarded after 10 slides.

Gill's hematoxylin II [9] has the advantage of ripening very quickly, not precipitating and lasting up to 1 year. As the solution ages, it may stain more quickly. Mix together 710 ml distilled water, 250 ml ethylene glycol, 4 g hematoxylin (CI 75290), or 4.72 g if the trihydrate is used. When dissolved add 0.4 g sodium iodate. When dissolved add 35.2 g aluminum sulfate ($Al_2(SO_4)_3\ 18H_2O$). When dissolved add 40 ml glacial acetic acid. Mix 1 h and filter. The ratio of hematoxylin to aluminum sulfate is very important! Adjustments will be needed to be made to the aluminum sulfate if the water of hydration differs from that given. Two grams of citric acid can substitute for the glacial acetic acid. For a counterstain make 0.05 % Eosin Y (CI 45380) in 70 % ethanol with 100 μl of glacial acetic acid added to each 100 ml of stain.

Other necessary materials include dry ice and 2-methylbutane (isopentane) for freezing tissue, chloroform, xylene, staining jars (Coplin) or dishes. For paraffin embedding an oven (preferably with vacuum attachment) capable of consistent 60 °C temperature, wax such as ParaplastPlus for embedding, plastic molds for embedding tissue (Peel-A-Way), and a heated water bath and a drying plate are useful.

A "brain blocker" of the type sold by Knof and other vendors may be useful for slicing the brain in a reproducible manner.

3 Methods

3.1 Fixation

How the tissue is exposed to the fixative will be dictated by the experiment. For optimal morphology of the CNS perfusion of fixative through the vascular system is best. The perfusion solutions, first a buffer to wash blood out of the vasculature then the fixative, are driven by gravity or by a small pump. It is important that the perfusion pressure not exceed the normal blood pressure of the animal. In a system driven by gravity, pressure can be approximated by converting the inches of water above the animal to millimeters of mercury. Two reservoirs, one for buffer and one for fixative, are connected with tubing and a Y connector and purged of air bubbles (Fig. 1).

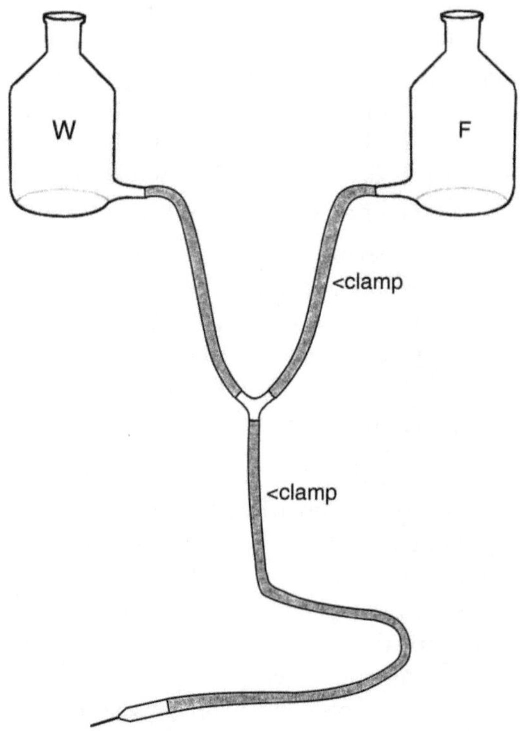

Fig. 1 A simple apparatus for perfusing fixative by gravity. One bottle (W) holds washout buffer and the other (F) buffered formalin fixative. Clamps are initially placed as indicated. First the clamp common hose is removed to facilitate washout, and then the clamp on the hose leading from the fixative is moved to the hose on the washout bottle. The bottles are known as "reservoir bottles" or "aspirator bottles with tubulation" and can be purchased from scientific supply vendors

The procedure, performed wearing gloves and under a fume hood to minimize exposure to formaldehyde vapors, is as follows. An assistant should be available.

1. The animal is deeply anesthesized and secured to firm surface, ventral side up. Have a pan or other receptacle in place to collect the perfusate which will be mixed with the animal's blood and may need to be disposed of as hazardous waste.
2. The chest is opened and the heart exposed.
3. Forceps are used to stabilize the heart while a cannula, usually a syringe needle whose inside diameter is the same as that of the aorta, is inserted a few millimeters into the left ventricle just to the right of the apex. A 15- or 16-gauge needle works well for a rat, 18 or 20 gauge for a mouse.
4. An assistant uses sharp scissors to make a small cut in the right atrium to allow the escape of blood.
5. The assistant starts the flow of a small amount, 5–20 ml depending on the size of the animal, of buffer at room temperature to wash most of the blood from the vascular system. It is neither possible nor necessary to wash all of the blood out so do not waste time trying.
6. The flow is quickly changed to fixative at room temperature. Perfusion of a volume twice the weight of the animal should be accomplished in about 10 min. A common error is to perfuse at too slow a rate either by using a cannula that is too small or using a rate of flow that is too slow.
7. After perfusion the CNS may be removed immediately or the entire animal may be placed in the refrigerator for several hours or overnight and the CNS removed the next day. The area of interest should be removed and sliced into pieces 5–10 mm in thickness and immersed in fixative for at least another 24 h but preferably several days.

The duration and temperature of fixation is the subject of some debate. While formaldehyde penetrates tissue rapidly, the reactions that actually fix and harden tissue proceed slowly, requiring several days. Furthermore, while fixation in the cold minimizes autolysis, it also slows the reaction of formaldehyde with tissue components. Thus, fixation for 48 h at room temperature fixative is probably the minimum, and a week is better. The common complaint that CNS tissue is too soft can usually be solved by longer fixation. Changing the concentration of formaldehyde in the fixative will not speed up the reaction.

For more detail on the chemistry of fixation, *see* refs. 1–5. For information on the relationship of fixation to immunostaining, *see* refs. 1, 6.

Fixation by immersion: There may be instances in which perfusion of the animal is impossible for various reasons, often because some fresh tissue is needed for other analyses. In this case the tissue in question is removed immediately after death, sliced no more than 5 mm thick, and immersed in a volume of fixative 10–20 times the volume of the tissue. Tissue should remain in fixative for at least 48 h, several days will give better results.

Embryonic tissues are often prepared by flooding the abdomen of a deeply anesthesized pregnant female with fixative or by removing the embryos and immersing them in fixative. It is possible to perfuse mouse embryos *in utero* as early as E10 [7]. The common complaint that embryos are difficult to fix or too soft after fixation is usually due to inadequate time in fixative. As mentioned above formaldehyde is known to fix tissue slowly so a minimum of several days in fixative will insure optimal results. Alternatively, a small amount of glutaraldehyde, 0.2 %, can be added to the fixative. Glutaraldehyde is a much more powerful cross-linker than formaldehyde and will fix more quickly, but staining may be affected.

Post-fixation processing: If tissue is to be frozen for sectioning it needs to be cryoprotected with sucrose and the formaldehyde washed out, see Sectioning. For embedding in paraffin wax, both the sucrose and the formaldehyde need to be washed out with buffer or water. Several changes of the same buffer in which the fixative was made 30–60 min per change are sufficient. Washing removes excess formaldehyde and sucrose that may interfere with subsequent processing. Tissue in the last buffer wash is stored at 4–6 °C until the next step. The buffer needs to be changed every 3–5 days to prevent growth of microorganisms. Stock solutions of buffer need to be filtered periodically for the same reason.

3.2 Sectioning

To make the tissue firm enough to cut thin sections, the water in the tissue must be frozen or replaced with a hard substance such as paraffin wax. Decisions about how the tissue will be sectioned (cut) for microscopic observation were made during experimental design.

Frozen sections: The easiest and fastest way to cut sections is to freeze the tissue very rapidly and cut sections with a sliding microtome equipped with a freezing stage. Tissue that is to be frozen must be cryoprotected to prevent damage by ice crystals. The tissue is transferred from storage buffer to a freshly prepared solution of 10 % sucrose at 4–6 °C for a few hours until it sinks and then moved to fresh 20–30 % sucrose at 4–6 °C until it sinks; overnight is usually sufficient. The importance of rapid freezing cannot be overemphasized; tissue must be completely frozen in 20–40 s. Slow freezing will cause large ice crystals to form and leave holes in the tissue, ruining morphology. When using a freezing stage on a sliding microtome, the tissue is frozen from below and simultaneously surrounded by crushed dry ice. Once frozen tissue is allowed to equilibrate for 10–20 min. The tissue should remain frozen but not so hard as to cause cracking when cut. Trial and error with practice tissue will determine the correct settings for your particular instrument and specimen. Twenty micron sections are cut, collected with a small brush, and stored in buffer in 12- or 24-well tissue culture plates. Keep in mind that in sections of whole embryos tissues may separate. Sections should be mounted on slides as soon as possible to avoid the growth of microorganisms. If sections must be stored, they should be refrigerated, the buffer changed often, and care taken to avoid evaporation. Once mounted on slides, sections are dried overnight or longer then stained as desired.

Cryostat sections: Frozen sections can also be cut with a cryostat, a freezing cabinet containing a rotary microtome. Cryostat sections can be thinner, but more skill is required for good results. As mentioned above the importance of cryoprotection and rapid freezing cannot be overemphasized, tissue must be completely frozen in 20–40 s. Simply placing tissue in a cold cryostat is unacceptable. Some cryostats have an apparatus for rapid freezing, and one can experiment to see if this gives acceptable results. If not, tissue is placed on the object disc supplied with the instrument and rapidly chilled with 2-methylbutane (isopentane) previously cooled with liquid nitrogen or dry ice. Tissue is placed in the cryostat previously cooled to about minus 20 °C and allowed to equilibrate for 20–30 min. Sections are cut at 8–20 μm and mounted on chilled, coated microscope slides. Sections are then thawed to adhere the section to the slide. A careful worker can put multiple sections on one slide. Once mounted on slides, sections are dried overnight or longer then stained as desired.

Both frozen and cryostat sections have the advantage of avoiding the shrinkage that occurs during processing for paraffin embedding. They also have the potential advantage of preserving myelin which is dissolved out of tissues by the organic solvents used in processing for paraffin wax embedding. However, retaining the myelin may adversely affect some types of staining. Frozen and cryostat sections should be washed with buffer before staining to remove any remaining sucrose.

Paraffin sections: Tissue can be infiltrated with and embedded in paraffin wax and sectioned on a rotary microtome. This is more time-consuming but thinner sections can be obtained and blocks of paraffin-embedded tissue can be stored for many years. For the CNS a paraffin such as ParaplastPlus with a melting point of 56 °C works well. For tissue to be embedded in paraffin wax, the water in the tissue must be replaced with a solvent miscible with wax. The classic method is to dehydrate with a graded series of ethanols and replace the ethanol with several changes of an organic solvent such as chloroform or xylene. This organic solvent step is known as "clearing" since the tissue takes on a translucent or transparent appearance depending on its thickness. Then the organic solvent is replaced with several changes of melted wax in an oven at a temperature slightly above the melting point of the wax, usually 60 °C.

Finally, the tissue is oriented in a mold filled with molten wax and allowed to cool and harden, then sectioned on a rotary microtome at the desired thickness. Sections are floated on a warm water bath, picked up on coated microscope slides, and dried on a warming plate before staining. Note that cutting sections much thicker than 12–15 μm can be difficult with paraffin-embedded tissue. Historically paraffin processing has been preferred for permanence, but properly processed frozen and cryostat sections can yield excellent longevity.

Below is a sample time schedule for processing tissue about 5–8 mm thick that has been fixed and washed in buffer (or water). The volume of each solution should be at least 20 times that of the tissue. Tissue should be agitated frequently in each solution and drained briefly between solutions. Instructions for preparing graded ethanol solutions are in Subheading 2. Multiple pieces can be processed at the same time, but care must be taken not to mix up samples!

50 % ethanol several hours or overnight
70 % ethanol 1–2 h or overnight
95 % ethanol 1 h
100 % #1 ethanol 1 h
100 % #2 ethanol 1 h
100 % ethanol:xylene (chloroform) in a 1:1 ratio 1 h
Xylene or chloroform #1 1 h
Xylene or chloroform #2 1 h

Melted paraffin, 3 changes 30–45 min each in an oven at 60 °C. Do not overheat!

A vacuum oven is an advantage as lowering the pressure speeds replacement of the organic solvent by molten wax. Chloroform has a lower boiling point than xylene, but the volatility and odor of the former may be troublesome. In either case avoid breathing the vapors of these solvents.

Embed tissue in molten wax in a mold and allow to cool completely, usually overnight. Be sure to put a small paper label in the block so it can be identified later!

Periodically the first 100 % ethanol is discarded, #2 moved up to #1 and a fresh #2 is made. The same is done periodically to the #1 xylene. Note that used xylene must be collected and disposed of as hazardous waste.

3.3 Staining

Only stains certified by the Biological Stain Commission (BSC) should be used. The BSC tests and certifies each batch sent to them by the manufacturer and they or the manufacturer affixes a certification label to each bottle of stain. Each stain is also identified by a Color Index (CI) number. Since most stains are simple percentage solutions in water or buffer, it is most economical to prepare staining solutions in the lab rather than buying commercial preparations of uncertain composition.

Frozen or cryostat sections may be stained once they have dried enough to adhere to the slide; overnight is usually sufficient. If desired, frozen sections can be stained before being mounted on slides, with histochemical reactions or immuno-reagents, for example.

Before staining paraffin sections, the wax must be dissolved out with several changes of xylene ("dewaxing") and the sections hydrated ("brought to water") with a graded ethanol series that is the reverse of the series used for embedding. For example:

- Xylene #1 5 min
- Xylene #2 5 min
- 100 % ethanol #1 5 min
- 100 % ethanol #2 5 min
- 95 % ethanol 5 min
- 70 % ethanol 5 min
- 50 % ethanol 5 min
- Running water or several changes of distilled water 5 min
- Proceed to staining

While slides can be stained individually, staining is usually accomplished in batches of slides held in a rack and moved from container to container as the sections are dewaxed and hydrated if necessary, stained, rinsed, dehydrated with alcohols, cleared with xylene, and ultimately coverslipped with a permanent mounting medium. If the number of slides to be processed is small, glass Coplin jars have ridges to hold slides apart during processing.

The choice of stains for routine morphology is straightforward. While hematoxylin and eosin (H&E) is the primary choice for tissues other than the CNS, eosin does not reveal any significant morphology in the neuropil. In addition, many hematoxylin formulations are complex to prepare, and some will react with the myelin in frozen and cryostat sections which may not be desired. There are other stains that are easier to prepare and apply that do an excellent job on the CNS. The most common is cresyl violet (cresyl violet acetate, cresyl echt violet) which gives a purple color to nuclei and Nissl substance (cytoplasmic RNA) or thionine which gives a bluer color. A simple method that mimics hematoxylin and eosin is the Azure-Eosin method [1, 3]. If a hematoxylin stain is deemed essential, Gill's hematoxylin II [9] is relatively simple to prepare and use.

Cresyl violet or thionine staining: The stain is prepared as described in Subheading 2.

1. Bring paraffin sections to water as above; frozen or cryostat sections are rinsed in water.
2. Stain 2–10 min in 0.1 % cresyl violet or thionine.

3. Rinse in 2 changes of distilled water, 30 s to 1 min each.
4. Dehydrate quickly in a few dips in 70 % ethanol then 95 % ethanol, 15 s, then 2 changes of 100 % ethanol, 1 min each.
5. Clear in 2 changes of xylene and coverslip with DPX mounting medium.

Results: Nuclei and Nissl substance violet or blue-violet and neuropil colorless. Slight metachromasia (violet-red staining) may be seen in white matter if frozen or sections have not been treated with alcohols or xylene (*see* Notes below).

Some judgment is necessary so that the proper amount of stain is removed during dehydration. Too much time in lower ethanols removes too much stain; too little time leaves the background colored.

The background tends to be higher in frozen sections because lipid components have not been extracted by paraffin processing. Some workers pretreat such sections with 70 % alcohol or xylene to remove lipids and "clean up" the background; sections are then rehydrated for staining.

Several dozen slides can be stained in this amount of stain.

If a counterstain for immunostaining reactions is desired, a shorter time in cresyl violet or thionine should suffice. Either stain provides an attractive contrast to the DAB reaction product.

It is important to use fresh 100 % alcohols to remove all water from the sections before clearing in xylene and mounting the coverslip. If not, small bubbles of water will appear in the final preparation. In this case the coverslip will have to be soaked off with xylene and the dehydration and clearing repeated with fresh chemicals.

Azure A-Eosin B stain [1, 3]: This stain is easy to prepare from stock solutions and gives a color picture very similar to that of hematoxylin and eosin staining, blue nuclei and Nissl substance with a pink background. The staining solution is prepared immediately before use and discarded after one set of 10 slides. Prepare as described in Subheading 2.

1. Bring paraffin sections to water. Frozen or cryostat sections may benefit from having lipids removed by a 15 min treatment in 70 % ethanol, then wash in water.
2. Stain 1 h in the above.
3. Pour off the stain and dehydrate with 3 changes of acetone (not alcohol!), 60 s each with agitation.
4. Clear in 2 changes of xylene, 5 min each, coverslip with DPX.

Results: Nuclei and Nissl substance blue, neuropil pink, connective tissue and muscle varying shades of pink.

Gill's hematoxylin II [9]. This solution has the advantage of ripening very quickly, not forming precipitate and lasting up to 1 year. As the solution ages, it may stain more quickly. Prepare as described in the Materials section.

1. Bring paraffin sections to water. Frozen or cryostat sections may benefit from having lipids removed by a 15 min treatment in 70 % ethanol, then wash in water.
2. Stain 3–4 min in Gill's hematoxylin II.
3. Rinse in running water, 5 min.
4. Treat with ammonia water (one drop of ammonium hydroxide in 100 ml water), 30 s until the sections turn blue. Avoid overexposure to strong alkalis as this may loosen sections.
5. Rinse in running water, 3–5 min.
6. Dehydrate in 50 and 70 % ethanol, 5 min each.
7. Counterstain for 2 min in 0.05 % Eosin Y in 70 % ethanol containing 100 μl glacial acetic acid per 100 ml.
8. Finish dehydration with 95 % and 2 changes of 100 % ethanol, 5 min each.
9. Clear with 2 changes of xylene, 5 min each, and coverslip with DPX.

Results: Nuclei and Nissl substance blue, neuropil pink, connective tissue and muscle varying shades of pink.

All stained slides should always be stored away from strong light to minimize fading.

References

1. Kiernan J (1999) Histological and histochemical methods, 3rd edn. Butterworth-Heinemann, Oxford
2. Gray P (1954) The microscopist's formulary and guide. Blakiston, New York, NY
3. Lillie RD, Fullmer HM (1976) Histopathological technique and practical histochemistry. McGraw-Hill, New York, NY
4. Thompson SW (1966) Selected histochemical and histopathological methods. Charles C. Thomas, Springfield, IL
5. Fox CH, Johnson FB, Whiting J et al (1985) Formaldehyde fixation. J Histochem Cytochem 33(8):845–853
6. Polak JM, van Noorden S (2003) Introduction to immunocytochemistry, 3rd edn. BIOS Scientific Publishers Limited, Oxford
7. Abrunhosa R (1972) Microperfusion of embryos for ultrastructural studies. J Ultrastruct Res 41:176–188
8. Rentrop M, Knapp B, Winter H et al (1986) Aminoalkylsilane-treated slides as support for in situ hybridization of keratin cDNA's to frozen tissue sections under varying fixation and pretreatment conditions. Histochemical J 18:271–276
9. Gill GW, Frost JK, Miller KA (1974) A new formula for a half-oxidized hematoxylin solution that neither overstains nor requires differentiation. Acta Cytol 18:300–311

Chapter 28

Cryosectioning

Alexander I. Son, Katie Sokolowski, and Renping Zhou

Abstract

Cryosectioning, the sectioning of frozen specimens, has been an important histological tool for more than a century and continues to be extensively utilized today. However, the ability to produce high-quality sections is often a difficult process requiring extensive patience and experience. In this chapter, we have detailed an effective method for the embedding, mounting, and sectioning of frozen tissues, as well as have provided suggestions in producing high-quality sections.

Key words Cryostat , Microtome , Frozen sectioning , Free-floating sections , Paraformaldehyde fixation , Cryoprotection

1 Introduction

The cutting of frozen tissues into fine sections for histological analysis has been a mainstay of modern biological and medical research. Established in the late nineteenth century, this practice was originally adopted to provide physicians with quick and accurate pathological examinations of their patients, with the first well-cited paper detailing an effective and reliable technique appearing in 1905 [1–3]. Over the course of the next century, frozen sectioning techniques, or cryosectioning, would be implemented in a large variety of research applications and remain an integral tool for biological study today. While some of the technology for this technique has advanced since its invention, the general concept and methods have remained similar.

In practice, cryosectioning is a quick and effective histological technique used for characterizing the morphology and composition of tissues [1, 2]. Typically, fresh or fixed tissues are quickly frozen within an embedding material. Samples are subsequently cut into sections under freezing temperatures using a cryostat, a large refrigerated chamber with an internal microtome to slice frozen samples. The resulting specimens are then collected and

Renping Zhou and Lin Mei (eds.), *Neural Development: Methods and Protocols*, Methods in Molecular Biology, vol. 1018, DOI 10.1007/978-1-62703-444-9_28,

placed onto slides for further processing [3]. The tissues can be analyzed using several techniques including immunohistochemistry, immunofluorescence, in situ hybridization, or various other counterstaining methods.

Frozen sectioning has the distinct advantage over traditional histological techniques in the efficiency which tissues can be sectioned and analyzed, as samples can be processed rapidly while still producing high-quality specimens for study. However, while the process is simple in concept, the ability to obtain cryosections of superior quality is often a difficult and laborious process. Several factors, including tissue fixation, the distance of the anti-roll plate, air humidity, and the temperature outside of the chamber, amongst other considerations, can dramatically affect the sectioning process. Learning these various caveats requires extensive experience, persistence, and patience.

Here, we have outlined a general method for embedding and sectioning tissues for the cryostat. We have included instructions for obtaining sections at varying depths; we recommend that thinner sections (less than 20 μm in thickness) are retrieved through direct slide mounting, while thicker samples (more than 20 μm in thickness) are collected using the free-floating section method. In addition, several notes have been added to aid in alleviating some of the challenges one may face during the cryosectioning process.

2 Materials

2.1 Tissue Fixation and Embedding Components

1. Fixation buffer: 4 % paraformaldehyde solution in 1× phosphate buffered saline (PBS), pH 7.4, store at 4 °C (*see* **Note 1**).
2. 30 % sucrose solution in 1× PBS, store at 4 °C (*see* **Note 2**).
3. Dry ice, finely crushed.
4. Disposable embedding molds (Fisher Scientific).
5. Tissue-Tek Optimal Cutting Temperature (O.C.T.) Compound (VWR Labshop).

2.2 Tissue Sectioning Components

1. Cryostat and cryostat chucks.
2. Accu-Edge Low-Profile Disposable Microtome Blades (VWR Labshop).
3. Fine nylon brush (VWR).
4. Kimwipes (VWR).
5. Razor blades (VWR).
6. Superfrost Plus Slides (VWR).
7. 70 % ethanol (VWR).
8. Freeze spray (VWR).

2.3 Free-Floating Tissue Mounting Components

1. Multi-well plates (VWR).
2. 25 % glycerin/30 % ethylene glycol solution in 1× PBS, pH 7.4, store at room temperature.
3. Long glass pasteur pipette (VWR).
4. Petri dish (VWR).
5. Slides (VWR).
6. 1× PBS.
7. Fine-pointed paintbrush (VWR).
8. Coverslips (Fisher Scientific).
9. Clear-Mount mounting medium with Tris buffer (Electron Microscopy Sciences).

3 Methods

3.1 Tissue Embedding Preparation

1. Fix the tissue with 4 % paraformaldehyde solution in 1× PBS, pH 7.4 (*see* **Note 3**). After fixation, thoroughly rinse the tissue in 1× PBS (three times, 10 min per wash) to remove any residual paraformaldehyde solution. Place the tissue in 30 % sucrose solution in 1× PBS overnight at 4 °C. Keep the specimen within the sucrose solution until it has sunk to the bottom of the container (*see* **Note 4**).
2. Prepare an embedding mold for casting the tissue by adding O.C.T. Compound to the container at room temperature (*see* **Note 5**).
3. Prepare an adequate amount of dry ice for freezing the sample (*see* **Note 6**). Pulverize the dry ice into a powdery consistency (*see* **Note 7**).
4. Remove the tissue from the sucrose solution and briefly rinse excess sucrose with ddH_2O (*see* **Note 4**).
5. Submerge the sample completely within the O.C.T. Compound in the embedding mold, keeping in mind the desired orientation in which the tissue is to be sectioned (*see* **Note 8**) (Fig. 1).
6. Place the entire mold with the tissue and O.C.T. Compound into the dry ice and allow the block to freeze, taking care that the dry ice is only touching the plastic mold and not the contents within. Carefully observe the block during the freezing process. The embedding mold should be relatively level on the dry ice to prevent the tissue from shifting or changing orientation, as well as to allow the block to freeze evenly. The O.C.T. Compound will be opaque when completely frozen (*see* **Note 9**).
7. When uniformly frozen, samples are ready for cryosectioning. Alternatively, sections may be stored in −80 °C for extended periods.

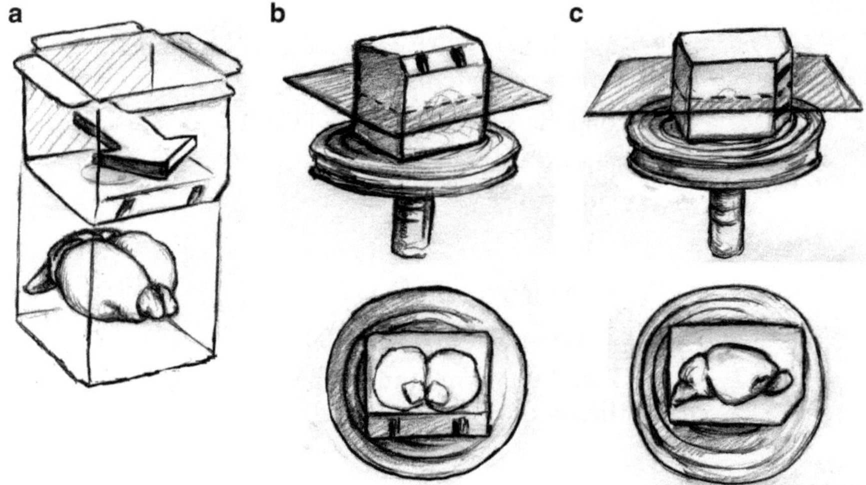

Fig. 1 Orientation of mouse brains in mold. (**a**) Orientation during embedding of specimen. Use the bevel and slots (b&s) to adjust the specimen in square molds and to maintain orientation after removing the embedded sample from mold. In this example, the embedded brain is oriented ventral-side-down with the rostral aspect towards the bevel and slots of the mold. (**b**) Orientation for coronal sectioning. Molds are applied onto the cryostat chuck using a small amount of O.C.T. Compound. For coronal sections, remove sample from mold and mount onto the chuck with the bevel and slots facing up. When chuck is placed into specimen holder of cryostat, the blade will cut coronal sections anterior to posterior. (**c**) Orientation for sagittal sectioning. For sagittal sections, mount sample onto the chuck with the bevel and slots touching chuck but still visible

3.2 Cryostat and Section Preparation

1. Set the cryostat chamber temperature. The default sectioning temperature is set around −20 °C (*see* **Note 10**).
2. Place the sample from the dry ice or −80 °C freezer into the cryostat chamber and allow the tissue block to acclimate for at least 15 min (*see* **Note 11**).
3. Organize materials for sectioning and place them within the cryostat chamber. Materials include cryostat chucks, slides, brushes, kimwipes, and razor blades (*see* **Note 12**).
4. Install a single microtome blade to be used for cryosectioning (*see* Fig. 2d). Thoroughly clean the blade with 70 % ethanol to remove any grease and dust that may impede sectioning. Dry the blade thoroughly and carefully place the blade within the knife carrier (*see* **Note 13**).
5. Cut or peel away the molding from the sample using a cold razor blade (*see* **Note 14**).
6. Place a small amount of room temperature O.C.T. onto the cold cryostat chuck within the cryostat chamber and quickly apply the mold onto the chuck. Allow the warm O.C.T. to freeze so that the sample is secured onto the cryostat chuck (Fig. 1).

3.3 Cryostat Sectioning

1. Place the cryostat chuck onto the microtome and secure it tightly within the holder. Angle the specimen so that the sample is perpendicular to the cutting blade (Fig. 2).

Fig. 2 Major internal components of the cryostat chamber. (**a**–**c**) The microtome mechanism used to hold and move the sample incrementally after each section. This includes the microtome specimen holder to hold the chuck (**a**), the adjustment knob to lock the chuck firmly in the specimen holder and alter the angle of the sample block (**b**), and the adjustment lock to hold the block in place during sectioning (**c**). (**d**–**g**) The stage that holds the cutting blade and holds the resulting sections. This mechanism includes a blade holder to firmly set the microtome blade (**d**), a stainless steel stage where tissue sections slides onto and rests (**e**), an anti-roll plate which the resulting section slides under and flattens (**f**), and a blade protector which swings forward to cover the blade and protect the user when handling the cryostat (**g**)

2. Set the desired cutting thickness. Initial sections may be cut thicker to trim excess O.C.T. Compound or to cut further into the sample before desired sections are saved.
3. Move the coarse feed so that the specimen is close to (but not touching) the microtome blade. Proceed to turn the handwheel continuously in the forward direction, as each turn will move the tissue block forward at the selected thickness (*see* **Notes 15** and **16**). The first several turns will yield no sections as the block moves incrementally to the blade.
4. Once the blade begins to cut into the sample, place the anti-roll plate down and continue to section (*see* **Note 14**). Adjust the anti-roll plate to the edge of the blade to ensure that

the section is properly sliding underneath and remains flat (*see* **Note 17**). Clean unwanted sections with a brush (*see* **Notes 18** and **19**).

5. When ready to save sections, turn the hand crank one revolution in an even motion to cut the section and then lock the crank into the secured position to prevent the block from moving while retrieving the section. Carefully lift the roll guard so as to keep the sections flattened.

3.4 Collecting Thin Sections on a Slide (See Note 20)

1. After cutting a section, place a cold slide face down to retrieve the section from the stainless steel surface (*see* **Notes 18** and **19**). The section should be flush and flat against the glass of the slide (*see* **Note 21**).
2. After the section has been retrieved onto the slide, turn the slide so the section is facing upward. Proceed to melt the section by placing a finger under the area of the slide where the section has been placed (*see* **Note 22**). Place the slide back into the cryostat chamber and continue with the next section.
3. After sectioning, slides may be kept in −80 °C indefinitely for extended storage (*see* **Note 23**). Prior to staining, sections should be air-dried at room temperature overnight or with a hair drier set on the lowest setting with indirect heat to thoroughly remove any excess moisture built on the slide and secure the sections onto the slide (*see* **Note 22**).
4. Slides are ready for processing. When finished, apply mounting medium and coverslip the slide.

3.5 Collecting and Mounting Thick Free-Floating Sections (See Note 24)

1. Prepare multi-welled plates filled with the 25 % glycerin/30 % ethylene glycol solution (*see* **Notes 25** and **26**).
2. After cutting a section, lift the anti-roll plate and collect tissue sections with a glass hook (*see* **Note 26**). Place the section in the well while removing O.C.T. from the tissue (*see* **Note 27**) (Fig. 3). Sections may be covered and stored at −20 °C (*see* **Note 28**).
3. Remove cryoprotectant and O.C.T. Compound from sections (*see* **Note 29**) in the wells. Proceed with desired staining (*see* **Note 30**).
4. When samples are ready for mounting, fill a 100 mm Petri dish with PBS and place a clean slide on the bottom of the dish facing upward. Place sections in the Petri dish (*see* **Note 31**).
5. Arrange the sections onto slide using a fine brush.
6. Draw off PBS slowly using a 5 or 10 mL pipette (*see* **Note 32**). Use increasingly smaller pipettes (P1000, then P100) to draw off excess PBS. Rearrange sections as needed.

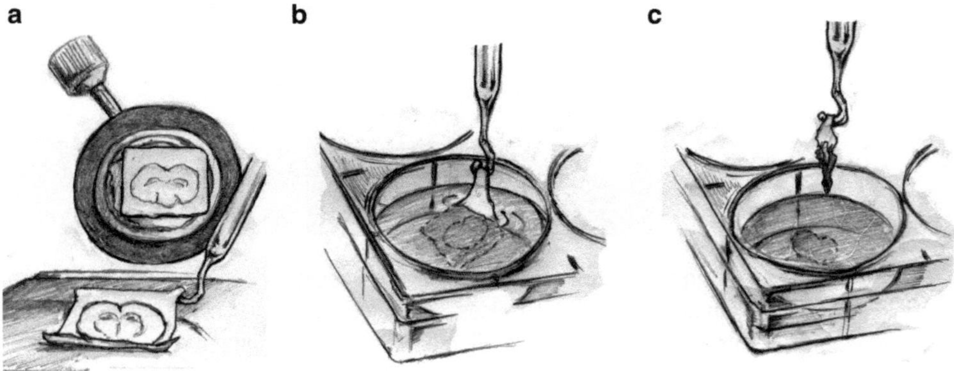

Fig. 3 Collection of free-floating sections. (**a**) Collection of sections from the cryostat. Section is collected using a glass hook (the hook is kept at room temperature) by touching only the edge of the O.C.T. and not the tissue section. (**b**) Submerging of sample into solution. Sample is carefully submerged into the glycerin solution while keeping one edge of the O.C.T. suspended in the air. (**c**) Removal of O.C.T. from the sample. Lift the glass hook immediately as the tissue section is fully submerged in the solution to remove the O.C.T. from the sample

7. Once sufficient PBS has been removed, the sections will remain attached on the slide when taken from the Petri dish. Rearrange sections as needed (*see* **Note 33**). Dry excess PBS from around sections with a dry paintbrush or kimwipe (*see* **Note 34**).
8. Apply mounting medium and coverslip.

4 Notes

1. 4 % paraformaldehyde solution is stable at 4 °C for up to 1 week. Some groups freeze their paraformaldehyde in −80 °C and store the solutions indefinitely.
2. Sucrose is a popular cryoprotectant for fixed frozen sections as it prevents the formation of ice crystals that can alter cellular integrity and result in a freeze–thaw artifact. The use of some level of cryoprotection for fixed tissue is absolutely necessary to maintain histological preservation. While 30 % sucrose is often used, lower, higher, or graded concentrations may be used depending on the tissue consistency, integrity, and fixation conditions. Tissues may be stored in sucrose for a short period of time.
3. Other fixatives, such as 10 % formalin solution (4 % formaldehyde) in 1× PBS, pH 7.4, may also be used.
4. If freezing unfixed tissue (fresh frozen tissue), skip this step.
5. The O.C.T. Compound should appear clear and viscous at room temperature.

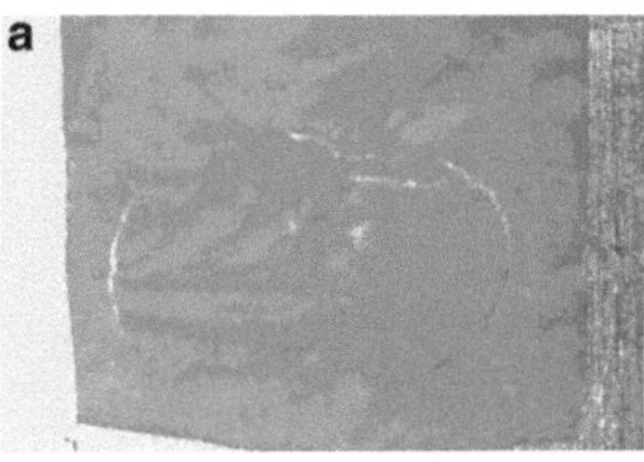
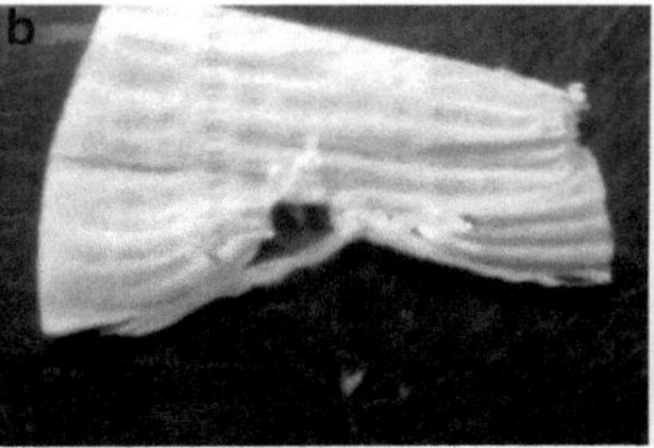

Fig. 4 Examples of cryosectioning cutting artifacts. (**a**) Typical brain section. Sections should remain flat without signs of tearing or crushing upon mounting on slides. (**b**) Example of tissue "scrunching," in which tissue that is too soft is crushed during sectioning. This is often a result of the temperature of the block being too warm and can be alleviated by decreasing the chamber temperature. (**c**) Example of tissue "shattering," in which the tissue is extremely brittle and becomes shredded during sectioning. This is due to the block being too cold and can be alleviated by increasing the internal chamber temperature

6. Freezing unfixed tissue (fresh frozen tissue) with liquid nitrogen is more optimal, though dry ice may still be used. However, be advised that the block is extremely hard, making it prone to splitting when freezing with liquid nitrogen.
7. Embedding molds freeze more thoroughly and quickly if the dry ice is in a powder consistency. Dry ice may be pulverized using a standard hammer or kitchen blender.
8. Note the orientation of the tissue in the embedding mold prior to freezing as the O.C.T. Compound becomes an opaque white color when frozen (Fig. 1). Under most circumstances, it is easiest to make the bottom face of the embedding mold as the front face in which the sections will be cut. For example, if cutting a brain as a sagittal section, it is preferable to lay the brain on its side against the bottom of the embedding mold.
9. Do not allow the mold to thaw after freezing. Thawing of frozen samples will disrupt cell membrane integrity and architecture, ruining the samples.
10. Optimal cutting temperature is dependent on several factors, including tissue composition and moisture of the outside environment, and often requires some experimenting for each user (Fig. 4). Tissue "scrunching" and crushing indicates that the temperature is too warm and that the chamber should be set at a cooler temperature (*see* Fig. 4b). Tissue "shattering" and flaking is an indication that the cryostat chamber is too cold and should be set at a warmer temperature (*see* Fig. 4c). Lipid- and fat-rich tissues such as the brain require temperatures colder than −20 °C, usually around −22 °C to −25 °C, while tissues with less fat such as the tongue require temperatures warmer than −20 °C, usually around −18 °C or −19 °C. Unfixed tissues, in comparison to fixed tissues, cut optimally in warmer temperatures. An environment in which the humidity

is higher often requires the chamber temperature to be set colder than usual, and one should take care to limit the amount of moisture within the chamber.

11. Embedded molds that are harder due to freezing in liquid nitrogen or storage under colder conditions may require longer time to acclimate, sometimes up to an hour. This step helps in preventing tissue from cracking or splitting while sectioning.
12. Be sure all materials are dried to prevent freezing in the cryostat chamber and that materials are adjusted to the cryostat temperature.
13. Microtome blades should be changed in regular intervals, often at the start of each new session. Difficulties in sectioning may be a result of the dulling of the blade, in which case the blade may be shifted to an unused region. Make sure the blade mechanism is secured tightly into the slot as this may also impair sectioning.
14. Excess O.C.T. Compound surrounding the tissue may be trimmed to aid in section and to straighten the sample. Typically, less O.C.T. Compound around the tissue is easier to section. However, allow for at least some embedding material to be around the sample to maintain even sectioning.
15. Turn the hand crank in one continuous and steady motion without stopping mid-revolution to obtain the most even sections. Jerking of the crank will result in uneven cuts, or tissue "chattering." Altering crank speeds (slower for more careful sectioning or faster to pass through more difficult areas of the tissue) may aid in sectioning and is to the discretion of the end user.
16. Ice and debris buildup can occur on the inside of the cryostat, particularly under humid conditions. Most machines have an automatic thaw setting to minimize ice formation. If any sort of resistance is felt while turning the hand crank, immediately stop all activity to prevent damage to the cryostat and proceed to clean the inside of the machine. The cryostat should be turned off and allowed to thaw, removing all traces of water and debris within the machine and allowing the mechanism to dry. Typically, this maintenance should be done about once every 6–8 weeks.
17. The anti-roll plate should be just within the lip of the blade, as the sample will be cut with the blade and slide underneath the anti-roll plate. The most typical reason for poor sectioning is a result of improper positioning of the plate. Cutting with the anti-roll plate too shallow will slice the tissue without flattening the section underneath the plate, crushing the specimen. Sectioning with the anti-roll plate too far out will cause the sample to nick the anti-roll plate and crush the tissue.

Protect the anti-roll plate from scratches and damage by keeping it raised while repositioning the tissue block for the next section.

18. The surface underneath the anti-roll plate should be kept clean at all times from debris or moisture and should be done regularly with a brush. The surface may also be cleaned with a cold kimwipe, though this should be avoided if possible. Be very careful of the microtome blade while cleaning the surface; always clean in the direction away from the blade (away from the user, towards the microtome) to prevent personal injury. A blade guard should be used during this process (*see* Fig. 2g).
19. The surface underneath the anti-roll plate must be kept cold to prevent sections from sticking onto the stainless steel surface. If sections are not sliding under appropriately and the anti-roll plate is properly adjusted, it may mean that the surface temperature is not cold enough. Close the cryostat door to cool the internal chamber or apply some freeze spray against the surface to cool the plate directly.
20. Collections of frozen samples directly on slides are optimal for samples sectioned at 8–20 μm.
21. When retrieving sections, carefully attempt to "pick up" the section against the slide without fully touching the stainless steel surface. Do not flatten the slide flush against the stainless steel surface as the sample will adhere to the stainless steel stage surface rather than attach to the slide. The slide may be warmed briefly with one's finger prior to retrieving the section to help attach the section to the slide. However, keep the slides cold to preserve tissue morphology.
22. Sections must be thoroughly melted onto the slide as insufficient melting will not secure the tissue completely onto the slide. Watch the surrounding O.C.T. Compound turn clear and wait an extra few seconds before returning the slide to the cryostat. If sections begin to detach during tissue processing, it may indicate that the tissue has been inadequately melted onto the slide or that the sections have been poorly dried.
23. In some instances, the tissue may be postfixed with either methanol/acetone or 4 % paraformaldehyde directly onto the slide at this point. This is typical practice for sectioning of unfixed tissue.
24. Collection of free-floating sections is optimal for samples sectioned at 20 μm or thicker.
25. Multi-well plates allow for the collection of multiple sets of serial sections.
26. Handle only the O.C.T Compound and not the section (Fig. 3). Touching the tissue directly may disturb the anatomical

and cellular morphology. Keeping the glass hook at room temperature will allow for better adhesion to the O.C.T. Compound.

27. The glycerin/glycol solution will instantly separate the section from the O.C.T. Compound. Pull the O.C.T away from the section and discard as soon as the section is submerged and while the hook is still attached to the corner of the O.C.T.
28. Multi-well plates are often provided with a plastic cover and provide protection from evaporation. For additional protection, seal the outside cover with paraffin.
29. Cryoprotectant can be removed through multiple washes in PBS with strainers.
30. Sections can be stained in any container with a relatively flat bottom (including clean wells, 2 mL tubes, or scintillation vials).
31. Sections may be transferred from one container to another using a hook fashioned from a small capillary tube or a paintbrush. The hook is useful for larger sections (5–10 mm) while a paintbrush is more useful when transferring very small sections (~2 mm).
32. Drawing off PBS too quickly will distort the arrangement of sections. If this happens, rearrange sections and draw off PBS slower or with an instrument with a smaller bore size.
33. Never arrange slides on a dry surface as this may damage tissue. Arrange sections only while the slide is damp.
34. Take care when coverslipping. If the slide is inadequately dried, sections may move upon coverslipping. However, sections should never dry completely and should remain moist, as insufficient moisture will cause the sections to crack.

References

1. Gal AA, Cagle PT (2005) The 100-year anniversary of the description of the frozen section procedure. JAMA 294(24):3135–3137. doi:294/24/3135 [pii] 10.1001/jama.294.24.3135
2. Dahlin DC (1980) Seventy-five years' experience with frozen sections at the Mayo Clinic. Mayo Clinic Proc 55(11):721–723
3. Wilson LB (1905) A method for the rapid preparation of fresh tissues for the microscope. JAMA 45:1

Chapter 29

The Golgi–Cox Method

Gitanjali Das, Kenneth Reuhl, and Renping Zhou

Abstract

One of the best neurohistologic methods to reveal the cytoarchitecture of the brain and detailed morphology of neurons with unsurpassed clarity has been the Golgi staining. It is based on the principle of metallic impregnation of neurons, allowing visualization in their entirety including cell soma, axons, dendrites, and spines. In this chapter, we describe the Golgi–Cox protocol standardized in our laboratory that can be used to study experimental effects of different genetic manipulations on spatial distribution of neurons, dendrite density, spine number and morphology to elucidate gene functions during development and in adult brain.

Key words Axon, Camillo Golgi, Cox, Dendrite, Golgi staining, Light microscope, Neuron, Spine

1 Introduction

The foundation of modern neuroscience was laid with the discovery of the "Black Reaction" by Camillo Golgi in 1873. This reaction was based on the principle of metallic impregnation of neurons, allowing complete visualization of the neuronal architecture. Though recent cytological advances in neuronal imaging have pushed this method to the backbench, nevertheless, none of them came close to revealing a complete overview of the brain morphology with as much clarity as the "Golgi Method." Thus, even after more than 139 years since its discovery, it is still used to study neuronal morphology and brain architecture after induced gene changes, neuropathology and regenerative interventions to get an idea about the effects in neuronal and brain morphology. The beauty of the method lies in one of its shortcomings: the Golgi protocol stains only a few hundred neurons out of the millions present. Thus, it is possible to visualize neurons and trace their path and connections in the brain against a pale yellow background with unsurpassed clarity. Without this ability to stain the few select neurons, everything would have looked like a large black blob on the brain section.

Renping Zhou and Lin Mei (eds.), *Neural Development: Methods and Protocols*, Methods in Molecular Biology, vol. 1018,
DOI 10.1007/978-1-62703-444-9_29,

Since its discovery, various modifications have been incorporated into the original protocol to obtain reproducibility, uniform coloration, and reducing the time taken for staining. The Golgi method can basically be categorized into two major steps: the first is the immersion of the tissue block into a solution of potassium chromate and potassium dichromate followed by the immersion in a silver nitrate solution for the formation of silver chromate crystals [1]. A modification of this step has been the incorporation of the metal mercuric chloride into the potassium dichromate solution [2] before treatment with a dilute sodium hydroxide or carbonate solution [3] to intensify the reaction product. Other variations introduced into the technique were the use of sodium sulfite [4, 5] and dilute ammonia [6] instead of the sodium hydroxide/carbonate step. Cox [7] introduced a significant modification to the Golgi's method by impregnating the sample with a solution of mercuric chloride and potassium dichromate mixed with potassium chromate that helped in improving the impregnation by reducing the acidity of the solution; this later came to be known as the "Golgi–Cox" reaction.

The chemistry underlying the Golgi staining is the formation of the black deposit permeating the cell cytosol as revealed by electron microscope and X-ray studies [8–10]. In general chromium salts bind to proteins in the cells. In the Golgi reaction, it is the formation of the silver chromate deposit product while in the Golgi–Cox method, firstly is the formation of the whitish mercuric chloride that is further transformed to the black mercuric sulfide deposit upon alkali treatment [11].

Golgi staining is unique in its ability to stain all components of the nervous systems viz., neurons, glia and blood vessels. Though it stained only a few neurons in a particular brain section nevertheless, it gave a detailed picture of the nerve cell body, its axon, dendrites, and spines that allowed Ramon Y Cajal to document the architecture of the brain and share the Nobel Prize in Physiology (1906) with Camillo Golgi.

The Golgi–Cox protocol standardized in our laboratory is detailed below along with important notes about things to keep in mind during the procedure.

2 Materials

2.1 Reagents

1. Gelatinization of Slides: (*see* **Note 1**) Gelatin: 5 g, Chromium Potassium Sulfate: 0.5 g, Ultrapure Water: 1,000 mL, Thymol: 0.75 g. Heat 500 mL of ultrapure water and dissolve 5 g gelatin in the water. Use a magnetic stirrer if needed. Once the gelatin is dissolved add 0.5 g Chromium Potassium Sulfate (CPS) and Thymol (preservative) and dissolve them in the

solution before making up the solution to 1 L. (This step must be performed under a fume hood, as CPS is a very dangerous toxic substance.) Next filter the solution through a piece of Whatman filter paper and let the solution cool down to room temperature. Now dip clean glass slides into the solution, taking care to avoid forming bubbles, as otherwise after coating there might be areas with non-uniform coating due to the presence of the bubbles. Next let the slides dry out in a clean environment overnight or at 37 °C in the oven. The dried slides can be re-dipped in the gelatin solution once more before using them for the Golgi protocol. Gelatin-coated slides should be always kept at 4 °C until use. The thymol present in the gelatin solution keeps it good for 1 month after preparation if kept at 4 °C.

2. Golgi–Cox solution: The Golgi–Cox solution is made from the following three stock solutions. Solution A: 5 % Potassium Dichromate in double distilled water. Dissolve the solution in a glass bottle using a magnetic stirrer under a fume hood. Solution B: 5 % Mercuric Chloride dissolved in double distilled water. The solution must be made in a glass bottle using a magnetic stirrer under a fume hood. Solution C: 5 % Potassium Chromate in double distilled water. Dissolve the solution using a magnetic stirrer under a fume hood. All the solutions should be covered with Aluminum foil both during and after preparation. The stock solutions can be stored in room temperature under light-tight conditions for a few months. Fresh working solution from the stock should be prepared every time brains are immersed in the impregnating solution. To make the working solution, mix the stock solutions in the following sequence: First, mix 5 volume parts of A with 5 volume parts of B. Then slowly add with continuous stirring: 4 volume parts of C diluted with 10 volume parts double distilled water. Let this final mixture sit in the dark for 2 days. The reddish precipitate in the Golgi–Cox solution that forms should be filtered out under a fume hood and the solution retained in a new glass bottle and should be stored under light-tight condition. This solution can be used till 1 week after preparation.
3. Sucrose Solution: Prepare a 30 % stock solution of sucrose in double distilled water. Sucrose solution can be kept at 4 °C for months without any bacterial growth.
4. Ammonia: Make 75 % ammonia solution with double distilled water
5. Sodium Thiosulfate: Make 1 % sodium thiosulfate solution in double distilled water.
6. Alcohol: Make fresh alcohol grades of 50, 75, 95 and 100 %.
7. Anesthesia: Mice were anesthetized with ketamine HCL/xylazine HCL solution (50 mg/kg, intraperitoneally).

2.2 Equipments

Glass Bottles, Glass Coplin Jars with cover (5–10 slide jars), Brush (size 10, with thick finer hairs for gentle tissue handling), Chemical Fume Hood, Gloves, Hotplate Stirrer, Forceps, Glass Coverslip, Funnel, Whatman Filter Paper (standard grade 1 with 11 μm pore size), Vibrotome, Paper Towels, Aluminum Foil, Razor Blades.

3 Methods

Precaution: Always wear gloves and perform the experiments under the chemical fume hood except when specifically not told.

The Golgi–Cox protocol is outlined in detail below.

3.1 Dissection

Deeply anesthetize the animal (as per NIH Animal Use and Care guidelines). Make sure the animal is anesthetized by pinching its paw with forceps. If the animal shows no response, carefully dissect out the brain as quickly as possible.

3.2 Golgi–Cox Impregnation

Quickly rinse the dissected brain in double distilled water to remove blood and place the brain (*see* **Notes 2** and **3**) into the Golgi–Cox solution. Leave the immersed brain at room temperature in the dark. After 6 h discard the Golgi–Cox solution according to the university hazard chemical disposal guidelines. Put fresh Golgi–Cox solution and leave the brain immersed in it for either [*]1 week (*see* **Note 4**; Fig. 1) or [$]2 weeks (*see* **Note 5**; Fig. 2) as per your experimental requirement and discard Golgi–Cox solution into the hazard chemical container. Wipe the brain very gently with a kimwipe and place the brain in a 30 % sucrose solution at room temperature. After 1 h place the brain into fresh 30 % sucrose solution and leave at room temperature for 8 h (*) or 3 days at 4 °C ($).

3.3 Sectioning the Brain

Mount the brain onto a vibrotome stage with glue and fill the vibrotome reservoir with 30 % sucrose solution. Set the speed of the vibrotome blade at 6 and the vibration/amplitude at 6. Cut brain sections at 250 μm and very gently pick the floating brain sections with the help of a brush and mount on the gelatin coated slides. Dry the section with the help of kimwipe and let the sections lie flat in a dark place for 3 days so that the sections are firmly glued to the gelatin slides.

3.4 Color Development

Wash the mounted sections two times 2 min each with double distilled water to remove traces of the impregnating solution. Place the mounted sections in a coplin jar filled with the 75 % ammonia solution for 10 min in the dark (*see* **Note 6**) at room temperature. Wash the sections with double distilled water six times 5 min each (*see* **Note 7**). Immerse the sections in 1 % sodium thiosulfate to fix the stain for 10 min at room temperature in the dark (*see* **Note 8**). Wash the sections six times 5 min each with double distilled water.

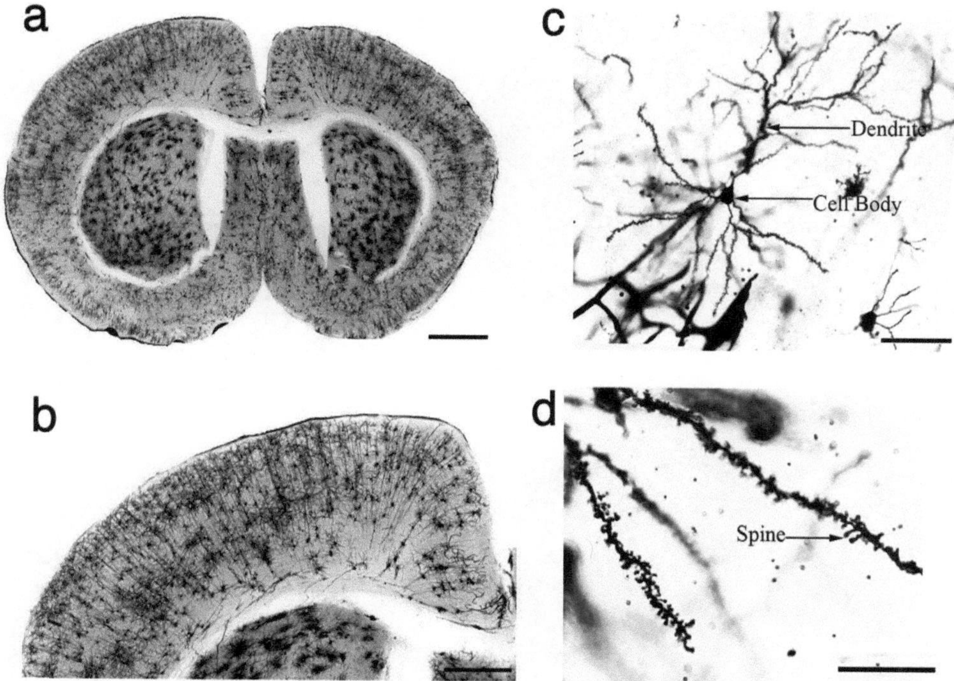

Fig. 1 (**a**) Representative image of a 1 week Golgi–Cox impregnated brain (scale bar = 500 μm). (**b**) Cortical neurons of a 1 week Golgi–Cox impregnated brain (scale bar = 100 μm). (**c**) Layer 1 cortical neuron showing dendrite branches of a neuron. *Arrows* indicate the neuron cell body and dendrite branching. (*Note*: all the dendrite branches of a specific neuron can be observed distinctly; scale bar = 50 μm.) (**d**) Spine of cortical neurons impregnated with Golgi–Cox stain. (*Note*: each spine head and spine neck is clearly visible and distinguished from surrounding spines; scale bar = 10 μm)

3.5 Dehydration

Process the slides through dehydration steps (*see* **Note 9**) of 50, 75, 95 % EtOH for 4 min each, and then process through two changes of 100 % alcohol 4 min each.

3.6 Coverslipping

Clean the sections three times 4 min each with xylene and leave the sections in fresh xylene for 2 h in the dark. Take out each slide carefully with a forceps and mount the slides with DPX/Permount (*see* **Note 10**) and let dry under the fume hood for 3 days before examining under the microscope (*see* **Note 11**).

4 Notes

1. Gelatinization of slides is a very important starting point in the protocol. This makes sure the brain sections stick to the slides during the entire staining process. If not done properly the sections will fall off during the staining. The stickiness provided by gelatin is further enhanced by addition of chromium potassium sulfate which gives a positive charge to the slides.

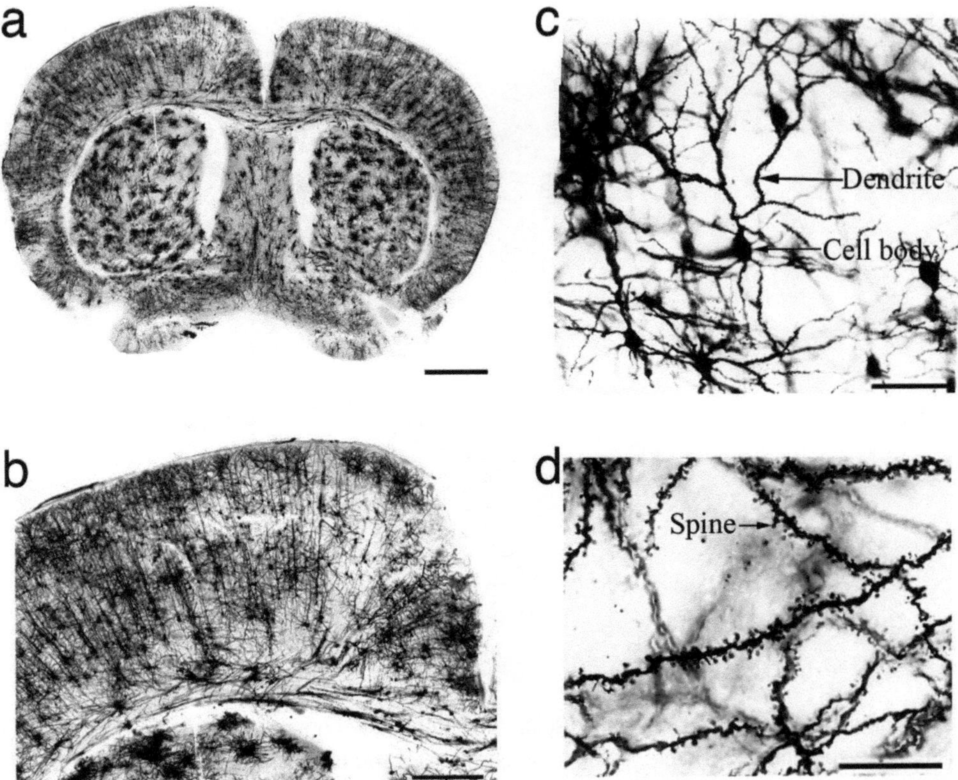

Fig. 2 (**a**) Representative image of a 2 week Golgi–Cox impregnated brain (scale bar = 500 μm). (**b**) Cortical neurons of a 2 week Golgi–Cox impregnated brain (scale bar = 100 μm). (**c**) Layer 1 cortical neuron showing dendrite branches of a neuron. *Arrows* indicate the neuron cell body and dendrite branching. (*Note*: due to more number of neurons getting stained by the Golgi–Cox solution, dendrite branches from neighboring neurons overlap each other and it is difficult to quantify dendrites of a specific neuron; scale bar = 50 μm.) (**d**) Spine of cortical neurons impregnated with Golgi–Cox stain. (*Note*: though spines are clearly visible there is overlap between neighboring spines; scale bar = 10 μm)

The latter tends to form a tight bond with the negative charge of the tissue section. Further, extra care needs to be taken for cutting such thick sections for Golgi staining as the brain sections are prone to lifting off the slide if they are not particularly well charged.

2. It is important to put fresh brains into the Golgi–Cox solution. Paraformaldehyde (PFA) or formalin perfused brains tend to cause problem for proper impregnation of the Golgi–Cox solution. In our experience we found that perfusion of the animal with 4 % PFA before putting the brains into the Golgi–Cox solution tends to give poor neuronal visualization in their entirety. More glial and vascular cells were also stained upon PFA perfusion. Further, the neurons that were visible did not show finer details of dendrite and spine number and morphology (Fig. 3).

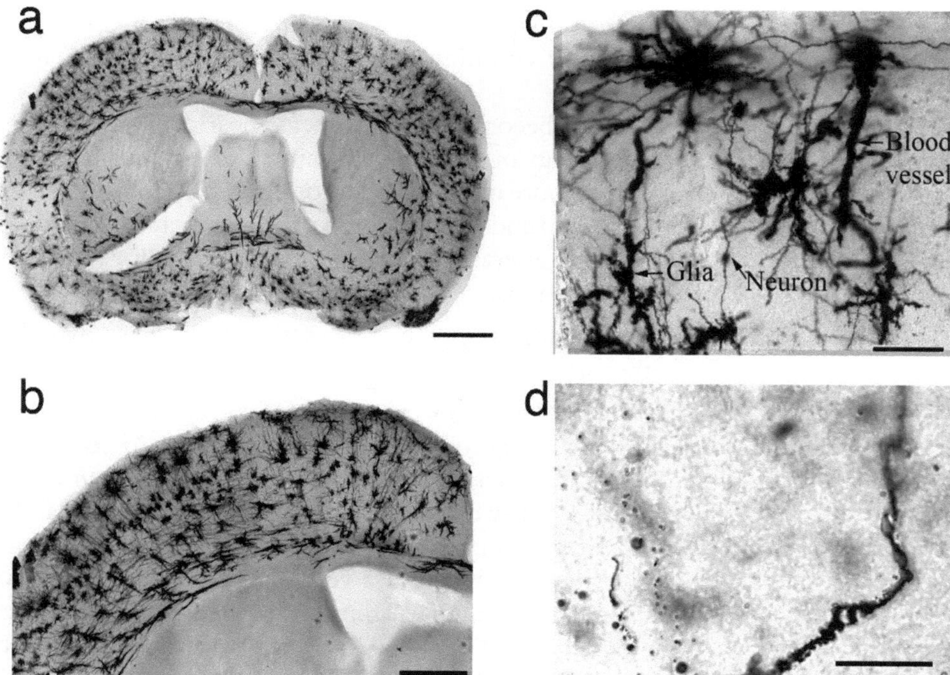

Fig. 3 (**a**) Representative image of a 4 % PFA perfused brain preceding impregnation of 2 weeks with the Golgi–Cox solution (scale bar = 500 μm). (**b**) Cortical neurons of PFA-perfused and 2 weeks Golgi–Cox impregnated brain. (*Note*: PFA perfusion impeded proper impregnation of the central areas of the brain with the Golgi–Cox solution; scale bar = 100 μm.) (**c**) Layer 1 cortical neuron showing dendrite branches of a neuron. *Arrows* indicate the neuron cell body, glia and blood vessels. (*Note*: due to PFA perfusion, fixation of tissues occurred rending it difficult to get properly stained neurons in their entirety; scale bar = 50 μm.) (**d**) Higher magnification image showing sparsely and poorly labeled neuronal process (scale bar = 10 μm)

3. It is very important that the brain samples immersed in the Golgi–Cox solution should not be more than 5 mm in thickness for proper impregnation of the deeper layers. So if using an adult brain, it is better to give a coronal cut in the middle before placing the brain blocks in the Golgi–Cox impregnation solution. Otherwise, though the Golgi–Cox solution may stain the cortical layers well it may have difficulty in staining the deeper subcortical regions such as the thalamus, striatum or septal areas uniformly (Fig. 3). This is explained by an autolytic process that starts after the animal is killed and the time that the impregnating solution takes to reach the deeper brain regions when using a larger brain cannot outmatch the autolytic process and so large brain pieces often display poor centrally impregnated regions
4. This step helps in counting number of dendritic branches per neuron. In our experience we found that incubating brain samples for 1 week in the Golgi–Cox solution impregnated neuronal populations sparsely but in their entirety. Thus, it is

easier to quantify the dendrites per neuron without making a mistake of counting overlapping branches from neighboring neurons.

5. This step becomes useful to get a clear overview of the actual morphology of the brain's architecture in a qualitative manner. Because more number of neurons gets stained with this impregnation method, it is easier to get a general overview of any structural alterations in the brain due to diseases or altered gene functions.

   (So depending on the goal of the study it is advised to use either step * or $ of the protocol).
6. Alkalinization of the Golgi–Cox impregnated sections make the stained neurons appear intensely dark in color against a light yellow background making the entire neuron visible with excellent clarity of finer details. These structures are also visible without alkalinization though with less clarity, appearing as ghosts peering through a yellow film. During the process of alkalinization, aqueous ammonia reacts with mercuric chloride present in the Golgi–Cox solution to produce metallic mercury (black) and mercury (II) amidochloride (white) a disproportionation reaction [12].
7. It is important to wash the brain sections thoroughly so that the extra black deposits formed by the ammonia reaction with the Golgi–Cox solution gets washed away. Frequent changes of water wash away the ammonia thereby stopping the reaction from proceeding further and keeping the background noise to the minimal.
8. Sodium thiosulfate acts as a fixer for the metallic black mercury deposits but care should be taken not to incubate in this solution for long periods as it also tends to bleach out the stain.
9. Dehydration is a very important step for slow removal of water (both free and bound form) from the tissue sections with organic solvent. This process avoids the formation of air bubbles after coverslipping. Further, care should be taken not to prolong the dehydration time as this will make the tissue hard and brittle affecting image integrity.
10. It is very important to coverslip the tissue after immediately after taking out of xylene. The tissue section should never be left to air dry after taking out of xylene as this makes the tissue hard and brittle affecting good image resolution.
11. A similar protocol is offered as a kit by FD Neurotechnologies, MD, USA.

Acknowledgment

This work has been supported by NIH grant RO1EY019012 and PO1HD023315 to R.Z.

References

1. Golgi C (1873) Sulla struttura della sostanza grigia del cervello. Gazz Med Ita Lombarda 33:244–246
2. Golgi C (1879) Di una nuova reasione apparentemente nera delle cellule nervose cerebrali ottenuta col bichloruro di mercurio. Arch Sci Med 3:1–7
3. Golgi C (1891) Modificazione del metodo di colorazione deli elementi nervosi col bichloruru Di mercurio. Riv Med Napoles 7: 193–194
4. Tal I (1886) Modificazione al metodo di Golgi nella preparazione delle cellule gangliari nel sistema nervoso centrale. Z Wiss Mikr 4:497
5. Cajal SR (1933) Elementos de tecnica micrografica del sistema nerviosa. Tipografia Artistica, Madrid, pp 103–128
6. Pal J (1887) Ein beitrag zur nervenfarbetechnik. Z Wiss Mikr 4:92–96
7. Cox WH (1891) Impregnation des centralen nervensystems mit quecksilbersalzen. Arch Mikr Anat 37:16–21
8. Fregerslev S, Blackstad TW, Fredens K, Holm MJ (1971) Golgi potassium-dichromate silver-nitrate impregnation. Nature of the precipitate studied by x-ray powder diffraction methods. Histochemie 25(1):63–71
9. Fegerslev S, Blackstad TW, Fredens K, Holm MJ, Ramón-Moliner E (1971) Golgi impregnation with mercuric chloride: studies on the precipitate by x-ray powder diffraction and selected area electron diffraction. Histochemie 26(4):289–304
10. Blackstad TW, Fregerslev S, Laurberg S, Rokkedal K (1973) Golgi impregnation with potassium dichromate and mercurous or mercuric nitrate: identification of the precipitate by x-ray and electron diffraction methods. Histochemie 36(3):247–268
11. Ramón-Moliner E (1970) The Golgi-Cox technique. In: Nauta WJH, Ebbesson SOE (eds) Contemporary research methods in neuroanatomy. Springer, New York, pp 32–55
12. Svehla G (1996) Reactions of the cations. In: Vogel's qualitative inorganic analysis. Longman Group Limited, England, pp 59–162

Chapter 30

Neuroanatomical Tract-Tracing Using Biotinylated Dextran Amine

Nikolai E. Lazarov

Abstract

Biotinylated dextran amine (BDA) is a highly efficient and powerful marker for bidirectional tracing of nerve pathways in a wide variety of species at the light and electron microscopic level. The BDA tract-tracing method can readily be combined with other anterograde or retrograde tracers for multiple neuroanatomical labeling studies to map the neuronal connectivity, or with immunocytochemistry for neurotransmitters and their receptors to reveal details of synaptic specializations within the multisynaptic neuronal circuits. Here, we describe an experimental protocol for anterograde and retrograde tracing using BDA. By applying BDA 10 kDa as an anterograde tracer, we demonstrate the existence of a direct bilateral nigro-trigeminal pathway in the rat.

Key words Anterograde tracing, Biotinylated dextran amine, Mesencephalic trigeminal nucleus, Multiple axonal labeling, Neuroanatomical connectivity, Retrograde tracing, Substantia nigra

1 Introduction

The neuronal tracer biotin dextran amine (BDA) is widely used to trace neuronal projections as it is well transported both retrogradely and anterogradely depending on its molecular weight. Moreover, extensive anterograde and retrograde BDA labeling occurs in all neuroanatomical pathways studied, and in animals of all ages regardless of the way of its application (*see* [1, 2] *for further details*).

BDA combines the advantages of amine-conjugated dextrans, namely, their sensitivity, and of biotinylated compounds, i.e., permanent labeling [3]. Dextran amines in their biotinylated form are commercially available in different molecular weights, ranging from 1,000 to 2,000,000 Da, but the most commonly used molecular weights in neuroanatomical studies are 3 and 10 kDa. Higher molecular weight biotinylated dextran amine (BDA 10 kDa) is an established, though not exclusive anterograde tracer [4], exquisitely labeling axons and their terminals. In contrast, lower molecular

Renping Zhou and Lin Mei (eds.), *Neural Development: Methods and Protocols*, Methods in Molecular Biology, vol. 1018,
DOI 10.1007/978-1-62703-444-9_30,

weight BDA (3 kDa) yields sensitive and detailed retrograde labeling of dendritic structures and neuronal cell bodies [3, 5] although it also does produce some anterograde labeling [6].

BDA possesses good water solubility, low toxicity, contains free amines and can therefore be fixed in place with aldehydes after labeling. Accordingly, BDA is dissolved in aqueous buffers, commonly introduced into the cells via an iontophoretic or pressure microinjection and after a sufficient period of time for the transport of the tracer, the tissue is fixed in formaldehyde and/or glutaraldehyde with no significant attenuation of labeling and finally visualized with the avidin biotin peroxidase complex (ABC) procedure. Following a standard or metal-enhanced diaminobenzidine (DAB) reaction, the electron-dense DAB reaction product can be easily detected by either light or electron microscopy. The resultant labeling can readily be combined with other morphological methods, such as other types of tract-tracing and/or immunocytochemical techniques for the demonstration of specific neuronal pathways [4, 7, 8].

BDA has been considered preferable for anterograde pathway studies [5]. Anterograde axonal tract-tracing is a neuroanatomical technique for tracing efferent projections from their source (the neuronal cell body or perikaryon) along the axonal tracts to their target termination point. We use the term "anterograde labeling" to describe the presence of relatively thick, straight or undulating labeled fibers, each of uniform thickness (i.e., no varicosities). In our tracing studies the term "terminal labeling" denotes the presence of thin, tortuous fibers with varicosities (boutons en passant and boutons termineaux) (Figs. 1 and 2). Here, we describe a detailed tract-tracing protocol for single labeling using the sensitive tracer substance BDA. In our laboratory this protocol is used to anterogradely trace axonal projections from different nuclei of origin (e.g., the substantia nigra, amygdala) to the mesencephalic trigeminal nucleus (*see* ref. 9). The procedure for retrograde labeling of neuronal cell bodies and their dendrites from their axon terminals with BDA 3 kDa and the subsequent processing is largely the same as for BDA 10 kDa used in our anterograde single-label studies (*see* **Note 1**). In principle, the present protocol applied for adult tissue can be employed for studies in developmental neuroscience too.

2 Materials

Animal housing, maintenance, and preparation should meet the national and institutional animal care guidelines for the use of experimental animals in research. No particular materials than the ones usually employed for neuroanatomical research are required. These include glass slides and coverslips, glassware for incubation, washing, and staining of the sections. A hot plate with a stirring bar, pH meter, and a vortex rotator are needed to prepare the

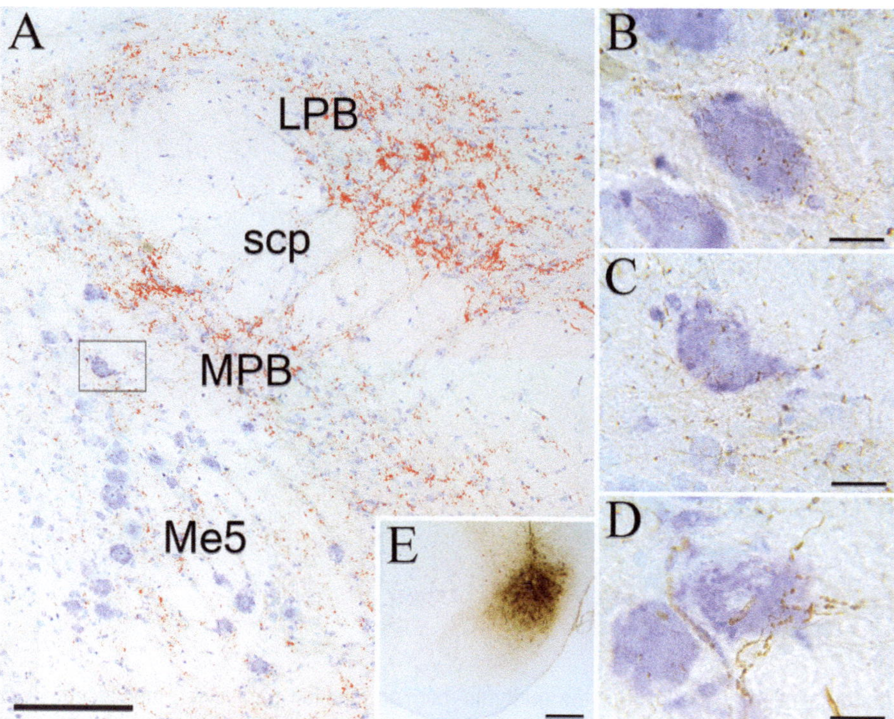

Fig. 1 Photomicrograph of a case where BDA was injected into the lateral substantia nigra (SN), compact part, and adjacent portions of substantia nigra, reticular part and substantia nigra, lateral part. (**A**) Low-power micrograph of the dorsolateral pontine tegmentum. Note the extensive terminal labeling into the parabrachial nuclear complex. The field of labeled fibers and terminals is denser in the lateral parabrachial nucleus (LPB) in comparison with the medial parabrachial nucleus (MPB). Light but distinct labeling is also visible in the caudal mesencephalic trigeminal nucleus (Me5). (**B**) Anterogradely labeled fibers and "en passant" boutons around and on the cell bodies of large pseudounipolar neurons in the ipsilateral (**B**, **C**) and contralateral (**D**) Me5. (**C**) Greater detail of the *outlined* region in (**A**). (**E**) The injection site in the SN. *scp* superior cerebellar peduncle. Scale bars = 100 μm in (**A**), 20 μm in (**B**–**D**), and 500 μm in (**E**)

reagents, and a rocking table shaker is also used in the staining procedure. The special equipment includes a stereotaxic apparatus, an ejection system for the tracer, a drill for opening the skull, a freezing microtome or vibratome for tissue sectioning, and light and electron microscopes for sample observation.

Prepare all buffer solutions used with distilled water. Adjust these at physiological pH, i.e., pH 7.2–7.4 and store them at room temperature (unless indicated otherwise). Attention should be paid to the preparation of the tracer and ABC working solutions while reagents used for tissue fixation and processing, including color development substrate solutions are standard for the ABC immunohistochemical procedure.

1. BDA solution: Dissolve 10 mg lyophilized powder (Molecular Probes™) in 100 mL appropriate buffer (either 0.01 M acetate buffer, pH 4.5 or 0.01 M sodium phosphate buffer (PB),

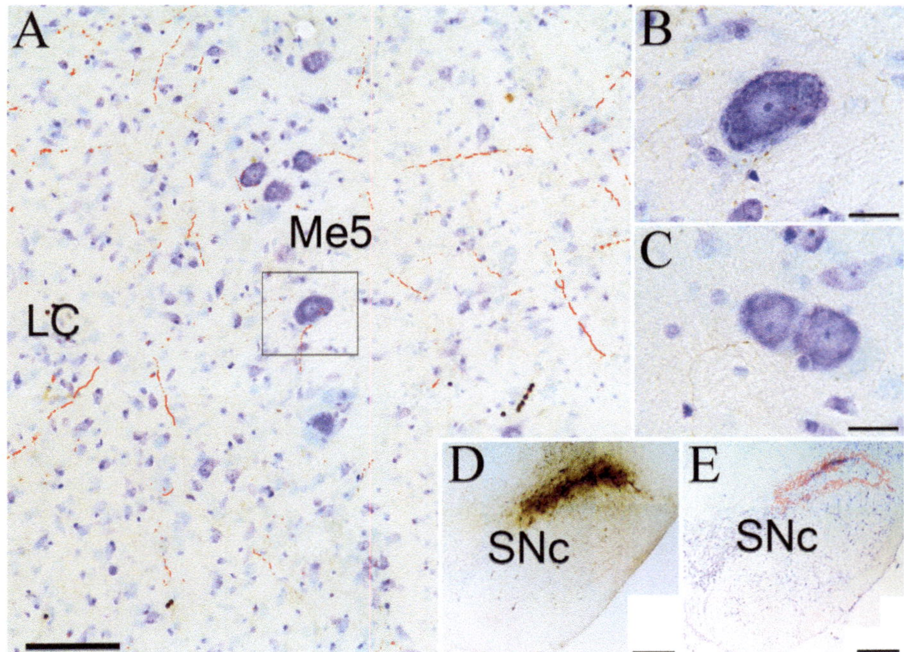

Fig. 2 A frontal section at the level of the midbrain–pontine junction part of the mesencephalic trigeminal nucleus (Me5) showing the distribution of axons and terminals anterogradely labeled with BDA following unilateral injection within the lateral substantia nigra, compact part (SNc). (**A**) Multiple BDA-labeled varicose fibers crossing the Me5 are seen in this section. Smoother labeled fibers and further terminal labeling is also present in the neighboring locus coeruleus (LC). (**B**) Higher magnification of the *boxed* area in (**A**) illustrating labeled varicosities in close proximity to a Me5 perikaryon. The two neurons in (**C**) receive both fine and coarse terminal labeling. (**D**, **E**) BDA-injection site and the adjacent Cresyl violet-stained section into the lateral SNc. Scale bars = 100 μm in (**A**), 20 μm in (**B**, **C**) and 500 μm in (**D**, **E**)

pH 7.3) (*see* **Note 2**). Gently rotate until all powder is dissolved and the solution is clear. Leave one aliquot in the refrigerator at 4 °C for current use and store the remaining aliquots in a freezer at ≤−20 °C. In our laboratory we prefer to make this solution fresh each time before use (*see* **Note 3**).

2. Vectastain ABC reagent: As recommended by the manufacturer (Vector Laboratories Inc, Burlingame, USA), add two drops of reagent A (avidin DH solution) and two drops of reagent B (biotinylated horseradish peroxidase) to 5 mL 0.1 M PB, mix thoroughly the vial on a rotator and allow the reagent to stand for at least 30 min before use.

3. DAB solution: Dissolve 5 mg of DAB (Sigma Chemical Company, St. Louis, MO, USA) in 10 mL of 0.05 M Tris–HCl buffer, pH 7.6 and after add 3.3 μL of H_2O_2 (*see* **Note 4**). Prior to use, filter the resulting solution through a 0.2 μm syringe filter or filter paper. The BDA substrate solution can be prepared as a stock solution and stored in a freezer until use.

Due to the fact that hydrogen peroxide (H_2O_2) evaporates over time and its stock (30 %) solution diminishes its concentration, add it to DAB solution immediately prior to use.

4. DAB-Ni incubation solution: Dissolve 5 mg of nickel ammonium sulfate (Sigma-Aldrich Corporation, St. Louis, MO, USA) and 10 mg of DAB (Sigma) in 25 mL of 0.05 M Tris–HCl buffer, pH 7.54. Immediately before use, add 10 μL of 30 % H_2O_2. Filter the resulting solution prior to use.
5. Stock 0.2 M SORENSEN's phosphate buffer (PB): Dilute 28.4 g anhydrous sodium phosphate dibasic to 1,000 mL distilled water (stock A) and 27.6 g sodium phosphate monobasic to 1,000 mL distilled water (stock B). Mix well 23 mL of stock A and 77 mL of stock B, and adjust pH of the working buffer to 7.3, pH 7.2–7.4 is also fine. Use phosphoric acid but not HCl to lower the pH if necessary. Store at 4 °C.
6. Working 0.1 M PB, pH 7.3: Dilute stock PB (0.2 M) 1:1 with distilled water. Recheck pH.
7. 0.01 M phosphate buffered saline (PBS): Make 10× PBS by mixing well to dissolve 10.9 g Na_2HPO_4 (anhydrous), 3.2 g anhydrous NaH_2PO_4, and 90 g NaCl in 1,000 mL distilled water. Adjust pH of this solution to 7.2 and store it at room temperature. Dilute the stock solution 1:10 with distilled water before use and adjust again its pH if necessary.
8. Heparinized saline: Add sodium heparin (10,000 IU) to 1 mL physiological saline (0.9 %) to make a final concentration of 500 IU/mL.
9. 4 % paraformaldehyde in 0.1 M phosphate buffer (PB), pH 7.4: Add 40 g paraformaldehyde (Merck, Darmstadt, Germany) to 500 mL distilled water, heat to 70 °C with stirring on a hot plate under a fume hood for several minutes (*see* **Note 5**), and add a small squirt of 0.1 M NaOH (about 1 mL). Once the solution has cleared (it should take 5 min or less), filter it with a side-arm flask, Buchner funnel, and Whatman No. 2 filter paper. Add 500 mL 0.2 M PB into the same container. Cool the solution in ice and store it at 4 °C for up to 2 weeks. This fixative solution is solely used in animal perfusion for light microscopy (LM).
10. 1 % glutaraldehyde and 1 % paraformaldehyde in 0.1 M PB, pH 7.4: Add 10 g paraformaldehyde to 1,000 mL 0.1 M PB, pH 7.4. Warm up to 60–65 °C while stirring. Add a few drops of 1 N NaOH until the solution clears. Continue to stir to entirely dissolve, and filter. Cool the solution to room temperature, adjust the pH to 7.4, and cool it further to 4 °C. Just before the perfusion add 20 mL of 50 % glutaraldehyde (Sigma Aldrich–Fluka) and mix well. This solution is used for animal perfusion, especially for electron microscopy (EM).

11. Sedative and analgesic cocktail: Mix 0.2 mg fentanyl and 5 mg diazepam in 1 mL physiological saline. These solutions are also commercially available as Hypnorm® (Janssen Pharmaceuticals, Oxford, England) and Valium® (Hoffmann-La Roche Ltd., Basle, Switzerland), respectively. One must pay attention that Hypnorm is available in Europe but must be imported as an investigational drug in the United States.
12. Sodium pentobarbital solution: Dissolve 50 mg sodium pentobarbital powder in 1 mL distilled water. Store the injectable solution at room temperature (*see* also **Note 6**). A trade name of the drug is Nembutal® (Sanofi, Libourne Cedex, France).
13. Atropine sulfate: Dissolve 500 μg crystalline powder of atropine sulfate monohydrate (BIOTREND Chemikalien GmbH, Cologne, Germany) in 1 mL distilled water. If necessary, dissolve it by warming up to 55–60 °C and then cool. The resultant solution is clear and colorless.
14. 0.05 M Tris–HCl buffer, pH 7.54: Dissolve 6.06 g Tris (Merck) in 500 mL distilled water. Add 38.9 mL 1 M HCl to adjust pH to 7.54 (practically to 7.6) and bring volume to 1 L with distilled water.

3 Methods

Carry out all procedures at room temperature unless otherwise specified. Perform the surgical procedure under deep anesthesia.

3.1 Surgical Procedure, Tracer Application and Perfusion of the Animals

1. Prepare a fresh BDA solution and fill the microinjection device (*see* **Note 7**).
2. For deep anesthesia on adult rats (250–300 g body weight) make one dose (0.3 mL) of sedative and analgesic cocktail and administer the mixture intraperitoneally. To reduce the mucous secretion in the tracheobronchial tree, prepare a dose (0.1 mL) of atropine sulfate and inject it subcutaneously.
3. Place the animal in a stereotaxic frame in a flat skull position.
4. Expose the skull, perform a craniotomy using a drill, and penetrate the dura with a needle to prevent breakage of the pipette tip.
5. Deliver BDA into neurons of the region of interest either iontophoretically (single injection) or by the pressure injection method (multiple injections) and insure that the injected tracer is absorbed into the tissue (*see* **Notes 7–10**). Determine the coordinates of the selected area from stereotaxic atlases of a given animal such as the atlas of the rat brain by Paxinos and Watson (*see* ref. 10).

6. After the injection finishes, rinse the surgical site with sterile saline, pack with gelfoam, close the wound in layers, suture the skin, and give the animal an injection of Ringer's lactate.
7. Allow the animal to recover for an appropriate period. Determine empirically the survival period depending on the distance to traced targets (*see* **Note 11**).
8. Deeply re-anesthetize the animal with sodium pentobarbital (Nembutal®) and perfuse it transcardially. Start perfusion with a vascular wash-out with heparinized saline injected into the heart (*see* **Note 12**).
9. After perfusion, quickly remove the brain, postfix it in the same fixative if necessary (*see* **Note 13**), wash in several changes of 0.1 M PB for 4 h, and soak in a series of 20 % sucrose-PB solutions overnight at room temperature until the brain sinks.
10. Cut serial coronal sections on a freezing microtome or a vibratome, collect them in a free-floating state in ice-cold 0.1 M PB and store the sections for up to several days at 4 °C until further processing for BDA labeling visualization. For long-term storage (−20 °C) transfer the sections in antifreeze solution.

3.2 Tracer Detection

Visualize the BDA staining immunohistochemically according the ABC procedure (*see* ref. 11 and **Note 14**) as follows:

1. The procedure starts with a preincubation step with 0.1 % bovine albumin for 30 min.
2. Then incubate the sections in the Vectastain® ABC reagent for 90 min at room temperature on a rocking table shaker (*see* also **Note 14**). Apply enough solution on the slide to cover the entire section.
3. Between the separate steps, rinse the sections for 30 min (3 × 10 min) in 0.1 M PB at room temperature.
4. The peroxidase reaction is then visualized with an appropriate substrate solution (*see* **Note 15**). When using DAB as a chromogen to reveal the ABC reaction, the precipitate produced has a brown color. Alternatively, the sections could be reacted with freshly mixed nickel-enhanced DAB (DAB-Ni) for 5–10 min, resulting in a blue-black precipitate.
5. Once the staining is completed and the desired stain intensity develops (*see* **Note 16**), rinse the sections in distilled water for 5 min, mount them onto chrome alum coated glass slides, and leave them to air-dry at room temperature overnight. Counterstain the sections with 0.025 % Cresyl violet, quickly dehydrate them through graded (70 %, then 90 %, then 100 %) ethanols, briefly clear in xylene (5 min each), and finally coverslip with a fast embedding medium, Entellan® (Merck, Darmstadt, Germany). Store the slides in a cool dark place.

6. View the slides by light microscopy. Figures 1 and 2 show the result of an anterograde tracing study to reveal the presence of a direct bilateral nigro-trigeminal projection in the rat.
7. For electron-microscopic tracer studies, the procedure is principally the same as above, except for the use of a suitable for EM fixative solution to preserve the ultrastructural details of BDA labeling. Other investigators also recommend a permeabilization step (1 h at room temperature in 0.3 % Triton X-100 in 0.1 M PB) in glutaraldehyde-fixed animals before the ABC procedure to increase the penetration of the ABC complex (*see* ref. 3). Penetration of the reagent, coinciding with good preservation of ultrastructure, can also be improved by using a brief freeze–thaw treatment with Triton X-100 (*see* refs. 12, 13).
8. The sections prepared for EM should be processed in the same way but without Triton X-100. Postfix the sections in 1 % OsO_4 for 1 h, then dehydrate them with graded concentrations of ethanol and propylene oxide and embed in resin. Cut ultrathin sections on an ultramicrotome, counterstain grids with uranyl acetate and lead citrate, and examine under an electron microscope.

4 Notes

Several technical problems may arise when designing an universal protocol for BDA tracing.

1. The procedure for retrograde labeling of neuronal cell bodies with BDA 3 kDa is largely the same, except that as suggested by Reiner et al. [3] an acidic solution of the tracer is used to enhance the preferentially retrograde nature of the labeling.
2. Application of injection techniques requires the tracer to be administered in a liquid form. BDA is supplied as a lyophilized powder and should be dissolved in appropriate aqueous buffers. We use 0.01 M PB, pH 7.3 as an injection vehicle for preparation of BDA 10 kDa and 0.01 M acetate buffer, pH 4.5 to dissolve BDA 3 kDa.
3. Due to the fact that the working BDA solution becomes unstable after long usage, we recommend for each injection to prepare a fresh BDA solution that will be iontophoretically or pressure-placed into the selected brain area.
4. Prepare DAB solutions in Tris–HCl buffer, pH 7.6. When making the chromogen solution, keep in mind that DAB is potentially carcinogenic in humans and harmful if inhaled or ingested and, therefore, it should be handled with care. To avoid direct skin contact, a gown and plastic gloves should be worn, and the procedure should be carried out under a fume hood using a facemask.

5. Additional special attention must be paid to the fact that vapors from all fixatives used are highly toxic, so they should be handled with plastic gloves and prepared under the fume hood.
6. Sodium pentobarbital easily precipitates and, therefore, it should not be added to acidic solutions. Besides, its aqueous solution is not very stable and should not be used if it contains a precipitate. We suggest adding 2 % benzyl alcohol or 10 % propylene glycol to enhance the stability of the injectable product.
7. For iontophoretic intracellular application of the tracer we use a silicon coated glass micropipette with inner tip diameter 25–40 μm. Fill the thin portion of the pipette by slow withdrawal of the syringe plunger, with backward rotation of the control knob (backfilling and aspiration). Perform the ejections of BDA with an iontophoretic power source using positive-pulsed driving current (3–5 μA, 2 Hz, 5 s on/off for 30 min). In order to prevent leakage of the tracer along the penetration track, reverse the injection current after the injection. To avoid tissue lesioning within the injection site and to prevent deposit formation in the pipette tip, we recommend applying low currents and long injection times. Smaller micropipette tip diameter and shorter current delivery periods are needed for smaller injections applied to small animals.
8. For manual pressure injections we use a "homemade" picospritzer consisting of a micropipette tip attached to a 1 μL Hamilton microsyringe. Other investigators have constructed a simple device for making multiple pressure microinjections into deeper structures of the brain (*see* ref. 14). Alternatively, the tracer can be delivered automatically and in this case, a motor-driven microinjector connected to a stereotaxic micromanipulator (Narishige, Japan) is used.
9. The suitable volume of the BDA solution for multiple injections is determined experimentally and must be optimized by the investigator for applications into different brain regions. We typically apply 0.05–0.5 μL of the tracer, with 0.01 μL steps per minute until the desired amount is reached. At the end of each injection hold the pipette in situ for 15 min to insure sufficient absorption of the injected BDA into the tissue and to reduce the possibility of its spread within the pipette track. Intracellular uptake of the tracer can be enhanced by co-injecting permeabilizing agents like 5 % Triton X-100. Finally, we verify the "effective" site of microinjection on coronal sections.
10. Iontophoretic application of BDA is the method of choice. Its major advantage is the application of the tracer alone, without an additional solvent. This makes intracellular filling feasible. Besides, the iontophoretic injection site is smaller than that produced by pressure microinjection (*see* ref. 14) and,

moreover, the application site may be precisely located by conventional electrophysiological techniques with the same electrode (*see* ref. 15). Nonetheless, pressure injections should also be considered as an adequate procedure for administrating the tracer.

11. BDA has a wide spectrum of survival time ranging from just a few days to 3 weeks. We have observed that labeling appears as early as 48 h and remains unchanged up to 14 days following injection due to the slow catabolism of BDA in brain tissue (see ref. 4). We have found that a survival time of 7–14 days is optimal in our studies of nigro-trigeminal or amygdalo-trigeminal connections in adult rats.

12. The transcardial perfusion is in fact performed by inserting a cannula through the ascending aorta, secured by a ligature. It may be advantageous to perfuse the animal first with physiological saline to remove the blood. In addition, immediately prior to perfusion, before the right atrium and the left ventricle are incised, we inject 0.5 mL heparin and 0.5 mL 1 % sodium nitrite into the right ventricle of the heart. As an alternative, heparinized saline (12 mg heparin per 1 mL physiological saline) can be used for the procedure. This step prevents clotting in the blood vessels. In principle, the temperature of the perfusate should not be below the body temperature of the experimental animal. We normally use ice-cold solutions for the perfusion and, thus, to avoid vasoconstriction that would impair the effectiveness of the procedure, we add a vasodilator such as sodium nitrite to the perfusate.

13. If necessary, the perfused brain can be postfixed prior to cryoprotection. For this purpose, immerse the brain in the perfusion fixative (without glutaraldehyde for EM) for at least 2–5 h (better overnight) at 4 °C.

14. Dextran amines like BDA are conjugated with biotin, so that they can be detected via high affinity binding of avidin with the ABC procedure. We recommend the use of the commercial avidin-biotin-horseradish peroxidase complex (ABC) Elite kit (Vectastain® ABC Kit, Vector Labs) for BDA visualization.

15. In our experience, the metal-intensified DAB reaction for the BDA stain is more sensitive than the standard "pure" DAB procedure but the use of DAB-Ni as a chromogen results in higher background staining.

16. Allow the chromogen reaction to develop in the dark at room temperature. Do not shake the slides while the color is developing. Inspect the chromogen development on a microscope at regular (usually 2–5 min) time intervals. Note that the progress of the reaction is crucial to avoid unacceptable increased degrees of background staining.

17. BDA 10 kDa has two minor drawbacks. The first one is its transport also in the retrograde direction (although not generally observed) (see ref. 4). This problem could be overcome by using a smaller (<40 mm) micropipette tip diameter and a lower (2 %) concentration of the tracer (*see* ref. 13). Moreover, this disadvantage could be turned into a useful tool to simultaneously reveal in a single-labeling study the collateral projections of retrogradely labeled neurons (*see* ref. 3).
18. Further, BDA 10 kDa is not be suitable for neuronal tracing in cases of application via muscle injections due to its potential uptake by fibers of passage, injured axons and even, though at a lesser degree, by intact axon terminals at the injection site (*see* ref. 7).

In summary, BDA is a highly sensitive tracer and a valuable tool for bidirectional mapping of topographical connections within the CNS of developing and adult mammals. Regardless of its minor disadvantages (*see* **Notes 17** and **18**), its compatibility with other neuroanatomical methods, ease of use, brevity of the labeling procedure, simplicity of visualization, and excellent labeling make BDA a neuronal tracer of choice.

Acknowledgments

This work was supported by a grant 1015945 from the Alexander von Hunboldt Foundation, Germany. The author greatly thanks Dr. Enrico Marani for protocol consultation.

References

1. Rajakumar N, Elisevich K, Flumerfelt BA (1993) Biotinylated dextran: a versatile anterograde and retrograde neuronal tracer. Brain Res 607:47–53
2. Vercelli A, Repici M, Garbossa D et al (2000) Recent techniques for tracing pathways in the central nervous system of developing and adult mammals. Brain Res Bull 51:11–28
3. Reiner A, Veenman CL, Medina L et al (2000) Pathway tracing using biotinylated dextran amines. J Neurosci Methods 103:23–37
4. Veenman CL, Reiner A, Honig MG (1992) Biotinylated dextran amine as an anterograde tracer for single- and double-labeling studies. J Neurosci Methods 41:239–254
5. Reiner A, Honig MG (2006) Dextran amines: versatile tools for anterograde and retrograde studies of nervous system connectivity. In: Zaborszky L, Wouterlood FG, Lanciego JL (eds) Neuroanatomical tract-tracing 3. Springer Science+Business Media, Inc., New York, pp 304–335
6. Medina L, Reiner A (1997) The efferent projections of the dorsal and ventral pallidal parts of the pigeon basal ganglia, studied with biotinylated dextran amine. Neuroscience 81: 773–802
7. Brandt HM, Apkarian AV (1992) Biotin-dextran: a sensitive anterograde tracer for neuroanatomic studies in rat and monkey. J Neurosci Methods 45:35–40
8. Lanciego JL, Luquin MR, Guillen J et al (1998) Multiple neuroanatomical tracing in primates. Brain Res Brain Res Protoc 2:323–332
9. Lazarov NE, Usunoff KG, Schmitt O, Itzev DE, Rolfs A, Wree A (2001) Amygdalotrigeminal projection in the rat: an anterograde tracing study. Ann Anat 193:118–126

10. Paxinos G, Watson C (1998) The rat brain in stereotaxic coordinates. Academic, New York
11. Hsu SM, Raine L, Fanger H (1981) Use of avidin-biotin-peroxidase complex (ABC) in immunoperoxidase techniques: a comparison between ABC and unlabeled antibody (PAP) procedures. J Histochem Cytochem 29:577–580
12. Wouterlood FG, Jorritsma-Byham B (1993) The anterograde neuroanatomical tracer biotinylated dextran-amine: comparison with the tracer *Phaseolus vulgaris*-leucoagglutinin in preparations for electron microscopy. J Neurosci Methods 48:75–87
13. Reiner A, Veenman CL, Honig MG (1993) Anterograde tracing using biotinylated dextran amine. In: Wouterlood FG (ed) Neuroscience protocols. Elsevier Science, Amsterdam, pp 1–14
14. Kratskin IL, Yu X, Doty RL (1997) An easily constructed pipette for pressure microinjections into the brain. Brain Res Bull 44: 199–203
15. Köbbert C, Apps R, Bechmann I et al (2000) Current concepts in neuroanatomical tracing. Prog Neurobiol 62:327–351

Chapter 31

Retrograde Tracing Technique for Neonatal Animals

Kengo Funakoshi, Akira Yoshikawa, and Yoshitoshi Atobe

Abstract

Tract tracing is a fundamental technique in neuroanatomy for examining fiber connections in the nervous system. After the introduction of horseradish peroxidase 40 years ago, many tracing substances have been used for neuroanatomical studies on various nervous systems. Here, we described retrograde tracing techniques using multiple fluorescent tracers, which make it possible to detect axonal collaterals. This technique is useful to study the development of axonal trajectories, as well as regenerative and compensatory mechanisms of animals that undergo neural damage at early stages.

Key words Axon collaterals, Retrograde tracer, Cholera toxin, Multiple tract tracing, Immunohistochemistry

1 Introduction

Neuroanatomical tract tracings are fundamental techniques for studies of the fiber connections in both the peripheral and the central nervous systems. Modern neuroanatomical tracing studies, based on the axonal flow, have their origin in an article describing the uptake and transport of the enzyme horseradish peroxidase (HRP) [1]. Depending on the direction of transport, tracing substances are divided into two groups, i.e., retrograde and anterograde tracers. Following injection into the target area, retrograde tracers are taken up by axon terminals, incorporated in transport vesicles, and then transported to the cell bodies. Therefore, retrograde tracing studies label the cell bodies and main stem dendrites of neurons that project axons to the target area. Anterograde tracers, by contrast, are taken up by the cell bodies and dendrites, transported from cell bodies into axons, and accumulate in the axon terminals and terminal boutons.

HRP was first introduced as a retrograde tracer transported by the peripheral nerves [1]. Transported HRP in the cell body is directly visualized using diaminobenzidine (DAB), which is converted to an

Renping Zhou and Lin Mei (eds.), *Neural Development: Methods and Protocols*, Methods in Molecular Biology, vol. 1018, DOI 10.1007/978-1-62703-444-9_31, © Springer Science+Business Media, LLC 2013

insoluble brown reaction product and water by the enzyme HRP in the presence of hydrogen peroxide [2]. After development of the HRP method, a modified retrograde tracing technique using HRP conjugates with the plant lectin wheat germ agglutinin (WGA; WGA-HRP) was introduced [3]. Subsequently, HRP conjugate with the β subunit of cholera toxin (CTB; CTB-HRP) was developed as a retrograde tracer, although it is also transported anterogradely [4, 5]. The general performance of WGA-HRP and CTB-HRP is much higher than that of HRP, so that retrogradely transported tracers fill the main dendrites as well as the cell bodies [6]. Availability of unconjugated CTB as a retrograde tracer was proved when it was visualized by immunocytochemistry using antibody against CTB [7]. This technique made it possible to identify the chemical nature of the labeled neurons by using double immunohistochemistry. CTBs conjugated to fluorescent molecule such as fluorescein isothiocyanate (FITC; FITC-CTB) are also available.

Retrograde tracing techniques using fluorescent dyes as tracers were developed, in parallel with the HRP and CTB methods. Evans Blue is one of the earliest introductions [8]. Subsequent studies revealed the availability of bisbenzimide (Bb), propidium iodide (PI), fast blue (FB), true blue (TB), granular blue (GB), nuclear yellow (NY), and diamidino yellow (DY) as retrograde tracers [9–11]. Fluoro-gold (FG), first introduced in 1986 [12], is now a widely used retrograde fluorescent tracer due to its excellent sensitivity and stability. Retrogradely transported FG labels the main dendrites as well as cell bodies. Immunohistochemistry using primary antibody against FG is used for observation under bright field microscopy.

One of the major advantages of fluorescent tracers is that the application of multiple tracers makes it easier to detect axonal collaterals [13]. Multiple labeling using fluorescent tracers is useful in studies searching for the collaterals of the corticospinal tract [14]. We injected FITC-CTB on one side of the spinal cord, and FB on the other side, to examine the axon collaterals of the corticospinal tract in neonatal rats with hemidecortication (Fig. 1) [15].

Dextran amines (DAs), which are biologically inert hydrophilic polysaccharides, are also widely used as tracers. DAs conjugated to biotin (BDA), or fluorescent molecules such as tetramethylrhodamine (RDA), and fluorescein (FDA) are commercially available. Although DAs of molecular weights ranging from 3,000 to 20,000 Da are transported both retrogradely and anterogradely, the low-molecular-weight form (i.e., 3,000 Da) of DA is preferentially transported retrogradely [16, 17]. Unlike FG, DAs are taken up even by axons damaged by the injection [18, 19]. DAs are also available for multiple retrograde tracing studies in combination with other tracers.

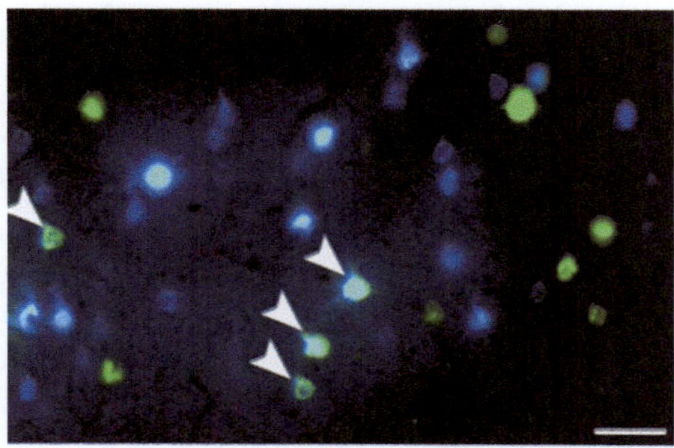

Fig. 1 Double fluorescence photomicrograph of the coronal section of the cerebral cortex. After FITC-CTB and FB were injected into the right and left side of the spinal cord, respectively, both FITC-CTB-labeled neurons (*green*) and FB-labeled neurons (*blue*) were observed in the right cerebral cortex. Some double-labeled neurons (*arrowheads*) were also observed. Scale bar = 50 μm. This figure is modified with permission from Yoshikawa et al. [2011] (Developmental Neuroscience © 2011, S. Karger AG, Basel)

2 Materials

Prepare all solutions using ultrapure water (prepared by purifying deionized water to attain a sensitivity of 18 MΩ cm at 25 °C).

2.1 Tracer Components

1. 2 % FG in distilled water (*see* **Note 1**).
2. 1 % FITC-CTB in 0.05 M phosphate buffer (PB), pH 7.4 (*see* **Note 1**).
3. 2 % FB in distilled water or saline (*see* **Note 2**).
4. 10 % 3 kDa FDA in 0.1 M phosphate-buffered saline (PBS), pH 7.4 (*see* **Note 2**).

2.2 Fixation and Sectioning Components

1. Fixative solution: 0.1 M PB containing 4 % paraformaldehyde, pH 7.4 (*see* **Note 3**).
2. Washing solution: saline containing 1 % heparin.
3. 0.1 M PB containing 15 % sucrose.

2.3 Immunohistochemistry Components

1. 0.1 M PBST: 0.1 M PBS containing 0.5 % Triton-X, pH 7.4.
2. Blocking solution: 0.1 M PBST containing 10 % normal donkey serum (NDS).
3. Diluent solution: 0.1 M PBST containing 1 % NDS, 1 % bovine serum albumin (BSA), and 0.1 % NaN_3.
4. Rinsing buffer: 0.025 M PBS, pH 7.4.
5. Tissue-Tek compound.
6. Moist chamber.

3 Methods

3.1 Preparation for Tracer Injection

1. For injection, a glass electrode (tip diameter: 50 μm) is pulled on a micropipette puller (*see* **Note 4**).
2. Aspirate an appropriate amount of solution containing the tracer substrate (i.e., 1 μl of 1 % FITC-CTB in 0.05 M PB or 1 μl of 2 % FG in distilled water) into a glass electrode using an injector (*see* **Note 5**). This operation should be done under a stereoscopic microscope.

3.2 Tracer Injection into the Animal

1. Anesthetize the animal with isoflurane (1.5 %, flow rate 2.0 l/min). In addition, 1 % lidocaine is injected into the skin covering the target area for local anesthesia.
2. Remove the musculature and bones, and expose the target areas of the central nervous system.
3. Fit an electrode containing the tracer solution into a pipette holder, and then insert an electrode into the target points using a micromanipulator (*see* **Note 6**).
4. Inject the tracer solution with an injector.
5. After injection, the skin opening is sutured with 5-0 silk and the animal is returned to its home cage.

3.3 Perfusion and Fixation

1. Four days after injection, re-anesthetize the animal deeply with isoflurane.
2. Open the abdomen and thorax so that the heart is well exposed, and cannulate the aorta via the left ventricle using a large gauge needle or catheter.
3. First perfuse the vasculature with washing solution.
4. Next perfuse the vasculature with 50 ml of fixative solution.
5. Brains and spinal cord are removed immediately, postfixed in the fixative solution for 1–2 h at 4 °C, and incubated in 0.1 M PB containing 15 % sucrose overnight at 4 °C for cryoprotection.

3.4 Preparation of Sections

1. Embed the tissue in O.C.T. compound (Tissue-Tek O.C.T. compound), and freeze it on the surface of liquid nitrogen or in 2-methyl-butane cooled by liquid nitrogen.
2. Frozen tissue is bound to a chuck and cut serially into sections of appropriate thickness in a cryostat at −20 °C.
3. The frozen sections are placed onto a cold gelatin-coated glass slides.
4. Warm the backside of the slide with the palmer surface of the fingers to thaw the sections onto the slides.
5. Store the mounted sections at −20 °C.

3.5 Immunostaining

Immunostaining can be used to enhance fluorescence intensity or to change colors.

1. Dry the sections for 30 min at room temperature and then postfix them in the fixative solution for 15 min at room temperature.
2. Incubate the sections with 0.1 M PBST for 60 min, and rinse four times for 5 min each with rinsing buffer.
3. Apply 50–100 μl of blocking solution to the section, and allow it to remain in place for 10 min.
4. Tip off the blocking solution, and apply 50–100 μl of primary antibody diluted with the diluent solution (*see* **Note 7**). Incubate the slide in a moist chamber for 12–24 h at 4 °C (*see* **Note 8**).
5. Rinse four times for 5 min each with rinsing buffer, and apply 50–100 μl of secondary antibody diluted with the diluent solution (*see* **Note 9**). Incubate the slide in the moist chamber for 2 h at room temperature.
6. Rinse four times for 5 min each with 0.025 M PBS, and mount antifade reagent onto the slide.
7. Cover with a cover slip and observe the stained slices with an epifluorescence microscope.

4 Notes

1. Tracer solutions are stored at 4 °C.
2. FB can be substituted for FG as a tracer that is visualized using an ultraviolet excitation filter. FDA can be substituted for FITC-CTB as a tracer that is visualized using a blue excitation filter.
3. Fixative solution is freshly prepared and refrigerated at 4 °C.
4. Glass electrodes (borosilicate glass; O.D. = 1.0 mm, I.D. = 0.75 mm, 10 cm length; item: B100-75-10, fire polished, Sutter Instruments, Novato, CA, USA) are pulled using a micropipette puller (P-97/IVF: Sutter Instruments). When the tracer solution is too viscous to be aspirated, the tip of the electrodes should be broken off using a surgical knife. The diameter of the broken tip should be 100–200 μm.
5. Although both an oil pressure injector and an air pressure injector are available, the air pressure type (IM-9C, Narishige, Tokyo, Japan) is recommended due to its ease of handling.
6. To insert the electrode into the brain or spinal cord, penetrate the dura mater with the electrode. Avoid making an incision in the dura to prevent leakage of the cerebrospinal fluid.

7. The following primary antibodies are recommended: goat antibody against CTB (1:1,000, No. 703; List Biological Laboratories, Campbell, CA, USA) and guinea pig antibody against FG (1:1,000, NM-101; Protos Biotech Corporation, New York, NY, USA).
8. To test staining specificity, negative control sections are incubated with diluent solution without the primary antibody. Other staining processes are the same as those for normal sections.
9. For secondary antibodies, various products such as Cy2 (FITC)-conjugated donkey anti-goat IgG and Cy5-conjugated donkey anti-guinea pig IgG are available.

Acknowledgments

We are grateful to Dr. A. Takeda and Mrs. M. Kobayashi for their assistance.

References

1. Kristensson K, Olson Y (1971) Retrograde axonal transport of protein. Brain Res 29: 363–365
2. Graham RC, Karnovsky MJ (1966) The early stages of absorption of injected horseradish peroxidase in the proximal tubules of mouse kidney: ultrastructural cytochemistry by a new technique. J Histochem Cytochem 14: 291–302
3. Gonatas NK, Harper C, Mizutani T, Gonatas JO (1979) Superior sensitivity of conjugates of horseradish peroxidase with wheat germ agglutinin for studies of retrograde axonal transport. J Histochem Cytochem 27:728–734
4. Trojanowski JQ, Gonatas JO, Gonatas NK (1981) Conjugates of horseradish peroxidase (HRP) with cholera toxin and wheat germ agglutinin are superior to free HRP as orthograde transported markers. Brain Res 223:381–385
5. Trojanowski JQ, Gonatas JO, Steiber A, Gonatas NK (1982) Horseradish peroxidase (HRP) conjugates of cholera toxin and lectins are more sensitive retrograde transported markers than free HRP. Brain Res 231:33–50
6. Horikawa K, Powell EW (1986) Comparison of techniques for retrograde labeling using the rat's facial nucleus. J Neurosci Methods 17: 287–296
7. Luppi PH, Sakai K, Salvert D, Fort P, Jouvet M (1987) Peptidergic hypothalamic afferents to the cat nucleus raphe pallidus as revealed by a double immunostaining technique using unconjugated cholera toxin as a retrograde tracer. Brain Res 402:339–345
8. Kuypers HGJM, Castman-Berrevoets CE, Padt RE (1977) Retrograde anoxal transport of fluorescent substances in the rat's forebrain. Neurosci Lett 6:127–135
9. Kuypers HG, Bentivoglio M, van der Kooy D, Castman-Berrevoets CE (1979) Retrograde transport of bisbenzimide and propidium iodide through axons to their parent cell bodies. Neurosci Lett 12:1–7
10. Castman-Berrevoets CE, Lemon RN, Verburgh CA, Bentivoglio M, Kuypers HG (1980) Absence of callosal collaterals derived from rat corticospinal neurons. A study using fluorescent retrograde tracing and electrophysiological techniques. Exp Brain Res 39:433–440
11. Keizer K, Kuypers HG, Husiman AM, Dann O (1983) Diamidino yellow dihydrochloride (DY. 2HCl); a new fluorescent retrograde neuronal tracer, which migrates only very slowly out of the cell. Exp Brain Res 51:179–191
12. Schmued LC, Fallon JH (1986) Fluoro-Gold: a new fluorescent tracer with numerous unique properties. Brain Res 377:147–154
13. van der Kooy D, Kuypers HG (1979) Fluorescent retrograde double labeling: axonal branching in the ascending raphe and nigral projections. Science 204:873–875

14. Castman-Berrevoets CE, Kuypers HG (1981) A search for corticospinal collaterals to thalamus and mesencephalon by means of multiple retrograde fluorescent tracers in cat and rat. Brain Res 218:15–33
15. Yoshikawa A, Atobe Y, Takeda A, Kamiya Y, Takiguchi M, Funakoshi K (2011) A retrograde tracing study of compensatory corticospinal projections in rats with neonatal hemidecortication. Dev Neurosci 33:539–547
16. Fritzsch B (1993) Fast axonal diffusion of 3000 molecular weight dextran amines. J Neurosci Methods 50:95–103
17. Kaneko T, Saeki T, Lee T, Mizuno N (1966) Improved retrograde axonal transport and subsequent visualization of tetramethylrhodamine (TMR)-dextran amine by means of an acidic injection vehicle and antibodies against TMR. J Neurosci Methods 65:157–165
18. Brandt HM, Apkarian AV (1992) Biotin-dextran: a sensitive anterograde tracer for neuroanatomic studies in rat and monkey. J Neurosci Methods 45:35–40
19. Todorova N, Rodziewics GS (1995) Biotin-dextran: fast retrograde tracing of sciatic nerve motoneurons. J Neurosci Methods 61:145–150

Index

A

B

C

Renping Zhou and Lin Mei (eds.), *Neural Development: Methods and Protocols*, Methods in Molecular Biology, vol. 1018, DOI 10.1007/978-1-62703-444-9, © Springer Science+Business Media, LLC 2013

N

O

P

R

S

T

U

V

W

X

MIX
Papier aus verantwortungsvollen Quellen
Paper from responsible sources
FSC® C105338

If you have any concerns about our products,
you can contact us on
ProductSafety@springernature.com

In case Publisher is established outside the EU,
the EU authorized representative is:
Springer Nature Customer Service Center GmbH
Europaplatz 3, 69115 Heidelberg, Germany

Printed by Libri Plureos GmbH
in Hamburg, Germany